Beyond Kyoto: Addressing the Challenges of Climate Change 2009

Science Meets Industry, Policy and Public

IOP Conference Series: Earth and Environmental Science
Volume 8

Aarhus, Denmark
5 - 7 March 2009

ISBN: 978-1-5108-0203-2
ISSN: 1755-1307

Printed from e-media with permission by:

Curran Associates, Inc.
57 Morehouse Lane
Red Hook, NY 12571

proceedings.com

Some format issues inherent in the e-media version may also appear in this print version.

Printed by Curran Associates, Inc. (2015)

Institute of Physics
Dirac House, Temple Back
Bristol BS1 6BE UK

Phone: 44 1 17 929 7481
Fax: 44 1 17 920 0979

techtracking@iop.org

Additional copies of this publication are available from:

Curran Associates, Inc.
57 Morehouse Lane
Red Hook, NY 12571 USA
Phone: 845-758-0400
Fax: 845-758-2634
Email: curran@proceedings.com
Web: www.proceedings.com

Table of contents

Volume 8

Beyond Kyoto: Addressing the Challenges of Climate Change
Science Meets Industry, Policy and Public
5–7 March 2009, Aarhus, Denmark

Accepted papers received: 9 November 2009
Published online: 7 December 2009

PREFACE

POLICY, LAW AND SOCIETY

MONITORING AND ANALYSING THE RISKS

IOP Conf. Series: Earth and Environmental Science **8** (2009) 011001 doi:10.1088/1755-1315/8/1/011001

Preface: Beyond Kyoto – the necessary road

By Professor Ellen Margrethe Basse

The Beyond Kyoto conference in Aarhus March 2009 was organised in collaboration with other knowledge institutions, business and authorities. It brought together leading scientists, policy-makers, authorities, intergovernmental organisations, NGO's, business stakeholders and business organisations. The conference was a joint interdisciplinary project involving many academic areas and disciplines.

These conference proceedings are organised in central and recurring themes that cut across many debates on climate change, the climatic challenges as well as the solutions. In front there is a short presentation of the conference concept.

Part I of the proceedings focuses on issues related to the society – covering climate policy, law, market based instruments, financial structure, behaviour and consumption, public participation, media communication and response from indigenous peoples etc.

Part II of the proceedings concerns the scientific knowledge base on climate related issues – covering climate change processes *per se*, the potential impacts of projected climate change on biodiversity and adaptation possibilities, the interplay between climate, agriculture and biodiversity, emissions, agricultural systems, increasing pressure on the functioning of agriculture and natural areas, vulnerability to extreme weather events and risks in respect to sea-level rise etc.

Concert Hall Aarhus was one of the venues at the *Beyond Kyoto* conference.

The conference proceedings committee consists of four professors from Aarhus University: Jens-Christian Svenning, Jørgen E. Olesen, Mads Forchhammer and Ellen Margrethe Basse.

Aarhus University's Climate Secretariat has had the overall responsibility for coordinating the many presentations, as well as the practical side of arranging the conference and supporting the publication of papers.

As Head of the Climate Secretariat and Chair of Aarhus University's Climate Panel, I would like to thank everyone for their contribution. This applies both to the scientific and the practical efforts. Special thanks to Project Manager Henrik Dalgaard for his excellent editorial services and to stud.mag. Nanna Katrine Lüders Kaalund for her practical assistance with the proceedings.

The European Commission under the Regional Development Fund has funded the conference and the publication of the proceedings.

The 7 Aarhus Statements on Climate Change

By Professor dr.jur. Ellen Margrethe Basse[1], Professor Jens-Christian Svenning[2], Professor Jørgen E Olesen[3], Professor Flemming Besenbacher[4], Professor Jeppe Læssøe[5], Associate Professor Marit-Solveig Seidenkrantz[6] and Professor Lene Lange[7]
[1]Head of the Climate Panel and Secretariat, Aarhus University
[2]Department of Biological Sciences, Faculty of Science, Aarhus University
[3]Faculty of Agricultural Sciences, Aarhus University
[4]iNANO, Faculty of Science, Aarhus University
[5]Department of Curriculum Research, Danish School of Education, Aarhus University
[6]Centre for Past Climate Studies, Department of Earth Sciences, Faculty of Science, Aarhus University
[7]Department of Biotechnology, Chemistry and Environmental Engineering, Aalborg University

E-mail: emb@adm.au.dk

Abstract. More than 1,000 prominent representatives from science, industry, politics and NGOs were gathered in Aarhus on 5.-7. March 2009 for the international climate conference "*Beyond Kyoto: Addressing the Challenges of Climate Change*". Thematically, *Beyond Kyoto* was divided into seven areas of particular interest for understanding the effects of the projected future climate change and how the foreseen negative impacts can be counteracted by mitigation and adaptation measures. The themes were '*Climate policy: the role of law and economics*' ,'*Biodiversity and ecosystems*', '*Agriculture and climate change*', '*Nanotechnology solutions for a sustainable future*', '*Citizens and society*' and '*The Arctic*'. The main responsible scientists for the seven conference themes and representatives from the think-tank CONCITO delivered "*The 7 Aarhus Statements on Climate Change*" as part of the closing session of the conference. The statements were also communicated to the Danish Government as well as to the press. This article is the product of the collective subsequent work of the seven theme responsibles and is a presentation of each theme statement in detail, emphasizing the current state of knowledge and how it may be used to minimize the expected negative impacts of future climate change.

The international climate conference "*Beyond Kyoto: Addressing the Challenges of Climate Change*" was part of the official preparations for the United Nations Framework Convention on Climate Change (UNFCCC) Conference of the Parties (COP15) to be held in Copenhagen from December 7th to 18th 2009. It was organised by Aarhus University in collaboration with many other expert institutions, authorities, and enterprises. The conference aimed specifically at creating dialogue and knowledge transfer between the sciences, industry, and civil society to portray and communicate the current knowledge base to political decision makers, and was very successful in doing this.

Thematically, *Beyond Kyoto* was divided into seven areas of particular interest for understanding the effects of future climate change and how the foreseen negative impacts can be counteracted by mitigation and adaptation measures (Table 1).

Theme 1 *Climate policy: the role of law and economics*	This theme focused on how to design a regulatory architecture, including legislative and economic instruments, addressing environmental issues and competitive concerns.
Theme 2 *Biodiversity and ecosystems*	The focus of this theme was on the importance of the climate-biodiversity relationship, not just in terms of the challenges and adaptation possibilities for nature conservation in the 21st century, but also for the future climate itself, notably the scope for mitigation of greenhouse gas emissions by the management of tropical forests and other natural areas.
Theme 3 *Agriculture and climate change*	The theme was concerned with the impacts of climate change on agriculture and forestry, and how management can be adapted to the changing climatic conditions and improved to reduce greenhouse gas emissions.
Theme 4 *Nanotechnology solutions for a sustainable future*	Each of the five sessions in this theme consists of a mixture of talks on nanotechnology and renewable energy, hydrogen technology, solar energy, wind energy, as well as catalysis and biofuels.
Theme 5 *Citizens and society*	This theme focused on what drives changes in the social practices and individual actions that causes global warming and how to promote a more sustainable development in terms of reducing GHG emissions.
Theme 6 *The Arctic*	The sessions in this theme focused on climatic changes and their consequences in the Arctic, integrating central aspects of other themes to synthetically address the climate change issue for a single geographical region.
Theme 7 *Integrated Energy Solutions*	Renewable energy and energy efficiency are an important part of sustainable solutions and elements for mitigation of climate change and were the main issues of this theme.

Table 1

Each theme was further divided into 2-5 sessions selectively focusing on important areas within each theme. The main responsible scientists for the seven conference themes and representatives from the think-tank CONCITO delivered "*The 7 Aarhus Statements on Climate Change*" as part of the closing session of the conference. The statements were also communicated to the Danish Government as well as the press. Subsequently, the statements have been presented on a number of occasions. Below, we present each theme in detail with respect to current status of knowledge and how this may be used to counter future problems caused by the projected changes in climate.

Theme 1. Climate policy: the role of law and economics

The focal point of Theme 1 was the importance of the legal, economic, and infrastructural instruments. The different perspectives and possible instruments regarding the post-Kyoto regime were among the issues covered by this theme (table 2). The first session focused on the legal challenges of climate change. Session 2 discussed the concept of investment under the post-Kyoto regime in both developing and developed countries. Session 3 focused on important aspects of governance of the energy sector – with a view to transforming this sector from a conventional to a renewable energy basis (focusing on institutions). Session 4 dealt with various aspects of regulatory systems and instruments, economic analysis and the impact of policy-based commitments. Finally, infrastructure and the management of smart growth was the central issue discussed during session 5.

Session 1	*Legal aspects of long-term cooperative action*
Session 2	*Investment and financial flows to address climate change*
Session 3	*Making room for renewable energy in power systems*
Session 4	*Regulation, economic analysis and impact of policy-based commitments*
Session 5	*Infrastructure and the management of growth*

Table 2

The discussions during this theme resulted in a recommendation for the stronger involvement of all sectors in the decision-making as well as in the implementation of decisions. Climate change is one of the biggest challenges that have yet faced mankind – so we must all help to find the solutions. It was concluded that far greater efforts are needed to ensure the involvement of the private sector and civil society in climate solutions. Challenges in respect to climate change are above all one of leadership, co-ordination, and collective action.

The involvement of governments should lead to the establishment of new and stronger institutions as well as international law and regulations that ensure efficiency with regard to mitigation and adaptation as well as sustainable development. At the heart of the challenge is the need for institutions that provide coherent and credible signals about future risks and opportunities and to ensure that the institutional framework is sufficiently resilient to maintain its function over a very long term. The institutions must focus on the needs of the present without compromising the ability of future generations to meet their own needs [1]. The development of new policy instruments, including eco-taxes, cap-and-trade regimes and other market-based instruments, voluntary agreements, performance standards and informational devices such as labels, has grown as part of the climate policy. Most of the existing agreements and instruments are formulated in highly general and abstract terms. International regulations should be formulated in a way, which can be accepted and understood by the public. The lack of precise objectives is a major difficulty in measuring achievements. The need is not for less regulation, but for better regulation [2].

It is well known that participation of both the private sector and civil society at international, national, and local levels is crucial with respect to the effectiveness of the legal framework for climate-change mitigation and adaptation [3, 4, 5, 6]. The involvement of the civil society is crucial in both negotiating and implementing international treaties and protocols on climate issues. The regulatory design must ensure both human rights and the efficiency of global agreements. Public participation in decision-making legitimizes decisions, and, if properly employed, may also improve their quality.

The transformation into a non-carbon society faces many challenges, for instance with regard to financing facilities. In the post-Kyoto-regime the private sector is expected to be a much larger source of funding than the public sector, making it critical that policies promote the flow of such funding. Public funds are also important to catalyze the generation of private funds, but the private sector has to deliver the vast majority of the required investments in climate change mitigation and adaptation. Consequently, we need to ensure stronger incentives and to reduce the barriers for private investors in these fields. Hence, financial mechanisms and institutions should be framed to facilitate further private investments. Future financial mechanisms should as a minimum have clear and focused objectives that are implemented in a more strategic and holistic way. This action could be taken immediately. The UNFCCC established the legal framework as long ago as 1992.

Figure 1. *From left is* John D. Hofmeister *(Founder and CEO, Citizens for Affordable Energy),* Lauritz B. Holm-Nielsen *(Vice-chancellor, Aarhus University),* Dr. Gro Harlem Brundtland *(Special Envoy on Climate Change for United Nations Secretary-General Ban Ki-moon) and* Sir David King *(Professor, bbSmith School of Enterprise and the Environment) in a full up auditorium during the opening session at the Beyond Kyoto conference. Photo: Lars Kruse/AU-photo.*

Theme 2. Biodiversity and ecosystems

Theme 2 focused on the importance of the climate-biodiversity relationship, not just for nature conservation in the 21st century, but also for the future climate itself. During session 1 the participants discussed the scientific basis for predicting the impact of future climate change on biodiversity and ecosystem functioning. During session 2 the possibilities for mitigation of greenhouse gas emissions by the management of natural areas were discussed. The need for sustainable management of ecosystem services and biodiversity under climate change also was emphasized in session 3. Session 4 dealt with the ways in which climate-driven changes in agriculture will affect natural ecosystems and biodiversity, focusing on the possibilities for efficient agricultural adaptation without imposing additional negative pressure on ecosystems and biodiversity. The possibilities for minimizing losses of biodiversity induced by climate change by the adaptive management of landscapes and seascapes were the focal point in session 5. Finally, during session 6 the talks focused on the role of the global environmental conventions, notably the Convention on Biological Diversity (CBD) in protecting biodiversity and ecosystems under climate change.

Session 1	*The scientific basis for predicting the impacts of future climate change*
Session 2	*Mitigation by management of natural areas*
Session 3	*Sustainable management of ecosystems and biodiversity*
Session 4	*Theme 2 -3 jointly (theme 3: session 3):* *Climate change and impacts of agriculture on ecosystems and biodiversity*
Session 5	*Adaptive management of land- and seascapes*
Session 6	*The role of global environment conventions in protection of biodiversity and ecosystem in a phase of climate change*

Table 3

The concluding statement on the biodiversity and ecosystems theme emphasized the need to include biodiversity as an integrated part of the general climate change mitigation and adaptation efforts.

Currently, biodiversity is under negative pressure from land-use, biological invasions, and pollution. As a result, numerous species are threatened with extinction; this is true, for example, for 25% of the roughly 5,500 mammal species in the world [7]. Over the next 100 years the changing climate will constitute an additional pressure, with potentially severe impacts [8, 9]. Climate is already affecting ecosystems and biodiversity globally, causing changes in the functioning of ecosystems, species abundances, and species ranges [9]. With an increase in global mean temperature of just >1.5-2.5°C, 20-30% of the species studied may experience an increased risk of extinction [9].

Although major losses are unavoidable if climate change is not strongly curtailed beyond business-as-usual expectations, the climate-change impact will depend greatly on interactions with other pressures, notably land-use [8, 9]. Key adaptation strategies are described in the following two main points.

- One key adaptation strategy would involve decreasing other pressures on biodiversity (habitat loss etc.) to increase resilience to climate change [9]. It is therefore crucial that climate change mitigation and adaptation measures, especially in land-use (e.g. biofuel crops), are implemented in ways that alleviate rather than place additional negative pressure on biodiversity.

- A key facet of climate-change-integrated conservation planning is to actively improve the possibilities for the maintaining or re-establishing viable, broadly distributed, and genetically diverse populations in the face of climate change by a variety of means such as expanding reserve systems, designing reserve systems that are resilient to climate change, climate change-off-setting local management (controlled burning to reduce fuel loads etc.), captive breeding, assisted migration (translocation), engineering new habitats etc. [9. 10, 11, 12].

Figure 2. *Theme 2 focused on the importance of the climate-biodiversity relationship, not just for nature conservation, but also for the future climate itself. Notably, forest conservation including Reduced Emissions from Deforestation and Degradation (REDD) represents a major and cost-effective climate change mitigation opportunity. One of the key points mentioned in the statement was that biodiversity needs to be an integrated part of the general climate change mitigation and adaptation efforts. Photo: Thea Kristiansen.*

Biodiversity and in particular forests also constitute a major part of the solution to the climate change problem [13, 14]. Notably, forest conservation including Reduced Emissions from Deforestation and Degradation (REDD) represents a major and cost-effective climate change mitigation opportunity [14]. Emissions from land-use change, especially tropical deforestation contribute about 20% of total anthropogenic greenhouse gas (GHG) emissions [9]. At the same time, forests including old-growth and tropical forests also constitute very important carbon sinks [15, 16, 17, 18]. Consequently, an important opportunity for climate change mitigation is the conservation of forests, notably carbon- and biodiversity-rich tropical forests [9, 19, 20, 21]. By reducing pressure on tropical forests, deforestation and degradation is also likely to have direct benefits for biodiversity [22]. Afforestation and reforestation can also contribute importantly to carbon sequestration [9] and reduce pressure on forest biodiversity [23]. To realize this key opportunity for climate change mitigation it is absolutely crucial that economic structures that provide the incentives for forest conservation are implemented.

One of the key points mentioned in the statement was that biodiversity needs to be an integrated part of the general climate change mitigation and adaptation efforts. Climate-change adaptation and mitigation in other sectors may have positive, neutral, or negative impacts on biodiversity [23], and synergies could be promoted by formulating integrated policies cross-linking major UN conventions, notably the UNFCCC and the Convention on Biological Diversity [9].

Theme 3. Agriculture and Climate Change

Theme 3 focused on the contribution of agriculture to mitigating greenhouse gas emissions and, at the same time, the need for adapting agricultural practices to climate change to meet the increasing demands for food, feed, fibre, and fuel. The theme was therefore concerned with the impacts of climate change on agriculture and forestry, and how management can be adapted to the changing climatic conditions and improved to reduce greenhouse gas emissions and sequester carbon. The focus was on the effects of changes in both mean climate and inter-annual variability, and on different responses in different regions of the world (Table 4). These response differences are a key factor in the vulnerability of agricultural food production systems. They also mean that adapting agricultural production to changing conditions will take very different routes in various parts of the world, depending not only on soil and climatic conditions, but also on the capacity to adapt as determined by the economic, social, and technological status of a given region. The discussions during session 1 and session 2 were related to climate-change impacts on agriculture and forestry in developing countries and intensive agricultural systems in developed countries, respectively. In a global context, some resources are readily available to most countries; however, others are unevenly distributed given current available technologies. Session 3 primarily considered the ways in which climate-driven changes in agriculture will affect natural ecosystems and biodiversity, focusing on the possibilities for efficient agricultural adaptation without imposing additional negative pressure on ecosystems and biodiversity. Session 4 focused on the role of bioenergy and biofuels, which are both renewable energies that will play a key role in achieving a sustainable energy system, whilst session 5 focused on the possibilities for reducing agricultural GHG emissions.

Session 1	*Impacts and adaption in developing countries*
Session 2	*Impacts and adaption in intensive agricultural systems*
Session 3	*Theme 2 -3 jointly (theme 2: session 4):* *Climate change and impacts of agriculture on ecosystems and biodiversity*
Session 4	*The role of bioenergy and biofuels*
Session 5	*Reducing agricultural GHG emissions*

Table 4

The primary aim of agricultural mitigation is the reduction of greenhouse gas emissions (methane and nitrous oxide), and the increase of soil carbon storage [24]. Globally agriculture contributes to CO_2 emissions by, e.g. deforestation, which can only be reduced if the productivity of existing agricultural land is improved through better crops and better soil and water management. Theme 3 recommended increasing agricultural productivity and reducing the use of water and land. An effective implementation of such a strategy urgently requires a substantial increase in research and innovation as well as in education and knowledge-transfer to farmers.

Figure 3. *Droughts are projected to become a more frequent and longer lasting under climate change, and this will have severe impacts on food production and food security, in particular in developing countries. It may also compromise some of the ambitions for significantly expanding biofuel production due to lack of water for supporting the biomass production. Photo: Jørgen E. Olesen.*

To reduce the dependency of society on fossil fuels, there is a high demand for increasing the use of bioenergy and bio-fuels [25]. These renewable energy sources are attracting more and more attention globally, not least because they have the potential to address concerns about climate change, the environment, energy security, and the opportunity to modernise energy services in developing countries. It was the opinion of the participants that this can only be achieved by developing new, highly productive bioenergy cropping systems that make use of land, which is not suitable for food and feed production, and which does not currently have high biodiversity and/or high carbon stocks. In addition, there is a need to better integrate ethical concerns regarding effects of increasing biofuel production on food prices and on biodiversity.

Adaptation in agriculture to increased climate variability involves increasing the resistance and resilience of production systems [26]. This can be done by improving soils, water use, crops, and cropping systems. There is a need to ensure that soils are managed to harvest and store rainwater more effectively. Irrigation water must be used more efficiently, and crops must be made more drought resistant through breeding programmes. In addition, cropping systems must be designed to conserve more water and provide better protection from the consequences of droughts, heat waves, and floods [9].

Figure 4. *A picture of the worlds largest experimental biogas plant, situated at Research Centre Foulum, Aarhus University.*

The greatest challenge facing agriculture during the 21st century is probably how to feed the increasing not only an increasing population, but also an increasing number of more wealthy people on the Earth, while maintaining high-quality soil and water resources [27]. Climate change significantly adds to this challenge by reducing the quality of soils and the availability of water in many regions through the increasing variability in temperature and rainfall [28]. The already large contribution of agriculture to global greenhouse gas emissions will increase unless more effective and climate-friendly farming systems are adopted [9].

The challenges facing agriculture are therefore three-fold: to increase production, to reduce emissions, and to adapt to a warmer and more variable climate. There is a particular need to increase agricultural productivity, reduce greenhouse gas emissions, and adapt to climate change in developing countries [29]. These challenges require an urgent and substantial increase in the focus of research, innovation, transformation of knowledge and education at all levels across all sectors related to agriculture. To do so calls for capacity building in many cases, which requires the focus not only of national and local governments, but also of international donors and the international research community. There is a further need internationally to promote the development of cropping systems and technologies that deliver highly productive systems for combined food, feed, and bioenergy production [30, 31]. Such actions require a collaborative effort across private and public research institutes and across many research disciplines.

Agricultural, food, and energy policies should be better integrated at international, regional, and local levels. They should seek to use, develop and maintain the available resources more effectively (land, soil, water, nutrients, agrochemicals, energy, genetic diversity, research capacity, information systems, infrastructure, culture) in a climate change and variability context. There is a need for flexible policies that focus on the multidimensional roles of agriculture, with priorities varying among regions depending on local resources, capabilities, and demands [9].

Theme 4. Nanotechnology solutions for the future

The application of modern nanotechnology and its contribution to ensure both energy efficiency and better utilisation of sustainable technology were covered by theme 4. One of the familiar challenges covered by the sessions was the growing political and economic pressure for an increase in the utilisation of renewable energy sources and the technological development needed to make this happen on a large scale. New sustainable energy supplies that can meet the increasing demands for energy will require major scientific and technological breakthroughs. Presently, one of the main challenges is to convert the maturing nano-science into nanotechnology and pave the way for a new generation of companies with a nano-technological foundation. The aim of the theme was to facilitate the link between science and industry that is crucial for the transfer of new scientific breakthroughs into commercial products. The scientific presentations focused on general principles and possible applications rather than on detailed fundamental studies. Presentations from the industry focused on research and development with less emphasis on commercial aspects.

It was pointed out in the statements made at the conference that nanotechnology will lead to the next industrial revolution, providing major new technological breakthroughs. The most urgent need is to stimulate maturation and implementation of the many new, promising nanotechnologies in developing sustainable energy technologies and improving the energy efficiency of current systems.

As clearly stated by the Fourth Assessment Report of the IPCC [9], a business-as-usual approach to research and development will not result in the fast implementation of renewable energy sources that is mandatory if society is to cope with the predicted increase in energy consumption worldwide, while simultaneously reducing emissions of greenhouse gasses from the energy sector [9]. In general, the challenge of clean energy sources involves improving renewable energy, energy efficiency, reducing the price per kWh produced, and increasing the scale of systems to that of large commercial systems. Nanotechnology provides new scientific and technological breakthroughs in the synthesis, characterisation, and production of new functional nano-materials with tailored properties optimised for applications within the fields of (for instance) solar energy, biofuels, fuel cells, catalysts, wind energy and hydro-energy, as well as for energy storage and improving the energy efficiency of existing systems.

The increase in the world population and the increase in GNP (not least in Asia) are expected to result in a 200% increase in world energy demand by 2050. If this energy demand is to be supplied by increasing the use of fossil fuels alone, it will most probably lead to unacceptable and potentially catastrophic CO_2 levels in the atmosphere. The development of clean energy technologies, such as solar energy, wind energy, biofuels, new energy storage systems and a new worldwide grid transmission system is therefore unavoidable [9]. Nanotechnology offers the tools needed to design and develop new functional nano-materials for sustainable energy systems to cope with increasing energy demand. Hence, investments in nanotechnology research aimed at developing new nano-structured energy materials could create immense economic and environmental benefits.

Investment in nano-science and nanotechnology is evolving rapidly, and nano-technological applications have already started to emerge at research laboratories within areas such as solar energy, biofuels, fuel cells, catalysts, wind energy, hydro energy and energy storage. However, there is a need for significant improvements with regard to efficiency, stability, scaling-up to mass production, and reductions in the cost per kWh produced. If clean energy systems are going to revolutionise the energy supply, the investments involved should have a commercial payback time of 5-10 years. This means that public funds must be made available in the transition period until renewable alternatives have matured to a level that ensures economic viability. Increased investments in research in and the development of new functional nano-energy materials and renewable energy technologies must be the top priority for any government wishing to address the challenges of increased energy consumption and climate change. Funding should stimulate both research and development, both at the fundamental level at research institutions and in the industry with a particular emphasis on maintaining the connection between the two.

Theme 5. Citizens and society

Theme 5 focused in particular on how to strengthen the knowledge transfer between business, science, public policy, and administration. The discussion focused on the potentials and obstacles for the involvement and social learning of citizens in process of governance. It is well-known that the mitigation of, and adaptation to, climate change will require actions and interactions between nations, governmental and non-governmental organisations, industry in general, individual companies, and ultimately every single individual as a consumer and citizen. Understanding the complexities and mechanisms of the relationship between citizens, society and climate change is fundamental to ensure the effectiveness and efficiency of regulatory actions. This relationship involves a variety of aspects that were addressed during this theme (Table 5).

Session 1	*Corporate social responsibility in an age of climate change*
Session 2	*Low energy buildings and user behaviour*
Session 3	*Human health and disease*
Session 4	*Communication and climate change*
Session 5	*Participation, learning, and socio-cultural change*

Table 5

Session 1 focused on corporate social responsibility in an age of climate change. Low-energy buildings and user behaviour were the issue dealt with during session 2. Human health and disease were discussed during session 3. Communication and climate chance were discussed during session 4, and finally the focus was placed on participation, learning and socio-cultural change as important social and cultural dimensions of the climate challenges.

The discussions during Theme 5 concentrated on socio-technical systems, competence development, learning, processes of social and cultural change, communication, the corporate responsibility of enterprises, and mechanisms and tools for facilitating learning and socio-cultural changes. The statements recommended that the agreement to be adopted at the COP15 in Copenhagen should include the commitments of the Aarhus Convention as well as other incentives to support the participation and empowerment of citizens.

The legal rights of citizens to information and participation constitute a necessary precondition for addressing the challenge of climate change. The Aarhus Convention has proved that it is possible to persuade government and local authorities to commit themselves to an agreement on these rights (cf. www.unece.org/env/pp). The Convention underpins participatory rights in the context of environmental governance. It emphasises linking the public with the public authorities in a democratic context, as well as establishing a new process for public participation in the negotiation and implementation of international agreements at national level. The UNFCCC also focuses on the importance of promoting public access to information, on training scientific, technical, and managerial personnel, and on developing and implementing educational and public awareness programmes.

Furthermore, there are good examples of companies working with social corporate responsibility and from the education sector that prove that citizen participation can help to promote ownership and social learning, thereby guiding people's actions in meeting the challenge of climate change [32, 33, 34].

Mass media communication also plays an important part in the task of education [35, 36]. However, in this field, there is a need to improve the public-service contribution to the involvement of citizens in deliberation, and innovative efforts to find socially robust solutions regarding the climate-change challenge.

To involve more companies, activate the whole education sector, and influence the public-service media, there is a need for both schemes and incentives. Hence, in addition to an agreement regarding legal rights to information and participation, the forthcoming global protocol on climate change should also include commitments regarding supportive schemes and incentives.

Theme 6: The Arctic

This theme covered the well-known particular vulnerability of the Arctic to climate change in terms of both ecosystems and socio-economic aspects. Extending from the Arctic Climate Impact Assessment [37], the theme pointed out changes and consequences for the Arctic following projected future changes in climate and brought together central aspects of previous themes related to a single geographical region. In contrast to the tradition of providing a kaleidoscopic presentation of causes and consequences, the focus on the Arctic provided a synthesis of an entire climate-driven 'cause-effect-consequence' spectrum involving all aspects of natural, cultural, human and socio-economic sciences.

Session 1	*in the Arctic – past, present and future*
Session 2	*Arctic ecosystem changes and feedbacks*
Session 3	*No session 3*
Session 4	*Use of natural resources*
Session 5	*Human health and socioeconomic aspects of a changing environment*

Table 6

In session 1 the discussion addressed the climate of the Arctic in a past, present and future perspective in order to place the present climate development in a broader context. In session 2 the discussion focused on Arctic ecosystem changes and feedbacks. Use of natural resources was the issue discussed in session 4. Finally, human health, living conditions, and other socio-economic aspects of a changing environment were the issues discussed under session 5 (Table 6).

Studies of past climate show that the Arctic region has clearly always been subject to significant climate change and a need for adaptation. It was also stated that adaptation to fast changing conditions is essential at both cultural and ecosystem level. Indeed, an even faster level of change may be necessary for the arctic societies than for other geographical regions, because here the way of life and economy is based almost entirely on available natural resources. The Arctic is a region of extreme and rapid climate changes, significantly exceeding the global average [e.g. 9, 38,], with fragile ecosystems and indigenous people heavily dependent on natural resources. Its natural and traditional cultural setting is under pressure from climate change, a growing industrial interest, and enhanced exploitation of non-renewable resources. The development may promise increased wealth, but also encompasses a significant risk of degradation of ecosystems and existing social structures [9].

There are far too many examples of economic development in the Arctic in which local people have suffered a complete break-down of social structures resulting in poor health and loss of identity leading to alcoholism and abuse [9]. Therefore, the conference statements related to theme 6 recommended ensuring the sustainable development of the Arctic region in collaboration with the local communities. Participatory processes should be used in order to address the needs and concerns of local people whenever decisions are made to begin any kind of economic development in the Arctic. People in the Arctic often have traditional lifestyles such as reindeer herding, hunting, and fishing, often based on a non-monetary economy [9]. Participatory processes involve developing tools and mechanisms, which enable local people to address the inevitable social and economic changes in a world dominated by climate change.

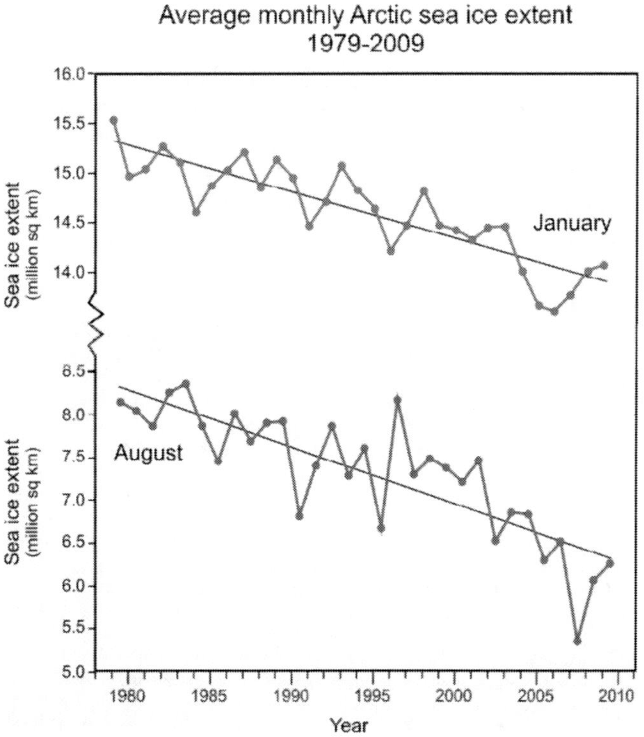

Figure 5. *Theme 6 focussed on the impact of climate change in the Arctic region. One of the major changes in recent years is the decreasing extend of Northern Hemisphere sea ice. The above figure shows the changes in winter (January) and summer (August) sea-ice extend of the Northern Hemispheres from 1979-2009 [39].*

Actions are needed to address these concerns and to ensure that the management of natural living resources is sustainable and based on the best available knowledge of the consequences of climate change. In many areas this depends upon the development of relevant institutions as well as research and education facilities [9].

Likewise, it is important to focus on the environmental and socio-economic impact of an increased extraction of non-renewable resources such as oil, gas and minerals as well as on increased shipping in vulnerable areas [9]. New technologies should be developed and environmental risk assessments as well as socio-economic studies would be needed in this process. Particularly fragile areas should be protected against non-sustainable exploitation. A particular effort should be made to understand the interaction between climate, ecosystems, natural-resource utilisation and socio-economic consequences and development. A solution could be to implement a temporary moratorium on further exploitation of non-renewable resources until these issues are resolved.

In order to improve our understanding of human impact and to distinguish it from natural climate variability, while augmenting our appreciation of non-linearity in climate effects and possible "tipping points", i.e. significant changes in ecosystems, we need an enhanced understanding of all elements constituting the climate system such as ocean currents and atmospheric conditions through the interdisciplinary study of the past climate record in order to better explain and predict present and future climate changes. In conjunction with this, we need to establish a firm understanding of how the structure and function of ecosystems react to short- and long-term changes in climate and, in particular, how resilient arctic species and, hence, the ecosystems in which they are embedded are to significant short as well as long-term changes in climate.

Partnerships between governments, scientists, industry, environmental NGOs as well as local and indigenous organisations are needed in order to identify problems and find solutions. The initiative to create these partnerships must be taken immediately after the COP15 Conference, embracing both perspectives of short- and long-term changes and consequences.

Ideally, a legally binding treaty protecting the Arctic Ocean against non-sustainable exploitation should be implemented following the agreement at COP15 in Copenhagen in December 2009.

Theme 7. Integrated energy solutions

Focus on renewable energy and energy efficient technologies were identified as the two most important areas where actions and efforts are needed in order to achieve mitigation of climate change and provide sustainable solutions for the future (Table 1)

According to the International Energy Agency (IEA), electricity generation from renewable energy sources (excluding hydro) has grown from an almost invisible 0.6% globally to close to 2.6% over the last 30 years [25]. However, in the same period, consumption has tripled, with gas and coal supplying most of the new capacity, and IEA foresees that the consumption will continue to grow for the next many years. Unless radical action is taken, fossil fuels will continue for years to be the backbone of global energy. During the last decade a wealth of new renewable energy technologies as well as energy efficiency technologies has been developed to reduce the dependence on fossil fuels, but by far the majority of the new technologies have not been implemented as part of the energy system. Consequently, the final statements of Theme 7 recommended that there is an urgent need for timely and targeted incentives for stimulating the introduction and implementation of new energy technologies.

The papers given in the meeting highlighted the following points: New energy technologies require cross-disciplinary collaboration. Integrated collaboration between biologists, biotechnologies, and engineers is especially essential. Further, the importance of the value adding effect of close collaboration with ICT expertise should not be underestimated, especially with respect to gaining and exploiting the synergy achieved by implementing integrated energy systems, embracing energy from many resources as well as incentives for introduction of energy saving instruments.

The challenge within the area of bio-energy is to integrate into it a biomass conversion concept, so that it can contribute to the global efforts of developing alternatives to the entire spectrum of products currently obtained from oil and natural gas. Biomass conversion in biorefineries can provide not just bio-energy, but also production of biomaterials, bio-chemicals, bio-fertilizer, nutraceuticals, food and feed ingredients etc. The conceptual frame for such an integrated approach to technologies for a fossil-free society is the bio-refinery. The basic two elements in a bio-refinery are the feedstock and the agents for converting the biomass polymers into the products needed. The sustainable feed stock resources, consisting primarily of lignocellulosic materials from plant cell walls, will come from crop residues, from side streams from the agroindustry, and from municipality waste. The agents that make the biomass conversion into valuable and value added products possible are microorganisms and microbial products, primarily enzymes. With such an approach, the bio-refinery technologies can be developed without threatening the production of food and feed.

However, the most efficient short cut to energy solutions is energy savings. A series of new Danish industry initiatives have achieved a significant competitive advantage in the European and the global market by developing and commercializing energy efficient and environmentally friendly technologies, processes and products. A political action is now called for from both industry and from the research environment. Regulatory and economic instruments and clear signals that Denmark and Europe fully and wholeheartedly support the clean-tech revolution could have a strong beneficial effect. This would especially be true, if it became a central pillar for the EU policies within all four sectors: industry, agriculture, environment, and commerce. Denmark is ready to take a concrete step in implementing this vision: Danish municipalities have already very advanced plans for fossil-free solutions and energy-saving technologies and mechanisms. A central component here is the plans for CO_2-neutral combined power and heat production already from 2030. The second important component is the construction of new energy-saving and energy plus houses as well as energy conscious renovation of the older houses and infrastructures.

Based on the presentations and discussions in Theme 7 it became clear that there are ample technologies, but too slow momentum in the current implementation, emphasizing an urgent need for action. As a consequence it was proposed to establish an international fund to .... and gain political support to create optimized market conditions, with clear incentives for introduction of fossil-free solutions and mechanisms for lowering greenhouse gas emissions. It is important that society establish framework conditions, which create markets that will pull the new technologies into use [9]. This will in itself stimulate investments in zero carbon emission technologies.

It was the overall recommendation from Theme 7 to establish a new global carbon tax system, which would put tax on use of fossil fuels and hereby creates incentives for use of renewable resources and energy-saving technologies and instruments. Such a global tax system could directly stimulate the development, commercialization, and use of new technologies and would furthermore aid building up a more stable momentum for the clean-tech revolution, as it would be independent of the fluctuation in oil prices. Another important advantage of an international fossil-fuel carbon tax as compared to a CO_2-quota cap and trade mechanism is that it is effective as a means to ensure the polluter pays principle the profit can be used to help the developing countries. It makes it attractive for the developing countries to build their future development on renewable and/or smart clean-tech technologies.

Figure 6. *The Aarhus University campus. Photo: Lars Kruse/AU-photo*

Reference list

[1] World Commission on Environment and Development 1987 *Our Common Future* Oxford University Press p 8

[2] Massai L and Montini M (eds.) 2007 *The Kyoto Protocol and Beyond: Legal and Policy Challenges of Climate Change* T.M.C. Asser Press

[3] Stokke O S, Hovi J and Ulfstein G (eds.) 2005 *Implementing the Climate Regime. International Compliance* Earthscan

[4] Depledge J 2005 *The Organization of Global Negotiations. Constructing the Climate Change Regime* Earthscan

[5] Lundqvist L J and Biel A (eds.) 2007 *From Kyoto to the Town Hall. Making International and National Climate Policy Work at the Local Level* Earthscan

[6] Sullivan R (eds.) 2008 *Corporate Response to Climate Change. Achieving Emissions Reductions through Regulation, Self-Regulation and Economic Incentives* Greenleaf Publishing Ltd.

[7] Schipper J, Chanson J S, Chiozza F, Cox N A, Hoffmann M, Katariya V, Lamoreux J, Rodrigues A S L, Stuart S N, Temple H J, Baillie J, Boitani L, Lacher T E, Mittermeier R A, Smith A T, Absolon D, Aguiar J M, Amori G, Bakkour N, Baldi R, Berridge R J, Bielby J, Black P A, Blanc J J, Brooks T M, Burton J A, Butynski T M, Catullo G, Chapman R, Cokeliss Z, Collen B, Conroy J, Cooke J G, da Fonseca G A B, Derocher A E, Dublin H T, Duckworth J W, Emmons L, Emslie R H, Festa-Bianchet M, Foster M, Foster S, Garshelis DL, Gates C, Gimenez-Dixon M, Gonzalez S, Gonzalez-Maya J F, Good T C, Hammerson G, Hammond P S, Happold D, Happold M, Hare J, Harris R B, Hawkins C E, Haywood M, Heaney L R, Hedges S, Helgen K M, Hilton-Taylor C, Hussain S A , Ishii N, Jefferson T A , Jenkins R K B, Johnston C H , Keith M , Kingdon J, Knox D, Kovacs K M, Langhammer P, Leus K, Lewison R, Lichtenstein G, Lowry L F, Macavoy Z, Mace G M, Mallon D P, Masi M, McKnight, M W, Medellin R A, Medici P, Mills G, Moehlman P D, Molur S, Mora A, Nowell K, Oates J F, Olech W, Oliver W R L, Oprea M, Patterson B D,

Perrin W F, Polidoro B A, Pollock C, Powel A, Protas Y, Racey P, Ragle J, Ramani P, Rathbun G, Reeves R R, Reilly S B, Reynolds J E III, Rondinini C, Rosell-Ambal R G, Rulli M, Rylands A B, Savini S, Schank C J, Sechrest W, Self-Sullivan C, Shoemaker A, Sillero-Zubiri C, De Silva N, Smith D E, Srinivasulu C, Stephenson P J, van Strien N, Talukdar B K, Taylor B L, Timmins R, Tirira D G, Tognelli M F, Tsytsulina K, Veiga L M, Vie J C, Williamson E A, Wyatt S A, Xie Y and Young B E 2008 The status of the World's land and marine mammals: diversity, threat, and knowledge *Science* **322** 225-230

[8] Sala O E, Chapin F S III, Armesto J J, Berlow E, Bloomfield J, Dirzo R, Huber-Sanwald E, Huenneke L F, Jackson R B, Kinzig A, Leemans R, Lodge D M, Mooney H A, Oesterheld M, Poff N L, Sykes M T, Walker B H, Walker M and Wall D H 2000 Global biodiversity scenarios for the year 2100 *Science* **287** 1770-1774

[9] IPCC 2007 *the Fourth Assessment Report of the Intergovernmental Panel on Climate Change*

[10] Hannah L, Midgley G F and Millar D 2002 Climate change-integrated conservation strategies *Global Ecology and Biogeography* **11** 485-495

[11] Skov F and Svenning, J-C, 2004 Potential impact of climatic change on the distribution of forest herbs in Europe *Ecography* **27** 366-380

[12] Hoegh-Guldberg O, Hughes L, McIntyre S, Lindenmayer D B, Parmesan C, Possingham H P and Thomas C D 2008 Assisted colonization and rapid climate change *Science* **321** 345-346

[13] Pacala S and Socolow R 2004 Stabilization wedges: solving the climate problem for the next 50 years with current technologies *Science* **305** 968-972

[14] Dinkel J, Enkvist P-A, Nauclér T and Pestiaux J 2009 *Pathways to a low-carbon economy. Version 2 of the global greenhouse gas abatement cost curve* McKinsey & Company

[15] Lewis S L, Lopez-Gonzalez, G, Sonké, B, Affum-Baffoe, K, Baker, T R, Ojo, L O, Phillips, O L, Reitsma, J M, White, L, Comiskey, J A, Djuikouo, M-N, Ewango, C E N, Feldpausch, T R, Hamilton, A C, Gloor, M, Hart, T, Hladik, A, Lloyd, J, Lovett, J C, Makana, J-R, Malhi, Y, Mbago, F M, Ndangalasi, H J, Peacock, J, Peh, K S H, Sheil, D, Sunderland, T, Swaine, M D, Taplin, J, Taylor, D, Thomas, S C, Votere, R and Woll, H 2009 Increasing carbon storage in intact African tropical forests *Nature* **457** 1003-1006

[16] Luyssaert S, Schulze E-D, Börner A, Knohl A, Hessenmöller D, Law B E, Ciais P and Grace J 2008 Old-growth forests as global carbon sinks *Nature* **455** 213-215

[17] Phillips, O L, Aragao, L E O C, Lewis, S L, Fisher, J B, Lloyd, J, López-Gónzalez, G, Malhi, Y, Monteagudo, A, Peacock, J, Quesada, C A, van der Heijden, G, Almeida, S, Amaral, I, Arroyo, L, Aymard, G, Baker, T R, Bánki, O, Blanc, L, Bonal, D, Brando, P, Chave, J, de Oliveira, A C A, Cardozo, N D, Czimczik, C I, Feldpausch, T R, Freitas, M A, Gloor, E, Higuchi, N, Jiménez, E, Lloyd, G, Meir, P, Mendoza, C, Morel, A, Neill, D A, Nepstad, D, Patiño, S, Peñuela, M C, Prieto, A, Ramirez, F, Schwarz, M, Silva, J, Silveira, M, Thomas, A S, ter Steege, H, Stropp, J, Vasquez, R, Zelazowski, P, Dávila, E A, Andelman, S, Andrade, A, Chao, K J, Erwin, T, Di Fiore, A, Euridice, H, Keeling, H, Killeen, T J, Laurance, W F, Cruz, A P, Pitman, N C A, Núñez Vargas, P, Ramírez-Angulo, H, Rudas, A, Salamão, R, Silva, N, Terborgh, J and Torres-Lezama, A 2009 Drought sensitivity of the Amazon rainforest *Science* **323** 1344-1347

[18] Phillips OL, Lewis S L, Baker T R, Chao K-J and Higuchi N 2008 The changing Amazon forest *Philosophical Transactions of the Royal Society* **B 363** 1819-1827

[19] Dinkel J, Enkvist P-A, Nauclér T and Pestiaux 2009 *Pathways to a low-carbon economy. Version 2 of the global greenhouse gas abatement cost curve* McKinsey & Company

[20] Kindermann,G., Obersteiner,M., Sohngen,B., Sathaye,J., Andrasko,K., Rametsteiner,E., Schlamadinger B Wunder S and Beach R 2008 Global cost estimates of reducing carbon emissions through avoided deforestation *Proceedings of the National Academy of Sciences* USA **105** 10302-10307

[21] Jackson R B, Randerson J T, Canadell J G, Anderson R G, Avissar R, Baldocchi D D, Bonan G

B, Caldeira K, Diffenbaugh N S, Field C B, Hungate B A, Jobbagy E G, Kueppers L M, Nosetto M D and Pataki D E 2008 Protecting climate with forests *Environmental Research Letters* **3** 044006.

[22] Laurance W F 2007 Have we overstated the tropical biodiversity crisis? *Trends in Ecology & Evolution* **22** 65-70

[23] Paterson J S, Araújo M B, Berry P M, Piper J M and Rounsevell M 2008 Mitigation, adaptation, and the threat to biodiversity Conservation Biology **22** 1352-1355

[24] Smith P, Martino D, Cai Z, Gwary D, Janzen H H, Kumar P, McCarl B, Ogle S, O'Mara F, Rice C, Scholes R J, Sirotenko O, Howden M, McAllister T, Pan G, Romanenkov V, Schneider U, Towprayoon S, Wattenbach M and Smith J U 2008 Greenhouse gas mitigation in agriculture *Philosophical Transactions of the Royal Society* **B 363** 789-813

[25] IEA 2008 *Energy Technology Perspectives 2008 - Scenarios and Strategies to 2050* International Energy Agency

[26] Howden S M, Soussana J F, Tubiello F N, Chhetri N, Dunlop M and Meinke H 2007 Adapting agriculture to climate change *Proceedings of the National Academy of Science* **104** 19691-19696

[27] Cassman K G, Dobermann A, Walters D T and Yang H 2003 Meeting cereal demand while protecting natural resources and improving environmental quality *Annual Review of Environment and Resources* **28** 315-358

[28] Tubiello F N, Soussana J F and Howden S M 2007 Crop and pasture response to climate change *Proceedings of the National Academy of Science* **104** 19686-19690

[29] Mertz O, Halsnæs K, Olesen J E and Rasmussen K 2009 Adaptation to climate change in developing countries *Environmental Management* **43** 743-752

[30] FAO 2006 World agriculture: towards 2030/2050 *Global Perspective Studies Unit* FAO Rome

[31] Searchinger T, Heimlich R, Houghton R A, Dong E A, Fabiosa J, Tokgoz S, Hayes D and Tun-Hsiang Y 2008 Use of U.S. Croplands for Biofuels Increases Greenhouse Gases Through Emissions from Land-Use Change *Science* **319** 1238-1242

[32] Wals A E J (eds.), 2007 *Social Learning - towards a sustainable world* Wageningen Wageningen Academic Publichers

[33] Reid A, Nikel J, Jensen B B and Simovska V 2008 *Participation and Learning - Perspectives on Education and the Environment Health and Sustainability* Springer

[34] Carlsson M and Læssøe J (forthcoming 2009) Learning and participation in developmental projects directed towards sustainable development in conference centres *Stevenson, Robert and Justin Dillon: Environmental Education: Learning, Culture and Agency* SensePublishers

[35] Baker E C 1998 The media that citizens need *U Penn Law Review* **147**

[36] Carvalho A 2007 Ideological cultures and media discourses on scientific knowledge *Public Understanding of Science* **16** 223-243

[37] ACIA 2004 *Impact of an Arctic Warming*. Arctic Climate Impact Assessment, Cambridge University Press.

[38] AMAP 2009 *Update on Selected Climate Issues of Concern*. The Arctic Monitoring and Assessment Programme, Oslo

[39] The National Snow and Ice Data Center, University of Colorado Boulder, USA; ftp://sidads.colorado.edu/NOAA/DATASETS/G02135

Beyond Kyoto: Addressing the Challenges of Climate Change IOP Publishing
IOP Conf. Series: Earth and Environmental Science **8** (2009) 012001 doi:10.1088/1755-1315/8/1/012001

Balancing the Risks: Choosing Climate Alternatives

C R Payne

Lecturer, University of California, Berkeley, School of Law
Director, Global Commons Project
Associate Director, Center for Law, Energy and the Environment
University of California, Berkeley, CA, 94720, USA

E-mail: cpayne@law.berkeley.edu

Abstract. Very aggressive reductions in greenhouse gas emissions are needed over the next ten years to avoid a "planet on fire." Current sub-national, national and international policy assumes that carbon sequestration, biofuels, nuclear power, ocean fertilization, atmospheric aerosols, and other such technologies, which heretofore have been considered too novel or too dangerous to use, will have to be deployed at large scale, globally. Moving forward with promising technologies that might preserve us from the consequences of global warming will be difficult because they also pose potential hazards, promise uncertain benefits, and in some cases are already burdened with restrictive legislation and poor public image. The lack of a rational process of risk assessment and public decision making is likely to lead to a poor long-term outcome. Moreover, the standard administrative and political processes used to assess such risks can take years, time that we do not have. Principled and practical policymaking demand citizens participate in the decision to develop and use these novel technologies. Environmental assessment, horizon scanning, and new research on human and organizational factors suggest techniques to improve technology development decisions.

1. Defining the problem

"All scientific work is liable to be upset or modified by advancing knowledge. That does not confer upon us a freedom to ignore the knowledge we already have, or to postpone the action that it appears to demand at a given time" [1].

Very aggressive reductions in greenhouse gas emissions are needed over the next ten years to avoid a "planet on fire." A great challenge for policymakers today is to introduce low carbon technologies on a fast-track without sacrificing environmental and human health or equity. Democratic, environmental and human rights principles require processes that weigh a range of values and expert opinion.

It is now believed that global average temperature increase must be held to no more than 2°C to 2.5°C above pre-industrial levels to avoid widespread, intolerable impacts on human well-being [2]. The ultimate objective of the framework climate change treaty is "stabilization of greenhouse gas concentrations in the atmosphere at a level that would prevent dangerous anthropogenic interference with the climate system … within a time frame sufficient to allow ecosystems to adapt naturally to climate change, to ensure that food production is not threatened and to enable economic development to proceed in a sustainable manner" [3]. Climate change goals already adopted by some governments are stringent. For example, California has committed to reduce greenhouse gas emissions by 80% by 2050, and replace 10% of liquid transportation fuels with low-carbon fuels by 2020 [4].

Developed and developing countries alike will have to adopt radical solutions to achieve these emissions reductions. Current sub-national, national and international policy assumes that carbon sequestration, nuclear power, ocean fertilization, atmospheric aerosols, and other such technologies, which heretofore have been considered too novel or too dangerous to use, will have to be deployed at large scale, globally [5]. Yet it will be difficult to move forward with promising technologies that might preserve us from the consequences of global warming because they also pose potential hazards, promise uncertain benefits, and in some cases are already burdened with restrictive legislation and poor public image.

Elected officials, private investors, academic research scientists, nongovernmental organizations and government regulators will drive the choice of which technologies are pursued. They wield the instruments of innovation and dissemination: tax credits, grants, venture capital, intellect, public support, and regulatory carrots and sticks. They will research the effects and effectiveness of each new technology. They will weigh its economic, ecological, social and cultural value. The suggestions here are addressed to them, with the goal of improving technology policy choices by suggesting different approaches to consultation that will bring in new information and new actors.

These policy professionals and scientific experts should not and cannot act independently of the public. International law and domestic law have established the importance of public participation, particularly in matters of environmental risk [6]. By ratifying the Aarhus Convention, 42 nations committed to provide public notice and opportunity for effective participation in decisions regarding proposed activities that may have a significant effect on the environment [7]. However, the Aarhus Convention does not dictate who must be consulted, nor does it detail the modes of consultation that are most appropriate for a given context [8].

The past supplies the experience of DDT, a marvelous new pesticide whose harmful effects were long neglected until better communication brought about a more balanced policy approach to its use. Section 2 of this paper describes this example of the interplay between policymakers, scientists and the public. In the short history of biofuels and carbon sequestration, two of the newest and highly controversial technologies targeted at global warming provide examples of the risks of rapid commercialization. This discussion, in Section 3, describes the limitations of science and regulation when even the nature of possible harmful effects of these new technologies is uncertain. Risk assessment and cost-benefit analysis are the standard policy tools for this kind of problem; section 4 critiques excessive reliance either on narrow expertise or on the will of the general public. Rather than try to compensate by shifting authority from one set of actors to another, in section 5, this paper identifies the following strategies to improve access to information:

- Environmental assessment,[1] made more accessible
- Horizon scanning.
- Role of Human and Organizational Factors

Each of these approaches takes advantage of new research and modern information technology.

2. The example of DDT

DDT (dichloro-diphenyl-trichloroethane) offers the classic demonstration of a technology solution with traumatic consequences. A new pesticide proved highly successful at addressing serious agricultural and public health problems. Public policy decisions to disseminate the technology did not fully take account of relevant, available scientific information about harmful effects of wide-scale use. After many years of global use, the human, ecological and financial costs of using the pesticide were

[1] In this paper the term "environmental assessment" will be used, recognizing that many different terms are used and that some have special meaning. For example, the regulations implementing the environmental review requirements in the United States use environmental assessment to refer to a preliminary review that determines whether a full analysis, called an environmental impact statement, is required.

finally acknowledged. Stakeholders who affected the policy processes included industry, government and academic scientists, journalists, nongovernmental organizations, and the general public. The role of information – what information was available, to whom, when, and with what credibility – was key to the decision making process.

During World War II, DDT was discovered to be extremely effective in controlling insects that are vectors for malaria, typhus and other diseases. Further study and use of DDT showed that it was low cost, less toxic to humans than the alternatives available at that time, and that it had the added virtue of being highly persistent in the environment so that fewer applications were required [9].

After many years of widespread DDT use, Rachel Carson's 1962 book, *Silent Spring*, exposed the devastating effects of the pesticide on bird and fish reproduction to the general public [10]. Her book was based on a well-developed but obscure body of research and observation, the work of ornithologists, foresters and naturalists. This, plus additional evidence indicating that DDT was a potential carcinogen and a possible threat to human reproduction, a dramatic fish kill attributed to pesticides, and a lawsuit by Environmental Defense Fund succeeded in getting DDT taken off the market in the United States in 1972 [11]. Just one measure of the persistent, deadly and expensive legacy of DDT for society and the environment is the lengthy litigation over responsibility for cleaning up toxic chemicals from the seabed caused by the Montrose Chemical Company's disposal of DDT-contaminated waste into the Pacific Ocean (the old saying used to go "the solution to pollution is dilution") [12]. The defendants agreed to pay $140.2 million to settle the lawsuit [13].

As Kehoe and Jacobson recount, DDT was not deployed without recognition of its possible downside. During and after World War II, the government did extensive testing on DDT safety. Government scientists and the popular press warned that DDT's wide-spectrum toxicity and persistence could be "a two-edged sword" and might "upset the whole balance of nature" [9].

Despite these concerns, US production of DDT expanded from less than 200,000 pounds in 1943 to nearly 130 million pounds in twelve months of 1955-56 [9]. During this time, government regulators were alert to human health impacts, but they failed to take account of harm to wildlife or developing resistant strains of mosquitoes, because those were issues outside their regulatory mandate.

A noteworthy point in the DDT story is that there was sufficient information available to warn decision makers long before *Silent Spring*. In Cambridge, England, a pesticide manufacturer voluntarily treated its DDT-contaminated wastes before releasing them into the river that flows by one of the world's most prestigious universities, Cambridge University. "Disposal of untreated pesticide wastes into the [River] Cam would have represented a highly visible threat to a natural resource amenity greatly valued by influential members of British society," as Kehoe and Jacobson observe [9].

The story could end with the recognition that DDT's benefits were offset by unacceptable harms and stand as a warning against incautious deployment of new technology. However, the value of DDT as a tool to fight malaria, a disease that kills over 1 million people annually mainly in Africa, has never been equaled by safer alternatives. As a result, it is still used in over 20 countries for that purpose, using indoor spraying only [14]. The intensity of the controversy over this is highlighted by the fact that DDT is now one of the "dirty dozen" synthetic and highly toxic chemicals regulated internationally under the Stockholm Convention on Persistent Organic Pollutants (POPs Convention); it can only be used to control insects which transmit diseases like malaria and in the production of another pesticide [15].

In retrospect, it appears that DDT use could have been targeted to the applications that would yield the greatest benefits and that the harms of DDT use could have been limited by measures like those taken in Cambridge, England. Why did this not happen? From this account, better communication between experts from the fields of public health, wildlife biology, local and national government and the public (who valued songbirds over DDT) could have led to better use practices for DDT.

3. Controversies over two new technologies: biofuels and carbon sequestration

3.1. An agriculture-based energy supply: biofuels

Energy from plants has promised to solve pollution, energy security and energy poverty problems with a single technology. Biofuels are commonly made from crops like sugarcane, sugar beets, corn, wheat, oil palm, rapeseed, or soy. Because carbon dioxide emitted by biofuels is theoretically drawn from the air when the feedstock plant was growing, biofuels have the potential to provide liquid transportation fuels that have no or low net greenhouse gas emissions. However, lifecycle studies of biofuels production have shown that there are significant direct emissions of greenhouse gas from the feedstock production and fuel manufacture [16]. Subsequent studies indicate that indirect greenhouse gas emissions from land clearing, in response to the increased agricultural demand from energy crops, may nearly double greenhouse emissions over 30 years [17].

This discovery was an unfortunate surprise for policymakers who had been strenuously promoting biofuels [18]. In addition to the realization that biofuels might not actually reduce greenhouse gas emissions, concerns range from food scarcity due to competition of staple crops with energy crops; increased stress on sensitive wetlands, forests and scarce arable land; competition for water to irrigate energy crops; and impacts on biodiversity to labor and human rights issues, resulting from a push to supply the world's transportation fuel [19].

Numerous laws and other incentives for biofuels production have been put into effect. The European Union took early steps in this direction, setting national indicative targets for member States [20]. California is adopting a the Low Carbon Fuel Standard that will require 10% reduction in carbon intensity of liquid transportation fuels sold in California by 2020, creating a demand for low carbon fuels [21]. The proponents of the regulation assumed that biofuels would be an important component of the low carbon fuel supply [22]. What is surprising is that the poor result of the life cycle analysis for corn ethanol was predicted but was overtaken by politics that supported the corn industry. Information about the foreseeable ecological and social problems of biofuels was neglected while significant private and public investments were made and government policies were put in place.

3.2. "Clean coal"

Carbon sequestration is a technology or natural process that removes carbon dioxide from the atmosphere. In natural processes such as plant growth, plants metabolize and retain carbon until the dead plant decomposes and re-releases the carbon to the atmosphere. The field of artificial carbon sequestration includes research on seeding the ocean with iron to stimulate marine organisms to take up carbon dioxide from the atmosphere and geological sequestration, where carbon dioxide is injected into geological formations underground. This is commonly done with small volumes of carbon dioxide as a technique to manage oil and gas extraction, but the experience of sequestering large amounts of carbon dioxide underground in order to permanently remove it from the atmosphere has a limited history in Norway and a few other pilot projects elsewhere [23]. It is considered a useful technology to capture some of the carbon dioxide that is released when coal is used as an energy source. China, Germany, Poland and the United States are among the countries for which coal is an important component of their energy supply. Since coal remains a widely available and relatively inexpensive energy source, carbon capture and sequestration (CCS) is potentially a significant climate change technology [24].

Although US President Obama supports it, there is strong public opposition to CCS from some communities. Some believe that coal-dependent nations like the United States and Germany would make a faster transition from coal if CCS were not available. There are those in the environmental justice movement who are working to unburden communities of color from bearing a disproportionate share of dirty industries, with coal-fired power generation being a significant contributor to health problems. For both these groups, CCS solves only one of the urgent pollution problems associated with coal burning, while leaving coal-based energy as a threat to public and environmental health.

Another set of risks associated with CCS is insufficient information about the extent, type and nature of harm that could result from wide-scale deployment of sequestration technology. Possible risks include contamination of drinking water; accidental leakage of carbon dioxide; and seismic instability [25]. Elizabeth Wilson and colleagues suggest that the actual threat of harm is likely negligible:

"Once injected, evidence from natural CO2 reservoirs and from numerical models suggests that CO2 can, in principle, be confined in geological reservoirs for time scales well in excess of 1000 yr and that the risks of geological storage can be small" [25].

On the other hand, large-scale failure of CCS projects, in terms of leakage, will be more damaging from a climate change perspective than not attempting it at all, for the following reason. Capturing and sequestering carbon requires the use of energy – with the concomitant GHG release – over and above that required by power generation or other purpose. This is what is called the "energy penalty" of CCS. It means that, if sequestration fails, *more* carbon dioxide will be released to the atmosphere than would have been produced by the initial activity alone.

The US public has had little opportunity to contribute to formal decision making about whether and how to proceed with CCS research, pilot testing and full-scale deployment. So far, a similar lack of information and participation in decision making about CCS seems to exist in the European Union [26].

4. Difficulty deciding: the experts and the people

The lack of a rational process of risk assessment and public decision making for selecting and deploying low-carbon technologies is likely to lead to a poor long-term outcome. We may fail to adopt useful technologies and suffer unnecessary GHG emissions or we may be careless and adopt a technology with severe, unforeseen side effects. The interplay between experts, appointed or elected official decision makers, and the public on whose behalf the decision is being made is critical.

Although the general topic is how we choose climate change technologies, the question must be re-framed as "how do we choose the risks that we are willing to accept?" For example, how do we weigh the possibility that carbon sequestration may contaminate groundwater with the possibility that a failure to use the technology might contribute to global warming impacts like sea-level rise of over one meter? For decisions without the vast uncertainties of these alternatives, methods like cost-benefit analysis, ecological risk assessment and environmental assessment are used to identify and compare relevant information. Cost-benefit analysis generally entails identifying and comparing the pros and cons of an activity; a more formal approach requires quantifying and assigning a monetary value to the pros and cons. Required by several laws in the United States, cost-benefit analysis may not include any public consultation. Environmental assessment identifies the environmental impact of a proposed action, positive and negative, and generally requires a comparison with alternative approaches which should include taking no action at all and public notice and comment [27].

Decision making distortions affect experts and non-experts alike, as has been amply demonstrated by social scientists. Applying their insights to climate and technology risk, we may discover that we perceive the risks of climate change quite differently than we perceive technology risks. This in turn may incline us to recoil from one alternative and accept another, and as we will see, this might lead to poor outcomes. Clearer understanding of the factors that can distort interpretation or lead us to overlook important data will not guarantee a perfect result, but it will improve the quality of analysis and choice in these highly uncertain and complex environments.

Traditional risk assessment and risk management is based on the notion of experts as the chief source of guidance for decision makers. Risk assessment is considered the more neutral, fact-based task; risk management is understood to be a political decision about governance [28]. One prominent academic proponent of technocratic decision making argues that the public "think poorly about dangers" [29]. In his view, the general public's flawed perception leads it to pressure government officials in favor of particular policies; government officials, responsive to the electorate, then proceed to ignore real risks and to waste resources on other, less important ones.

In classic comparisons of public risk calculation it is observed that people are willing to drive automobiles but refuse to live near a nuclear power plant, although the respective mortality statistics suggest that the car is more dangerous [29]. Like Mr. Spock of the Starship Enterprise, nuclear engineers are prone to say "this is not logical," relying on a calculated probability of a nuclear accident for their confidence [30]. The populist weighting of risk derives from several influences. Salient, vivid examples of negative outcomes stick in people's minds and influence their belief that it is more frequent or dangerous than it is. Acts of commission are viewed more negatively than omission. Some outcomes, like radiation poisoning, seem more horrible and are more feared than others. And people are apparently more willing to accept risks that are voluntary and over which they have control, like choosing to drive, than those over which they have no control, like the malfunction of a nuclear power plant. Other factors include fairness, procedural inclusiveness, morality, familiarity, and remoteness in time [31]. Peter Sandman usefully summarizes the factors at issue as "hazard" and "outrage." He says:

> "Call the death rate (what the experts mean by risk) 'hazard.' Call all the other factors, collectively, 'outrage.' Risk, then, is the sum of hazard and outrage. The public pays too little attention to hazard; the experts pay absolutely no attention to outrage. Not surprisingly, they rank risks differently" [32].

Distinguishing between experiential and analytic approaches to risk perception and management adds nuance. The tendency in the literature has been to associate the general public with experiential decision making, which is described as emotional. Decision making by experts is often considered more analytic, and thus more rational than the experiential approach. Deference to expert opinion is common, and the analytic approach is perceived to be more reliable. Examples are the expert view that nuclear power and genetic engineering have social benefits that far outweigh harm to public and environmental health, in contrast to public anxiety about the suffering that would result from radiation leaks and the unknown effects of genetically modified food - "Frankenfoods." Winickoff et al. argue that "Social science and regulatory experience instead emphasize that value judgments and public participation play an important role in generating reliable and conclusive risk assessments, especially in new and contested risk situations" like genetically modified foods [28].

While experts may be less subject to distortions in their decision making from heuristics that use stored personal experience (representativeness and availability heuristics) or the greater weight that non-experts give to downside risks, they may be prone to overconfidence in their judgments [33]. On the one hand, they have strong motivations to appear confident and accurate to themselves and to colleagues [34], and even experts have constraints on their attention that lead to confirmation bias and vulnerability to the proverbial, unanticipated, "flying cow" – the problem that it is difficult to know what we do not know [35]. As Dimitrov remarks, scientific information is "a product of social processes among scientists and other social actors" [36].

On the other hand, Weber observes that climate scientists give significantly greater weight to the dangers of global warming than the public does, because the scientists not only give greater weight to statistics and models of climate effects, but they have more personal experience observing effects like melting glacial ice and submerged islands [37].

A way out of these psychological traps is suggested by "multiattribute utility theory," which emphasizes communication and trust. Salient characteristics are the inclusion of multiple goals, minimization of goal conflicts, and recognizing that communication is binding because repeated interactions build or destroy trust. Operationalizing these involves framing situations in terms of social goals triggering the powerful need for social affiliation and social approval. Weber recommends employing scenario analysis to put experiential and affective processing and aversion to uncertainty to good use in planning for uncertain climate change outcomes. Weber also recommends combining analytic and experiential processes to improve interpretation of climate model results and to motivate responsive actions. Weber further advises influencing actions and choices with strategic framing of information to prime cooperation.

Finally, in situations of high complexity or novelty, there simply is not enough information for scientists to be considered fully informed experts. At the cutting edge of research, "[h]euristics (rules-of-thumb) replace direct observation and synthesis in guiding the formation of scientific consensus and introduce the possibility of multiple outcomes" [38].

Even those who assume that objectively sound decisions will result from the expert's analytic approach, public participation in risk management decisions is often recommended and frequently legally mandated. In some instances, the public may actually be expert – it may have information and informed judgments to contribute. The "public" may include all citizens, or the identity of participants might vary according to context. A core justification for public participation in policy is the fundamental understanding of democracy [39]. Even Sunstein sees a role for informed public deliberation, despite his suspicion of the "intuitions of interest groups and populism" [29].

This discussion leads us perforce to conclude that neither experts nor the public can provide a "correct" answer to complicated technology choice questions. Working together, pooling the different kinds of information that experts (whether from the academy, industry or government) and the general population possess, all of the potentially relevant information can be made available. Understanding the biases and distortions that can be traps for expert and popular judgment, rational but responsible opinions can be formed. There are many processes that can facilitate cooperative investigation and decision making. Techniques for collecting relevant information, bringing it to the attention of appropriate decision makers, and avoiding problems like information overload are constantly being sought by government officials, business and academic researchers. Some of these approaches are discussed in the following sections.

5. Information strategies

Missing a key element of a problem is a recurrent issue that is likely to arise with new energy, geoengineering and other climate change-related technologies. Difficult trade-offs may need to be made. As we saw in the case of DDT, policymakers focused on DDT as a low-cost, high performance solution to public health and agricultural problems while failing to evaluate DDT as an environmental health risk itself. Similarly, by focusing on the benefits of biofuels for energy security and for the agricultural sector, the European Union, the US, and the State of California overlooked a suite of potential costs. In swift succession, the international media, advocacy groups, and academic researchers published reports on the greenhouse gas performance and food security impacts of large-scale biofuel development. As a result, the prospect of avoiding something like the DDT catastrophe was vastly improved. The biofuel story suggests that modern communications, the engagement of civil society and unbiased contributions of the research community can produce better public policy on technology development and deployment. This section identifies three strategies that bring a range of perspectives and disciplines to bear, encourage repeated re-assessment, and use a systematic method to ensure that these steps are taken.

5.1. Environmental Assessment

Environmental assessments could make a significant contribution to the decision making process regarding potentially dangerous energy, albedo and carbon sink technologies by providing both varied sources of information and public voice. An environmental assessment identifies the ecological, social and economic consequences of a proposed activity, plan or program. The public may be given the opportunity to review and comment on the assessment at various stages of the process. Where environmental assessment is a legal requirement, the environmental assessment must be taken into account in determining whether the proposed activity may go forward. Environmental assessment is unusual because it is one of the few formal processes that brings very diverse opinions and expertise – from environmental professionals, advocates, self-selected citizens and government agencies – to bear on proposed activities. An environmental impact statement for a large-scale project can be a truly impressive compilation of data, politics and analysis [39].

Environmental assessment is by now a well-accepted, familiar approach [40] and indeed it is mandated in many domestic legal systems and international treaties [27][41]. Both customary international law [42] and the UN Framework Convention on Climate Change (UNFCCC) require environmental assessment. The UNFCCC states that parties shall

> "Take climate change considerations into account, to the extent feasible, in their relevant social, economic and environmental policies and actions and employ appropriate methods, for example impact assessments…with a view to minimizing adverse effects on the economy, on public health and on the quality of the environment, of projects or measures undertaken by them to mitigate or adapt to climate change" [3][42].

Nonetheless, the practice of environmental assessment varies widely across jurisdictions [43]. The threshold of environmental impact that triggers assessment, the type of projects covered, the extent of public access to information and public participation, the extent of true consideration of environmental assessments before decision is made and the existence of mandatory mitigation of harmful impacts are sometimes adequate and in other cases insufficient to provide effective assessment or true environmental protection. A hurdle is frequently political will. Every country that mandates environmental assessment experiences conflict over reluctance of authorities and project proponents to provide full, timely disclosure of information and effective consultation with other stakeholders.

Research projects and larger government research and development programs are not always subject to environmental assessment, even though the early stages of technological development are precisely when important investment decisions are taken. For example, the Aarhus Convention states that certain research, development and testing activities are exempt "unless they would be likely to cause a significant adverse effect on environment or health;" this exemption might be interpreted quite broadly [7].

Although in the US the National Environmental Policy Act (NEPA) does require environmental review for any federal action that significantly affects the environment, including research activities, a recent decision reversed course on assessing the overall research and development program for carbon sequestration technologies and practices. In 2004, the US Department of Energy (DOE) announced its intention to carry out a programmatic environmental impact assessment of the federal Carbon Sequestration Program, which at the time included 80 separate projects,

> "The Proposed Action is for DOE to continue implementation of its Carbon Sequestration Program with a focus on moving toward GCCI [Global Climate Change Initiative] goals and to eventually help meet the requirements of the Framework Convention on Climate Change. To achieve these objectives, the Program needs to consider, evaluate, develop, and implement carbon capture and carbon storage technologies, including effective measurement, monitoring, and verification methods, over a longer-range planning horizon. The Program also needs to provide technological viability data for the GCCI 2012 technology assessment" [44].

The Department of Energy proposed to "evaluate the issues and impacts associated with the demonstration and deployment of technologies to implement the key elements of the Program: carbon dioxide capture; sequestration (geologic, oceanic, and terrestrial); [Measurement, Monitoring, and Verification]; and breakthrough concepts" [44]. The alternatives that would have been examined included alternative technologies or variations in the mix of technologies, variations in the implementation of sequestration methods and variations in implementation by geographic region, as well as a "no action" alternative that would have been a more limited continuation of existing research. The assessment would have looked at impacts from the processing, land use effects of running pipelines, health and safety issues from instantaneous or slow leakage of carbon dioxide, socioeconomic and cultural impacts, effects on water resources and ocean chemistry and potential hydrologic fractures due to carbon dioxide injection that could harm aquifers or cause seismic events. As part of the scoping phase, the public was invited to attend explanatory meetings held in eight different sites around the country, and to comment on the issues to be addressed in the environmental assessment.

In 2007 the US DOE decided to cancel the programmatic environmental impact statement in favor of a project by project approach [45]. There are no doubt good reasons to prefer individual project assessments, but those would have been done in any case under the previous plan. The opportunity to identify a broader range of issues, to compare across geographic sites, and to scan for cumulative effects was lost by choosing to evaluate projects individually. There is even the possibility that smaller projects might not rise to the level of significance and would be overlooked entirely. The US biofuels research program has similarly devolved responsibility for environmental review to the facility or project level. Consequently, cumulative impacts have been assessed by researchers but not through the NEPA process.

A systematic, publicly accessible, flexible, low-cost archive of all environmental assessments, linked to maps would address the project by project approach along with other problems that erode the utility of environmental assessment. Though environmental assessments are rich sources of information, they are generally lost from history. One proposal is to create an internet-based, geographic information system (GIS)-linked repository of all environmental assessments [46]. The first step would preserve the value of millions of dollars invested each year in data, analysis and public input by putting all environmental assessments on the internet, ideally in a digital, searchable format. With historical data readily available, it would be possible to go back and evaluate the efficacy of mitigation and the accuracy of projections made by project proponents and opponents. Linking assessments to geographic information would facilitate cumulative impacts analysis, allowing easy viewing of projects in the same watershed or airshed, or affecting vulnerable human populations or particular species of concern.

Using environmental assessment to support decisions about climate change technology can be helpful, but to be really effective there need to be several reforms. Environmental assessment should be done at the research and development stage, and it should cover research programs that, as a totality, may have a significant impact on the environment. If research moves to field testing, re-assessment is needed, and it is needed again if the technology is scaled up since effects of scale and deployment in varied environments are very likely to have impacts that were not revealed in the lab or in field testing. Carbon sequestration presents this issue: pumping carbon dioxide into geologic strata has been used in oil and gas fields for many years, but it is recognized that much larger scale use of the technology in many different geological formations may have different consequences. Environmental assessments should be made available electronically at no cost as described for researchers to draw on and to inform the public, political debate.

5.2. Horizon scanning

Eyes looking in many directions, reporting emerging risks, are the goal of the many practices that comprise horizon scanning. They are based on the assumption that strategic planning requires a multitude of perspectives. Horizon scanning is particularly intended to identify low probability-high-impact events. The city-state of Singapore became interested in horizon scanning when, focused on the threat of post-9/11 terrorism, it was unprepared for the SARS epidemic; in 2005, the Risk Assessment and Horizon Scanning program was created [47]. Great Britain, the Netherlands and Switzerland have become interested in applying the basic techniques of horizon scanning to improve strategic planning in diverse domains. Horizon scanning is not a single approach, but rather a grouping of practices drawn from many disciplines: information technology, modeling, artificial intelligence and social networking may be used to work with data and human informants [48].

5.2.1. Singapore. Singapore's Risk Assessment and Horizon Scanning (RAHS) program "aims to provide policymakers with anticipatory knowledge of the nature of potential upcoming issues so that risks may be minimized and opportunities maximized" [49]. The Singaporean approach initially focused on strategic shocks that could affect national security based on experience of the 1997 Asian financial shock, political crises and terrorism. When 15 years of experience in scenario planning left the government unprepared to deal with the public health risk posed by SARS, it started to experiment with additional methods to improve awareness of threats across a 360 degree horizon and sensitivity to faint warning signals. A hallmark of the Singapore program is its emphasis on non-linear thinking as a source of resilience.

RAHS has developed a "whole-of-government" approach. While it emphasizes cross-agency collaboration in government and network building across professional academic and business communities, RAHS also uses powerful computing tools to link participants, provide data and carry out analysis. Detecting emerging trends is aided by the technology and by consultation with other government analysts within a closed network. The elements of the RAHS system are:

- Data collection and information organization;
- Context and scenario development using models that are constructed to understand and represent situations with an emphasis on context and structure;
- Monitoring and detection that allows the matching of incoming data to previously identified patterns or contents of interest;
- Pattern discovery and evaluation tools allow for the automated processing of data through natural language processing engines, as well as the visualization of information in order to aid the serendipitous discovery of new patterns;
- Collaboration, in terms of data, models, monitors and methods [47].

Singapore has been aggressive in pursuing fresh approaches that can be added to the developing body of horizon scanning practice. The belief is that "[t]he resultant diversity will reduce the possibility of getting blindsided by reliance on one particular approach or by one way of thinking about the future."

Although Singapore has opened up RAHS for its universities to study financial markets, agricultural commodities, public service, ethnic conflict and other topics, the approach is so strongly influenced by national security concerns that it may have limited application to the problem of low-carbon technology planning. Also, its planning horizon is only two to five years, while the planning horizon for new climate change-related technology must be on the order of decades.

5.2.2. Great Britain. In Britain, the Chief Scientific Adviser, currently Professor John Beddington, leads the Foresight Programme and its Horizon Scanning Centre to tackle sustainable energy and other topics that require strategic planning and risk assessment in highly complex, novel contexts. In such situations existing knowledge is not a sufficient guide to good decisions. One of the goals of the Centre is "[t]o spot the implications of emerging science and technology and enable others to act on them" [50]. Centre projects include a best practices guide for horizon scanning and two major projects, the Sigma Scan which explores future issues and trends that will face British public policy in the next 50 years and a scan on Wider Implications of Science and Technology [51]. The best practices guide recommends horizon scanning as a tool that can be used to review a broad spectrum of information beyond usual sources that is suitable for short to medium time horizons. The biodiversity scan, discussed below, provides an indicative example.

5.2.3. Biodiversity scan. A group of British scientists used horizon scanning to examine future threats to biodiversity in the United Kingdom within the next forty years, seeking to identify issues that were not already part of the public policy agenda [52]. Their goal was to identify all issues of potential future importance, a first step in planning policy. They looked for weak or conflicting signals of issues that are just beginning to emerge but that might become threats or opportunities in the future. The insight that faint signals that are observed in multiple contexts may be significant is drawn from horizon scanning, so consultation was designed to identify issues that persisted across disciplines by consulting informants from the natural sciences, social sciences, journalism and the field of horizon scanning itself.

Their method was relatively simple: representatives of government departments and non-governmental organizations (NGO) were invited to join academics, journalists and horizon scanning experts to prioritize, score, comment on and suggest additions to a list of issues. The original list was compiled from past horizon scans and other relevant government and NGO reports. Each issue was characterized by its potential benefits and threats, research needs, the likelihood it would occur and the level of threat that it posed. The project director recommends repeating the scan at 3-year intervals.

The biodiversity horizon scan placed in the top 25 of 195 issues geoengineering, demand for biofuels and increase in coastal and offshore energy generation, all potential technology responses to climate change and the type of technology with which this article is concerned. Each was scored as high likelihood, high threat to biodiversity. The horizon scan report identifies needs for further risk assessment and research into environmental impacts of geoengineering; more analysis of biodiversity and land use for biofuels crops; and the use of cost-benefit analysis and life-cycle analysis for offshore and coastal renewal energy development.

The report on the biodiversity horizon scan suggests several useful points. With regard to the horizon scan itself, the issues identified were found to be highly dependent on the people involved, yet the "challenge is to identify people who can give a broad view of developments in other fields" [52]. The project found it helpful to address this by including a mix of disciplines and by using other horizon scans as source material. Participants repeatedly identified a need to revise risk assessment to better include Earth system and ecological factors. They also asked for better monitoring of the natural environment to provide better early warning signals of adverse impacts.

5.3. Risk and the role of Human and Organizational Factors

A University of California, Berkeley research program is developing new tools to integrate technology with human factors in the context of environmental disasters, using insights from their studies of infrastructure failure and high reliability organization design [53]. These contexts are highly demanding, because system failure can so easily translate into catastrophe. The interdisciplinary research project initially came together in response to the Hurricane Katrina disaster. The team reported a set of recommendations that address the role of human and organizational error, drawn from their analysis of what occurred in New Orleans but capable of being generalized to other circumstances [54]. Their approach is of particular interest because it addresses the expert's tendency to discount expertise in other domains, which Oppenheimer refers to as the "inhibitory effect of disciplinary walls" [55].

This research is seeking a systematic way of integrating qualitative and quantitative information to improve risk assessment and risk management in highly complex situations. Computer modeling and geographic information systems (GIS) are used to integrate information about individuals and social groupings with environmental data from sensors, field reports and remote sensing. The computer models that result will be able to identify broad trends [56].

This work has produced insights about valuable sources of information. Journalists and the electronic news media have an important role to play, as was also noted in the Biodiversity Horizon Scan described above. Although generally transparency is preferred, by providing access for anonymous informants to report, sensitive information can be collected. Drawing from the study of high reliability organizations, the group observed that by requiring incidents of failure to be reported—even if it had to be done anonymously—the reports could be scanned to identify patterns that could be used as early warnings of dangerous conditions.

Current work involves the adaptation of an assessment system that was designed for offshore oil platforms, commercial ships, oil and gas refineries, and pipeline systems to new problems, Quality Management Assessment System (QMAS), and System Risk Analysis System (SYRAS). Structures such as offshore oil platforms or levees are designed for factors like strength and capacity, but they may experience failures in operation or due to maintenance [57]. These failures are traceable to human error, with organizational roots. Reliability Assessment and Management for offshore structures builds on more than thirty years' experience with both qualitative and quantitative analyses to detect and address potential problems in advance, learn from failure, and intervene during incidents to return the system to safe function.

Interactive assessment and management during the build-up toward a dangerous situation has the most to contribute to the problem of managing climate change and technologies. Society has already committed to developing technologies, so assessment can no longer be considered prospective; instead, problems will need to be identified and addressed as they arise. The problems that are likely to arise with offshore structures and technology development differ significantly in the time frame during which problems arise: hours versus months or years. However, they share the characteristics of complexity, uncertainty and novelty.

The basic procedure for the qualitative analysis is to periodically run an assessment with a team of experts from relevant disciplines, including people working within the system and outsiders. They identify factors of concern; interview stakeholders; develop scenario analysis based on information about the system and interviews with key individuals which allow for anonymous discussion if desired; score performance of system components (operators, organizations, interfaces, etc.); and propose mitigation of factors of concern that have been identified. In a technology setting, the group of stakeholders might be expanded beyond a specific facility to the industry and the people it affects. For carbon sequestration, this could include communities residing near sequestration sites; project managers, owners and employees; local, state or provincial, and federal regulators; seismologists, chemists, public health experts and ecologists.

The quantitative assessment evaluates the system's probability of failure and provides a tool to measure the effects of mitigation. For infrastructure systems failure is an objective event; technology policy will not be susceptible to similarly concrete definition of failure. However, the analytical framework might shed some light on the dimensions of performance for low carbon technology. For example, the life cycle phases of design, construction, operation and maintenance are relevant to carbon sequestration facilities; they might be further divided into research and development, pilot testing and full-scale deployment of the technology and considered with an industry-wide scope. The criteria of serviceability (carbon is actually reduced), safety (no public or environmental health problems), durability and compatibility (expected costs) are translatable. So is the recognition that failures derive from both intrinsic and extrinsic causes. The first are "factors such as extreme environmental conditions and other similar inherent, natural, and professional uncertainties" while extrinsic causes are human and organizational factors.

A sensitive point in applying the QMAS/SYRAS system to the very different technology context is the mathematical link between the qualitative and quantitative assessments, which was calibrated with historical data about structural failure and success. An analogous calibration would be difficult to perform. This highlights a distinction between the complicated but perhaps knowable context of structural failures, the complex context of technology development, and the chaotic climate system. The reliance on historical patterns of system failure or success in the QMAS/SYRAS approach suggests that the system is more knowable, less uncertain than either projections of climate change or the development of new low carbon technologies.

Bea emphasizes that the most important factor in a QMAS/SYRAS review is the people carrying out the assessment. Good selection of the right people for the task is important, in terms of training, experience and motivation. Teams should include a variety of backgrounds and experience and should be trained in techniques to mobilize the necessary awareness, skills and knowledge when called for. Training and retraining to refresh skills are necessary for individuals and teams.

This system provides a number of tools and a systematic approach to analyzing complex interfaces between people, the environment and technology. Perhaps the computer models and calculations can be converted into a tool to analyze climate change technologies. In any case, the work that has been done to create and demonstrate QMAS/SYRAS underscores three points:

- the value of bringing a wide spectrum of expertise to bear on the assessment;
- the need to train the assessors in a common group process to improve their communication with each other; and
- the importance of iterative assessments and follow-up assessment of mitigation.

6. Conclusion

Successful adoption of innovative GHG emission-reduction solutions calls for innovation in policy analysis. This paper has suggested new risk assessment, participation and decision support tools that can be used to evaluate urgently needed technologies to manage the threat of dangerous climate change. Uncertainties about geo-engineering, carbon sequestration, biofuels, and other technologies need to be reduced where possible, accounted for through democratic processes, and reevaluated over time. The great epidemiologist, Sir Austin Bradford Hill, pointed out that although we can act only on the information we have, however imperfect, we have an obligation to make timely decisions in the light of the information we *do* have [1]. In recent years, environmental assessment, horizon scanning and human and organizational factors research have evolved to help identify the information that we have, to account for the uncertainties that remain, and to support decisionmaking by bringing information and analysis to the attention of policymakers.

Multiple perspectives are essential to provide information from relevant disciplines and to represent the gamut of stakeholders likely to be affected. While environmental assessment does this well by systematically evaluating the proposed activity from different perspectives, government agencies responsible for assessment should make better use of the internet to preserve studies and analyses that have been done and to communicate with stakeholders during the assessment process. Horizon scanning programs have been very successful in taking advantage of social networking and other information technologies to collect and integrate knowledge and opinions from research, government, and civic communities. Singapore's continuous innovation and its use of powerful computing are noteworthy in this regard. Projects like the British biodiversity scan show that it can be scaled up or down in terms of cost, complexity and number of participants. When horizon scanning is integrated into government practice, it can be turned into a more or less continuous process. The British approach, looking out forty years in some cases, takes a long-term view well-suited to the needs of climate policy and acknowledges the need for repeating horizon scans at frequent, regular intervals to monitor significant change.

As observations throughout research, pilot testing and deployment phases reveal new information about innovative energy, carbon sink and albedo technologies, and as the effects of climate change and other features of the planet alter over time, such regular re-assessment can be provided by repeated environmental assessments or horizon scans. The QMAS/SYRAS process, if suitably adapted, adds techniques to investigate and learn from failures. Here, the recognition that failures derive from both intrinsic and extrinsic causes calls attention to the changing external environment on the one hand and the human process of technology development and deployment on the other.

Timeliness is a key factor: risk assessment must be undertaken early in the technology development process, as decisions about allocation of resources and regulation should be taken in light of the best projections of benefits to be obtained and downside risks to be managed. In the case of DDT, it was only after years of widespread use that attention was paid to the environmental health impacts of the pesticide. The high level of attention paid by the international research, policy and media communities to the development of biofuels as a low-carbon liquid fuel enabled fairly early review and awareness of the need for correction in public policy. An example of the kind of alternatives review that would be helpful in the early stage of development is the original programmatic environmental impact assessment of the federal Carbon Sequestration Program. It would have provided exactly the kind of information that is needed to support the difficult choices of where to invest and what regulation will be necessary, including the important evaluation of cumulative effects of all the projects in the program.

Communication about risk with the public is a necessary (though not sufficient) step to gaining public acceptance. Jeffrey Sachs of Columbia University's Earth Institute identifies public acceptance and carefully crafted regulatory controls as two factors so critical to the success of carbon capture and sequestration technology that a failure on one or the other "could kill the technologies" [58]. Formal environmental assessment requires this kind of public participation, inviting the public to review and comment on the assessment and, in principle, taking the public comment into account in the final decision. Vigorous use of these strategies is necessary now.

Acknowledgments

This project was supported, in part, by the National Science Foundation (NSF) under EFGRI Grant No. 0836047. Any opinions, findings, and conclusions or recommendations expressed in this material are those of the author and do not necessarily reflect the views of NSF. I would like to express my appreciation to Robert Bea, Daniel Farber, and an anonymous reviewer for their helpful comments, and to the organizers of the "Beyond Kyoto" conference.

References

[1] Phillips C V and Goodman K J 2004 The missed lessons of Sir Austin Bradford Hill *Epidemiologic Perspectives & Innovations* **1** 3 (quoting Hill AB 1965 The environment and disease: association or causation? *Proc R Soc Med* **58** 295-300.)

[2] Raven P H, Bierbaum R M, Holdren J P, MacCracken M C and Moss R H 2007 *Confronting Climate Change: Avoiding the Unmanageable and Managing the Unavoidable* (Research Triangle Park, NC: Sigma Xi: and Washington DC: United Nations Foundation) xi.

[3] *United Nations Framework Convention on Climate Change*, Article 2 (1992).

[4] *Cal. Exec. Order No. S-3-05* 2005 (establishing 2010, 2020 and 2050 targets for California to reduce greenhouse gas emissions);
California Global Warming Solutions Act of 2006 (AB 32) Cal. Health & Safety Code §§ 38,500–38,599 (committing California to reduce greenhouse gas emissions to 1990 levels by 2020);
Cal. Exec. Order No. S-01-07 2007 (establishing the California Low Carbon Fuel Standard).
Cal. Air Resources Bd., *Climate Change Proposed Scoping Plan,* 2008.

[5] Holdren J P 2008 AAAS Presidential address: science and technology for sustainable well-being *Science* **319** 424, 429-31;
Crutzen P 2006 Albedo enhancement by stratospheric sulfur injections: a contribution to resolve a policy dilemma? *Climatic Change* **77(3-4)** 211-219 (Proposing to artificially enhance earth's albedo and thereby cooling climate by adding sunlight reflecting aerosol in the stratosphere);
Stafford N 12 July 2006 Should we flood the air with sulphur? *Nature* news@nature.com [viewed 29 June 2009] (describing negative responses to Crutzen's proposal).

[6] Bernasconi-Osterwalder N 2008 Transparency, participation and accountability in international economic dispute settlement *Sustainable Development in International and National Law: What Did the Brundtland Report Do To Legal Thinking and Legal Development, and Where Can We Go From Here?* eds Bugge H C and Voigt C (Groningen:Europa Law Publishing);
Payne CR 1998 Local regulation of natural resources: efficiency, effectiveness, and fairness of wetlands permitting in Massachusetts *Environmental Law* **28** 519.

[7] *Convention on Access to Information, Public Participation in Decision-making and Access to Justice in Environmental Matters*, 25 June 1998, reprinted in 38 *ILM* 517 (1999), http://www.unece.org/env/pp [viewed 29 June 2009].

[8] The guide to the convention advises on best practice: Stec S and Casey-Lefkowitz S 2000 *The Aarhus Convention: An Implementation Guide* (New York & Geneva : United Nations).

[9] Kehoe T and Jacobson C 2003 Environmental decision making and DDT production at Montrose Chemical Corporation of California *Enterprise & Society* **4(4)**.

[10] Carson R 1962 *Silent Spring* (Boston : Houghton Mifflin & Cambridge, MA : Riverside Press).

[11] *Federal Environmental Pesticides Control Act* (21 October 1972) (A US Environmental Protection Agency history of DDT further notes that by 1972 DDT use had decreased significantly due to increased insect resistance and availability of more effective alternatives as well as public concern over safety.);
EPA 1975 *DDT Regulatory History: A Brief Survey (to 1975)* (US EPA History Office).

[12] *Consent Decree with Montrose Chemical Corporation of California et al., United States and State of California* v. *Montrose Chemical Corporation,* No. CV 90-3122-R (C.D. Western

Div. Calif., March 15, 2001), http://www.darrp.noaa.gov/southwest/montrose/admin.html [viewed 1 July 2009].

[13] NOAA, General Counsel for Natural Resources, Office of Habitat Conservation, Office of Response and Restoration, Southwest Region, http://www.darrp.noaa.gov/southwest/montrose/index.html, [viewed 24 May 2009].

[14] World Health Organization 2006 Indoor Residual Spraying. Use of indoor residual spraying for scaling up global malaria control and elimination *WHO Position Statement Series 1112* (World Health Organization, Geneva); Jackson D C 2009 Malaria, DDT, and desperation in Uganda *Boston Globe* (Boston).

[15] *Stockholm Convention on Persistent Organic Pollutants* 22 May 2001 40 *I.L.M.* 532; *Provisional DDT Register pursuant to Paragraph 1, Part II of Annex B*, http://chm.pops.int/Programmes/DDT/DDTregister/tabid/456/language/en-US/Default.aspx [viewed 1 July 2009].

Division of Toxicology 2002 *Public Health Statement - DDT, DDE, and DDD* ed. US Agency for Toxic Substances and Disease Registry (Describing human and environmental health effects of DDT exposure).

[16] Farrell A E, Plevin R J, Turner B T, Jones A D, O'Hare M, Kammen D M 2006 Ethanol can contribute to energy and environmental goals *Science [Reports]* **311** 506-08.

[17] Searchinger T, Heimlich R, Houghton R A, Dong F, Elobeid A, Fabiosa J, et al. 2008 Use of U.S. croplands for biofuels increases greenhouse gases through emissions from land-use change *Science* **319** (5867) 1238-40.

Letter responses to Searchinger et. al.'s paper entitled Use of U.S. croplands for biofuels increases greenhouse gases through emissions from land-use change *Science Online* 2008, http://www.sciencemag.org/cgi/eletters/319/5867/1238 [viewed 1 July 2009].

[18] Commission of the European Communities 1 Oct 2007 *Report on the progress made in the use of biofuels and other renewable fuels in the Member States of the European Union* (Brussels) (Report acknowledges that indirect land use change could alter the otherwise favourable greenhouse gas emissions from biofuels).

[19] UN-Energy 2005 *The Energy Challenge for Achieving the Millennium Development Goals* (United Nations).

[20] Directive 2003/30/EC on the promotion of the use of biofuels or other renewable fuels for transport (OJ L 123, 17.5.2003, p. 42);

Directive 2003/96/EC restructuring the Community framework for the taxation of energy products and electricity (OJ L 283, 31.10.2003, p. 51).

[21] California Low Carbon Fuel Standard Regulation 2009 (California Air Resources Board Resolution 09-31).

[22] Payne C R 2009 Local meets global: the low carbon fuel standard and the WTO *N.C. J. Int'l L. & Comm. Reg.* **34(Spring)** 891-917.

[23] Intergovernmental Panel on Climate Change 2005 *Special Report on Carbon Capture and Sequestration*;

International Energy Agency 2007 *Near-Term Opportunities for Carbon Dioxide Capture and Storage* (Paris : International Energy Agency).

[24] Schiermeier Q 2008 Europe to capture carbon *Nature [News]* **451** 232.

[25] Philibert C, Ellis J, Podkanski J 2007 *Carbon Capture and Storage in the CDM* (Organisation for Economic Co-operation and Development, Paris : International Energy Agency).

Wilson E J, Johnson T L, et al. 2003 Regulating the ultimate sink: managing the risks of geologic CO2 storage *Environmental Science and Technology* 3476-3483.

World Resources Institute 2008 *CCS Guidelines: Guidelines for Carbon Dioxide Capture, Transport, and Storage* (Washington, DC).

[26] Anderson J, Chiavari J, De Coninck H, Shackley S, Sigurthorsson G, Flach T, et al. 2009 Results from the project 'Acceptance of CO2 capture and storage: economics, policy and technology (ACCSEPT)'. *Energy Procedia* **1** 4649-53.
Johnsson F, Reiner D, Itaoka K, Herzog H. 2009 Stakeholder attitudes on carbon capture and storage - an international comparison *Energy Procedia* **1** 4819-26.
[27] 42 U.S.C. § 4332(2)(c) *US National Environmental Policy Act of 1969*, Pub. L. No. 91-190, Sec. 83 Stat. 852 (1970) (codified as 42 U.S.C. §§ 4321-4347), [NEPA].
40 CFR 1500-1508, Council on Environmental Quality Regulations for Implementing the Procedural Requirements of NEPA, July 1, 1986.
U.S. Environmental Protection Agency 1998 *Guidelines for Ecological Risk Assessment*.
[28] Winickoff D, Jasanoff S, Busch L, Grove-White R and Wynne B 2005 Adjudicating the GM food wars: science, risk, and democracy in world trade law *Yale J. Int'l L.* **30** 99.
[29] Sunstein CR 2002 *Risk and Reason* (Cambridge).
[30] Kunreuther H and Slovic P 1996 The process of risk management: science, values and risk *The American Academy of Political and Social Science Quarterly* **545** 116.
[31] Kuran T, Sunstein CR 1999 Availability cascades and risk regulation *Stan. L. Rev.* **51**:683, 708-9.
[32] Sandman P M 1987 Risk communication: facing public outrage *EPA Journal* 21-22.
[33] Oppenheimer M, O'Neill BC, Webster M 2008 Negative learning *Climatic Change* **89**:155-72, 156.
[34] Latin, H 1988 Good science, bad regulations, and toxic risk assessment *Yale J. on Reg.* **5** 89, 129 ("the majority of interveners in regulatory proceedings are sponsored by affected industries or trade associations, which means the scientific performance of agency officials will regularly be monitored and challenged by industry scientists who advocate less conservative risk assessment practices. Agency bureaucrats, like other people, are sensitive to criticism and may deliberately or subconsciously seek to placate persistent critics.").
[35] Ossorio P N 2001 Pills, bills and research shills: physician-researchers' conflicts of interest *Widener L. Symp. J.* **8** 76, 88-91 (explaining the role of judgment in scientific research, the potential for unconscious bias, and illustrating the negative effect on expert judgment when professional reputation, financial gain, and other interests conflict).
[36] Dimitrov R S 2006 *Science & International Environmental Policy* (Lanham, MD : Rowman & Littlefield Publishers, Inc.).
[37] Weber EU 2006 Experience-based and description-based perceptions of long-term risk: Why global warming does not scare us (yet) *Climatic Change* **77**:103-20, 108.
[38] Atik J 1996-1997 Science and international regulatory convergence *NW. J. Int'l L. & Bus.* **17** 736, 748.
[39] Ebbesson J. 2007 Public Participation in *The Oxford Handbook of International Environmental Law* eds Bodansky D, Brunée J, Hey E. (Oxford: Oxford University Press) pp 682-703.
[40] Principle 17, Rio Declaration, 31 *ILM* 874 (1992); endorsed by UN General Assembly Res. 47/190 (1992) ("Environmental impact assessment, as a national instrument, shall be undertaken for proposed activities that are likely to have a significant adverse impact on the environment and are subject to a decision of a competent national authority").
[41] EC Directive 97/11/EC, 3 March 1997 amending Directive 85/337/EEC on the assessment of the effects of certain public and private projects on the environment OJ No. L 073, 14/03/1997 P. 0005.
Convention on Environmental Impact Assessment in a Transboundary Context, in force 10 September 1997, 30 ILM 802 (1991) (Espoo Convention).
1994 Energy Charter, Art. 19.
[42] Sands P 2007 *Principles of International Environmental Law, 2d ed.*

Bastmeijer K, Koivurova T. 2000 Introduction *Theory and Practice of Transboundary Environmental Impact Assessment* eds Bastmeijer K, Koivurova T, (Ithaca: Cornell University Press) pp 1-25, 3-8.

[43] Dalal-Clayton B, Sadler B 2005 *Strategic Environmental Assessment: A Sourcebook and Reference Guide to International Experience* (Earthscan).

[44] *Notice of Intent to Prepare a Programmatic Environmental Impact Statement for Implementation of the Carbon Sequestration Program* 2004 69 Fed. Reg. 21514.

[45] *Notice of Cancellation of Programmatic Environmental Impact Statement for Implementation of the Carbon Sequestration Program* 2007 72 Fed. Reg. 8363.

[46] Farber D. 2007 Bringing Environmental Assessment into the Digital Age. *Taking stock of environmental assessment : law, policy and practice* ed Holder J, McGillivray D (Routledge-Cavendish).

[47] Tan Hong Ngoh, Edna and Hoo Tiang Boon eds 2008 *Thinking about the Future: Strategic Anticipation and RAHS* (Singapore: National Security Coordination Secretariat, S. Rajaratnam School of International Studies).

[48] Helft M 2008 Google uses searches to track flu's spread *New York Times* (analysis of frequency of searches for terms like "flu symptoms" in Google and Yahoo may provide early warning of regional flu outbreaks).

[49] RAHS Singapore, http://rahs.org.sg/t2_home.html, [viewed 29 June 2009].

[50] Foresight Horizon Scanning Centre. http://www.foresight.gov.uk/Horizon%20Scanning%20Centre/index.asp [viewed 5 July 2009].

[51] Government Office for Science 2009 *The Sigma Scan* http://www.sigmascan.org/Live/Home.aspx [viewed 29 June 2009];
Foresight Horizon Scanning Centre undated *Wider Implications of Science and Technology (WIST) Programme: Report of Findings.*

[52] Sutherland W J, Bailey M J, Bainbridge I P, Brereton T, Dick J T A, Drewitt J and et al. 2008 Future novel threats and opportunities facing UK biodiversity identified by horizon scanning *Journal of Applied Ecology* **45(3)** 821-833.

[53] Center for Catastrophic Risk Management, at University of California, Berkeley. The author is an affiliate of the Center.

[54] Farber D A, Bea R G, Robers K, Wenk E, Inkabi K, 2007, Reinventing flood control, *Tulane Law Review* **81** 1-43.

[55] Oppenheimer M O'Neill, B C and Webster M 2008 Negative Learning. *Climatic Change* 89:155-172. (Oppenheimer describes how disciplinary silos may have delayed progress on understanding the process of ozone depletion and may still be affecting research on the effects of global warming on the West Antarctic ice sheet).

[56] Bea R G, Mitroff I, Roberts K H, Farber D and Foster H, 2008, *A new approach to risk: the implications of E3* (Berkeley : Center for Law, Energy and the Environment).

[57] Bea R G, 2002, Human and organizational factors in reliability assessment and management of offshore, *Risk Analysis* **22(1)** 29-45.

[58] Sachs J D 2008 Keys to climate protection: Dramatic, immediate commitment to nurturing new technologies is essential to averting disastrous global warming *Scientific American* April 2009.

IOP Conf. Series: Earth and Environmental Science **8** (2009) 012002 doi:10.1088/1755-1315/8/1/012002

Considering WTO law in the design of climate change regimes beyond Kyoto

Sanford E Gaines
Visiting Professor in Environmental Law, Aarhus University, Denmark

Abstract: This article describes the most important provisions of World Trade Organization (WTO) agreements that should be considered in designing laws and regulations under likely post-Kyoto climate change mitigation regimes. The Kyoto Protocol and the expected post-Kyoto international climate agreement depend on national measures to implement market-based mitigation measures. This market strategy promotes international exchanges of goods, investments, and services such as cross-border trading of credits for emissions reductions and transnational financing for projects that avoid emissions through the Clean Development Mechanism. Moreover, the United States and other countries, concerned over "leakage" of greenhouse gas (GHG) emissions through relocation of industry to other countries coupled with political worry over manufacturing competitiveness, have proposed national climate legislation containing border adjustments on imported goods or implicit subsidies for national producers, raising additional WTO considerations. The article assesses the likely effectiveness of such trade-related measures in achieving climate change mitigation goals and the potential trade policy infringements and trade distortions that they might bring about. Alternative strategies for achieving GHG mitigation goals in closer conformity with WTO law and policy will be suggested.

1. Introduction

The edifice of international trade law rests on classical economic theory, which teaches that trade of goods between two trading partners, where each has a comparative cost-of-production advantage in producing the good it exports relative to the good it imports, results in macroeconomic welfare gains for both trading partners. Most contemporary economists continue to espouse the central theory of comparative advantage and its core conclusion that there are welfare gains from trade for both parties to the exchange, noting that it is "one of the most basic ideas in economics" [1] [2] [3], though some stress certain limitations or qualifications [4].

In order to secure the gains from trade for all nations, and to prevent economic protectionist behavior widely believed to have aggravated the Great Depression and fostered international conflict, nations negotiated the General Agreement on Tariffs and Trade (GATT) after World War II as one of the pillars of the post-World War II economic order [5]. In 1994, a much larger group of nations recommitted themselves to the GATT, entered into other agreements on specific trade topics, and established the World Trade Organization (WTO) to oversee the agreements and administer an enforceable procedure for the settlement of trade disputes [6].

As described more fully in section 2, the GATT and the other WTO agreements rest on two foundational principles. First, only tariffs may be used to restrict imports of goods (and by mutual agreement tariffs on manufactured goods have been sharply reduced over the years). As a corollary to this first principle, various non-tariff barriers to trade are prohibited or strongly controlled. The second bedrock principle is that nations should not discriminate in their trade relations. Nondiscrimination has two aspects: most-favored nation (MFN) treatment and national treatment. MFN treatment means that nations should not discriminate between trading partners, but grant to all partners the same privileges and benefits they give to the "most-favored nation" with respect to like goods. National treatment means that nations should not discriminate between foreign goods and like national goods in their national taxes and regulations, but instead give foreign goods "national treatment."

For a policy matter like climate change, the GATT allows an important exception to these strong principles under certain conditions. Unilateral trade measures may be used where necessary to protect human health or to promote conservation of natural resources, provided that the measures do not result in arbitrary or unjustifiable trade discrimination or serve as a disguised restriction on trade.

Tracking the GATT language on environmental exceptions, the Kyoto Protocol and drafts of post-Kyoto agreement language expressly commit the governments, most of whom are also members of the WTO, to avoid any arbitrary or unjustifiable discrimination or disguised restrictions on trade in their national laws and regulations implementing their climate control obligations. Nevertheless, each nation has its own differentiated mitigation responsibility under the umbrella of the international climate agreement, its own greenhouse gas (GHG) emissions profile and capacity to enhance carbon sinks, and its own economic circumstances. These factors influence the national legislation and regulations that each country has adopted or will adopt to fulfill its climate change obligations. As catalogued by trade analysts too numerous to cite here but usefully cited and selectively discussed by Voigt [7], it is in the many details of national policy that the design of national law, intentionally or unintentionally, may affect international trade, investments, or services in ways that could contravene the WTO agreements and give rise to trade disputes. With careful attention, most risks of impairment to international trade can be avoided. On some issues, it is even possible that the post-Kyoto climate agreement or other changes to international laws and institutions to remove existing barriers to trade and investment could thereby engage international trade affirmatively in pursuit of the world's climate change mitigation goals.

2. The WTO regime

2.1. Basic GATT elements

The WTO legal elements of most importance are contained in the GATT [5]. Most other WTO agreements, which are legally separate from the GATT, contain their own formulation of core GATT law principles. The legal interpretation of many GATT provisions has been refined through the reports of trade dispute settlement panels and decisions on appeals of those reports by the WTO Appellate Body, a select group of international trade jurists.

The GATT is built around the idea that tariffs should be the sole mechanism for countries to restrict imports, with other provisions meant to protect the integrity of the tariff regime. Article II of the GATT sets forth the basic approach to tariffs and tariff negotiations. Article XI reinforces the "tariffs only" approach by forbidding any non-tariff "prohibitions or restrictions . . . [such as] quotas, import or export licenses or other measures" that would control importation or exportation of products. This is generally described as the prohibition of quantitative restrictions, including any ban (think of it as a zero quota) on importation of a product that remains a legal product in the country.

Article I of the GATT expresses the MFN principle. It requires that "any advantage, favour, privilege, or immunity" granted by a WTO member to a product imported from or exported to any other country "shall be accorded immediately and unconditionally" to the like products of all other WTO members. There is one important and common exception to MFN: GATT Article XXIV provides that countries may enter into customs unions (e.g., the European Union) or free trade agreements (e.g., the North American Free Trade Agreement), which give trade preferences to other countries in the customs union or free trade area.

The national treatment principle, set forth in GATT Article III, has the general objective of preventing taxation or regulation of imported products "so as to afford protection to domestic production." (Article III, paragraph 1, or III:1). The particular rules for national treatment relating to taxes are in paragraph 2 of the article (Art. III:2), and the rules relating to national laws and regulations are in Art. III:4.

Art. III:2 on national taxation has two sentences with different nuances. The first sentence forbids any domestic taxation of imported goods "in excess of" the taxes on the like domestic product. This has been construed to mean that tax rate must be exactly the same or lower for the imported product when the two products are closely similar [8]. The second sentence more broadly prohibits other differences in tax treatment that afford protection to domestic production. As an official explanatory note makes clear, the second sentence covers taxes on imported products where "a directly competitive or substitutable [domestic] product" is not "similarly taxed." Directly competitive or substitutable products is a much broader category than "like" products in the first sentence, but the difference in taxation under the second sentence needs to be significant enough to be likely to affect competition in the market place [8].

Article III:4 concerning laws, regulations and conditions of sale, is simpler than Article III:2. It provides that imported products "shall be accorded treatment no less favourable than that accorded to like products of national origin." A regulation that treats foreign products more favourably in some situations but less favourably in other situations has been held to violate this provision. Moreover, the treatment of the foreign good does not need to be less favourable in a particular case, but only needs to create the potential of less favourable treatment [9].

As noted in the section 1, there is an exception to all of these requirements in the GATT itself, GATT Article XX, which is available for trade measures in pursuit of many environmental policies, including almost certainly climate change mitigation. Article XX begins with an introductory policy paragraph generally known as the "chapeau," followed by lettered subparagraphs identifying the national policies for which the exception may be available. For a climate change, three of these subparagraphs are relevant: (b), (d), and (g). Following WTO interpretive practice, let's look at the subparagraphs first, then at the general policy statement of the chapeau.

Article XX(b) allows use of trade measures when "necessary to protect human, animal, or plant life or health." The word "necessary" has been interpreted very strictly to mean that it is an essential requirement and that other less trade restrictive alternative measures to meet the health protection goal are not reasonably available [9]. Even if climate change is a human (or animal) life or health matter, it is unlikely that any particular mitigation strategy will be deemed "necessary."

Article XX(d) more generally allows trade measures that are "necessary" to "secure compliance with [national] laws or regulations" that are themselves not inconsistent with the GATT. Again, the word "necessary" is taken to mean that there is no other effective way than a control at the border to secure compliance by an imported product with the domestic law.

Finally, Article XX(g) allows trade measures "relating to the conservation of exhaustible natural resources" if they are used in conjunction with domestic restrictions on production or consumption. This provision has been broadly construed by the WTO Appellate Body. It has held that the phrase "related to" means only that there should be a reasonable means-ends connection between the measure and the intended resource conservation outcome [10]; and "exhaustible natural resources" has been specifically interpreted to include good air quality [11]. So Article XX(g) would seem to be available, in principle, to allow trade restrictive measures in pursuit of national climate change mitigation programs [12, p 84] [13, pp 35-36].

The policy language of the chapeau has proven to be the main stumbling block when countries have tried to invoke Article XX in defense of trade measures. The chapeau puts conditions on the exceptions by imposing two requirements. First and foremost, the trade restriction should not be applied in a way that results in "arbitrary or unjustifiable discrimination." This language has been construed by the Appellate Body to preclude an Article XX defense in several situations that are common in environmental policy. For example, a trade measure cannot be prescriptive. It may not require the foreign producer or government to follow essentially the same specific policies or practices required in the country taking the measure [11]. The chapeau language has also been interpreted to require that the acting government must make a good faith effort to reach an international agreement on the issue before it invokes a unilateral trade measure [10][14]. The second requirement in the chapeau is that the trade measure should not constitute a "disguised restriction on international trade." In one report, however, the Appellate Body seemed to equate the idea of disguised restriction with the idea of arbitrary or unjustifiable discrimination, so it is not clear whether "disguised restriction on trade" has any meaning as a separate test [11].

2.2. Other WTO agreements relevant to climate mitigation
In recent decades, trade policy makers have developed other agreements to address trade issues beyond the traditional domain of trade in goods. With one exception noted below, these other agreements have been incorporated into the overall WTO Agreement [6] as annexes. Several of them have potential relevance in the climate change context.

2.2.1. Agreement on Technical Barriers to Trade (TBT Agreement) This agreement contains detailed provisions on the nature and scope of permissible government action to set standards or regulations for products in pursuit of legitimate objectives like environmental protection. In most instances there is some jurisdictional overlap between the TBT Agreement and the GATT; more often than not, when it has had a choice the WTO Appellate Body has applied GATT rather than the TBT Agreement to settle trade disputes [10].

2.2.2. Agreement on Subsidies and Countervailing Measures (SCM Agreement) Some government measures applied in pursuit of climate policy, especially those focused on the energy sector, may constitute a subsidy (or state aid in European terminology). The WTO SCM Agreement sets forth a rather complex series of factual findings and legal determinations to be made in deciding whether a government measure is a subsidy and, if so, whether it may affect international competition or otherwise distort trade. The SCM Agreement defines subsidy broadly to include any government arrangement that a) makes a financial contribution (for example, below-market interest on loans, government credit guarantees, or not collecting revenue otherwise due, as through special tax relief), and b) the financial contribution confers a benefit on one or more enterprises. Within that broad sweep, however, general government payments, contributions, or other assistance that benefits all or most enterprises, including infrastructure development or education, is perfectly permissible from the WTO perspective. The WTO only concerns itself with subsidies that are specific, in law or in fact, to a firm, an industrial sector, or some other select group of enterprises.

Among specific subsidies, the SCM Agreement creates three classes: prohibited subsidies; "actionable" subsidies; and non-actionable subsidies. The class of non-actionable specific subsidies has subsequently become largely inoperative; that leaves prohibited subsidies and "actionable" subsidies. Subsidies that are directly connected to export activity or that are contingent on the use of domestic rather than imported goods are prohibited. Moreover, any prohibited subsidy is legally deemed to be specific. Governments can be required to remove prohibited subsidies. Actionable subsidies are specific subsidies that create some economic injury or "serious prejudice," generally meaning an effect on terms of competition between domestic and foreign producers or an effect on the world price for a product. Although "actionable" subsidies can be challenged through WTO dispute settlement, the "action" that is authorized is for the government of adversely affected producers to adopt a "countervailing measure" to offset the economic effect of the subsidy. The typical countervailing measure is a surcharge on the normal tariff for that product, known as a countervailing duty (CVD). The SCM Agreement indicates the methodology and procedure governments must use in setting the value of the countervailing measure and for deciding when they must be modified or removed.

2.2.3. General Agreement on Trade in Services (GATS) Trade in services is increasingly important in the world economy. The GATS creates a structure for bringing national regulation of services provided across borders under many of the same disciplines as apply to trade in goods. The GATS differs from the GATT in two fundamental respects. First, it takes a sector-by-sector approach to services. Some service sectors, such as telecommunications, are now covered by separate international agreements, while others are left to national regulation. The second difference between GATT and GATS is that GATS follows an "opt-in" approach, meaning that a government is only obligated to observe the liberalizing principles of GATS for those service sectors that it chooses to include.

Those two critical distinctions make it difficult to generalize about specific rules applicable to specific service sectors. In general, however, countries have committed to move toward eliminating discriminatory practices in their internal regulation of services without giving up their right to set reasonable requirements for companies or individuals that provide services in their territory. For example, countries may continue to require local licenses for people providing services, whether they be financial analysts, insurance brokers, medical doctors, lawyers, or some other skilled practitioner.

2.2.4. Some other agreements Brief mention should be made of some other WTO agreements that may, in specific situations, have some bearing on climate change law and policy. Many climate change strategies depend, directly or indirectly, on foreign investment. There is a WTO agreement called the Agreement on Trade-Related Investment Measures, commonly known as TRIMs. As its name implies, this agreement targets investment laws or restrictions that link to trade, such as requirements for a foreign investor to use domestic product inputs. TRIMs might have some bearing in particular circumstances on climate initiatives such as foreign investments related to forest conservation or afforestation and the related trade in forest products.

Another agreement with potential application in some circumstances is the Government Procurement Agreement. Unlike the other WTO agreements, the Government Procurement Agreement is plurilateral, meaning that it applies only among the nations that are party to this particular agreement. Among those parties, the agreement establishes a general principle of nondiscrimination and then provides specific practices that governments shall observe at each of the common stages of government procurement, such as setting qualifications for contractors, issuing tenders for contracts, and selecting contractors. Government contracts for renewable energy facilities, for example, would presumably fall within the scope of this agreement.

2.3. Multilateral Environmental Agreements and International Trade

Multilateral environmental agreements (MEAs) such as the U.N. Framework Convention on Climate Change (UNFCCC) and the Kyoto Protocol are negotiated when science identifies the environmental consequences of certain human activities or manmade products. The MEAs then apply legal controls and other policy instruments to regulate the activities or the products causing environmental harm. If the activities or products directly involve international trade of specific goods, the environmental agreements have included provisions requiring each government to ensure that such international transactions are prohibited or regulated in order to fulfill the objectives of the environmental regime. It is beyond the scope of this article to go into detail on the relationship between WTO trade rules and trade restrictions embedded in the MEAs. For present purposes, it can be said that the WTO generally accepts the validity of trade measures incorporated directly into MEAs or required by MEAs as means to the environmental end. This acceptance is based in part on the recognition that when many and diverse countries agree to trade restrictions in a MEA, those restrictions are unlikely to result in improper discrimination in trade. The WTO also seems to appreciate that MEAs have equal status in international law with the WTO agreements [15].

Several important MEAs contain trade provisions, observed by the parties and enforced by the administrative apparatus of the MEA. For example, the Convention on International Trade in Endangered Species of Fauna and Flora requires participating governments to establish mechanisms to prevent or control the export or import of listed endangered species. The Montreal Protocol to the Vienna Convention on Substances that Deplete the Ozone Layer likewise calls on party governments to track imports and exports of such substances and to prohibit trade in them with nations not party to the protocol. The trade restrictions help assure limits on production and consumption of these substances. The trade restrictions under these two agreements have never been contested through the dispute settlement mechanisms of the WTO or any other trade agreement. In fact, the WTO has invited the secretariat staff of these MEAs to be official observers in the deliberations of the WTO Committee on Trade and Environment.

Climate change, however, presents an environmental problem of an entirely different scale and complexity. Even while "differentiated responsibilities" for climate mitigation are debated among industrialized, emerging, and least developed economies, there is consensus that climate change presents risks to all and that the location of sources and sinks does not matter to the atmospheric concentrations of greenhouse gases. On the other hand, there is no single strategy for reducing GHG releases. The many ways to control emissions, enhance carbon absorption, or push energy users toward climate neutrality leave governments many choices among mitigation strategies, choices that have major implications for how goods are produced, energy and transport are provided, and buildings are designed. The social and economic implications of climate mitigation strongly suggest that the principle of subsidiarity should apply, delegating the choice among effective legal measures to the national or even the local level. Even within the multilateral framework of a climate treaty, then, the specific measures to meet global goals will be adopted through national laws and regulations.

It is precisely when each nation makes unilateral choices about how to meet multilateral climate goals that the risk of conflict with international trade law increases. But with foresight and proper design national climate control measures that appropriately regulate trade may be permissible, shielded from WTO challenge to some extent by their connection to the multilateral climate change agreement. There are even ways to use trade liberalization in positive ways to promote climate change, such as reducing barriers to trade in green products and technology.

3. **Trade issues in climate policy**

3.1. The multilateral framework and the role of national law

The UN Framework Convention on Climate Change (UNFCCC) allows, and the Kyoto Protocol explicitly embraces, market-based strategies for achieving reductions in emissions of greenhouse gases (GHGs) and enhancement of carbon sinks. The Kyoto Protocol authorizes emissions trading between developed countries and investments by firms in developed countries in carbon mitigation projects in developing countries through the Clean Development Mechanism (CDM). In this way it fosters international commercial transactions in carbon mitigation.

Regardless of the nature of the transaction and accounting under Kyoto, individual government policies with respect to such international transactions may raise issues under the WTO trade and investment agreements. Recognizing this, Article 3.5 of the UNFCCC urges that: "Measures taken to combat climate change, including unilateral ones, should not constitute a means of arbitrary or unjustifiable discrimination or a disguised restriction on international trade." The Kyoto Protocol reiterates this commitment: "The Parties included in Annex I shall strive to implement policies and measures . . . in such a ways as to minimize adverse effects . . . on international trade . . . ."

Under the Kyoto Protocol and expected post-Kyoto frameworks, there are two broad classes of national climate legislation and regulation that may raise considerations of WTO law. One class includes those national measures necessary or appropriate to implement the treaty mechanisms, specifically emissions trading and the CDM (and, post-Kyoto, a possible system for reduced emissions from deforestation and forest degradation, or REDD). The other class includes all other national implementing measures, such as national energy and industrial policies, that will help meet the GHG emission reduction goals of the international regime. In this class, trade issues arise especially with respect to energy production or green technology subsidies and efforts to help energy-intensive industries cope with increased costs under national GHG emission reduction programs. The following sections take up each of these four major issues. In each case, WTO law need not be and should not be an impediment to sound climate policy. To reduce or avoid potential trade problems, however, climate policy makers would be well advised to consult with trade experts and design their climate mitigation strategies accordingly.

Even more controversial from the trade policy perspective, the United States Congress and some other legislatures are considering national policies, including trade measures, to prevent "leakage" of GHG emissions–relocation of GHG-emitting manufacturing from countries with emission caps to countries not constrained by such caps. Intertwined with leakage concerns is political anxiety over loss of industrial competitiveness when energy-intensive and carbon-constrained enterprises have to compete internationally with firms in countries not assuming any mitigation obligations. The twin considerations of leakage and competitiveness have generated proposals for imposing GHG emission conditions at the border on the importation of manufactured goods, taxing such importation, or subsidizing domestic business sectors vulnerable to international competition. These proposed defensive measures raise numerous issues under WTO law.

3.2. Emissions trading systems

The Kyoto Protocol establishes a so-called cap-and-trade system for GHG emissions. Each Annex 1 country undertook a national commitment to reduce GHG emissions within the implementation period; that is the emissions cap. The theory behind emissions trading is that within each country, or even between countries with caps, different GHG emitters have different emission control costs, so it is economically efficient to allow those with low control costs to reduce emissions by more than the required amount and have the opportunity to sell their excess reduction credits to the firms with higher reduction costs, who will reduce their actual emissions somewhat less, using the purchased credits to make up the difference.

Kyoto, especially as articulated and elaborated through subsequent conferences of the parties, has an elaborate classification of these transfers and how they are accounted for. In summary, each Annex 1 party—the developed countries with GHG mitigation obligations—has a quantity of "assigned amount units" (AAUs). If two Annex 1 countries enter into a transaction by "joint implementation," the transferred units are known as emission reduction units (ERUs), while a project between an Annex 1 country and a developing country under the CDM, once certified, generates certified emission reductions (CERs). Moreover, even within an Annex 1 country, mitigation credit may be given for certain land uses, land use changes, or forestry, and those mitigation units are called removal units (RMU).

The European Union emissions trading system (ETS), which became effective in 2005 and was revised significantly in 2009, exemplifies the cap-and-trade system [16]. The European Commission has prescribed rules specifying which source categories are included in the trading arrangement, the initial and revised free allocation of emission units to nations and source categories, a phased reduction in emission units over time, a system for auctioning new emission units, and rules on how emissions and reduction credits are calculated. Trading systems also typically incorporate certification or approval systems to verify the credits. Although the validity and enforceability of such systems are often called into question, there is a robust market of private firms offering certification and verification of emission reductions. Because emissions trading typically involves private transactions at every stage, from offer of sale to final verification, national rules and regulatory decisions governing emissions trading have the potential to discriminate in trading between emitters or service providers in one nation and those in another, which could then give rise to trade disputes in the WTO. Careful drafting can avoid trade-related complications.

For trade law purposes, it is useful to establish at the beginning that it is not the physical emissions themselves that are being traded, but the legal authorization to emit a specific amount of GHGs during a specified period of time, normally a calendar year [7 p 209]. Thus, although the matter has apparently never been subject to definitive interpretation through the WTO, it can be asserted with reasonable confidence that emissions trading is not a matter of trade in goods as regulated by the GATT [17]. It can be said with equal confidence that the various services associated with trading emission rights—brokerage, financing, legal, accounting, and auditing—potentially fall under the

GATS (the WTO services agreement) or applicable bilateral or regional trade agreements covering services. What remains indefinite as a matter of trade law is whether the emission units or credits or offsets themselves might also be considered "services." On the one hand, they are like services in that they are intangibles with economic value that are exchanged across borders, and should therefore be subject to trade law provisions against discrimination. On the other hand, they are perhaps more analogous to money or financial instruments [7 pp209-210]. Although financial services such as banking and investment advice are clearly services, the underlying financial documents or instruments they deal with are not deemed to be services.

With that analysis in mind, the key trade consideration for national emissions trading systems is that they should be designed and administered in such a way that foreign sources of credits or allowances and foreign providers of services associated with emissions trading should be allowed to participate in the same manner and subject to the same rules and requirements as domestic sources and service providers. That is not to say that emissions trading must be open to all comers. There are bona fide issues for each nation in accounting for actual emissions and emissions reductions that make it appropriate for an emissions trading regime to be restricted, perhaps to certain industrial sectors, to nations that have themselves imposed emissions caps, or to nations that have the administrative capacity to verify the emissions credits or emission reductions being traded. Even so, deliberate efforts to set up emissions trading so as to exclude certain sources or service providers on the basis of national origin should be avoided. Similarly, licensing and regulation of associated service providers should be determined only by bona fide concerns for the integrity of the system such as the government's ability to supervise and regulate the service providers.

3.3 Emission reducing projects in developing countries—CDM and REDD

Under the CDM or a possible successor to CDM based on the UN Reducing Emissions from Deforestation and Forest Degradation (REDD) regime, GHG emitters in Annex 1 countries can get credit against emission reduction goals for "certified emission reductions" (CERs) resulting from international investments to enhance carbon sinks or forestall emissions in a developing country. These CERs can then be used to offset GHG emission increases in another country, typically an Annex 1 industrial country. At the same time, the CDM system is designed to serve sustainable development policies in the country receiving the investment.

The CDM and REDD raise many of the same trade law issues as emissions trading, but with several variations. In general, the international transfers of financial and/or environmental services may be subject to WTO agreements or related regional trade agreements on investments and services. Moreover, business firms brokering or certifying such transactions may be subject to national regulation, just as with emissions trading, creating the potential for discrimination between local and foreign service providers in sectors that, for some countries, may be covered by commitments under the GATS.

One important difference from emissions trading is that CDM transactions almost always involve foreign investment in a capital project or new activity that is going to generate the "certified emission reductions." The investment element brings into play the possibility of trade discrimination in the recipient country's treatment of foreign investment. This discriminatory potential is unlikely to fall under the jurisdiction of the WTO. Instead, there is a separate network of bilateral and plurilateral investment agreements that will set the ground rules for investment regulation in many cases. Some of these agreements give the country receiving the investment broad latitude in deciding which investments to accept, but many other agreements establish, in principle, a right of investment subject only to nondiscriminatory national regulation [18].

Another variation from emissions trading is that because of the close connection in the developing country with sustainable development many CDM projects have direct government involvement at one end or both ends of the transaction. In a particular case, this could mean that the transaction might also be covered by the WTO agreement on government procurement.

The most important difference between CDM and emissions trading is that the CDM is governed by an international board within the UN system. Although the CDM board has been criticized for inefficiency and for conflicts of interest between board members and project proponents [18], the direct involvement of governments in CDM administration diminishes the possibility of discrimination between countries or between domestic and foreign business entities in their opportunity to participate in CDM. For example, as required by the detailed provisions of CDM and as institutionalized through practice, third-party certifiers are needed before the emission reductions can be duly "certified." The certifiers, known as Designated Operational Entities (DOEs), are governed by the CDM board. Any discriminatory treatment in the appointment of certifiers is thus more likely to be a result of political favoritism by the board than trade discrimination based on national origin as such. The public concerns expressed about the DOEs relate to competence, improper accreditation standards, and ineffective supervision, not discriminatory treatment in the trade sense [20].

3.4 Subsidies

National climate change policies in many countries include subsidies (state aid) or tax incentives to promote GHG reductions or augment carbon sinks. For example, so-called "green" subsidies to promote development and commercialization of non-carbon "renewable" energy sources are widely deployed as an incentive to accelerate the transition to carbon-free energy. Meanwhile, pre-existing programs in many countries still subsidize the production of one or more fossil fuels. All such subsidies and tax policies are, in principle, reviewable under the WTO agreements on subsidies or government procurement. In particular, subsidies that benefit specific industries or firms could lead to trade disputes over products or technologies that compete in international markets, such as wind turbines.

Subsidies to producers of fossil fuels are common practice throughout the world. The government support takes many forms, including special production tax benefits, provision of rail and road infrastructure, and promotion or protection of energy and other facilities that consume fossil fuels. Indeed, in many countries, petroleum production and refining is a state enterprise. Perhaps because they are so pervasive, these subsidies seem to be taken for granted as the accepted way of doing business. They have never given rise to a trade dispute.

Subsidies for renewable or alternative energy products and systems have also proliferated in recent years as governments seek to build their non-carbon energy portfolios as quickly as possible. As a trade policy matter, the particular form of state support may determine whether or not it raises trade concerns. For example, many states in the United States have adopted so-called "renewable portfolio standards," state law requirements for electric power providers to obtain a certain ratio of their electricity from renewable sources. This government mandate clearly benefits energy systems and products that qualify as "renewable" over alternative energy sources such as fossil fuels or nuclear power. But these standards do not prefer one technology or product over another, so they do not distort the market within the rather broad category of renewable energy. They also do not discriminate between domestic and foreign providers of renewable energy products and services. In fact, in the United States some of the most prominent beneficiaries of this form of state intervention have been foreign firms such as producers of solar photovoltaic panels and wind turbines. Similarly, tax breaks for purchasers of green products such as compact fluorescent lights and low-emitting automobiles benefit those producers, from any country, most capable in manufacturing such products. In the United States, the firms that have benefited most have been European (for CFLs) and Japanese (automobiles), not US producers. It could be said that these subsidies to purchasers have stimulated trade, not restricted it.

Green goods and green technologies is one context in which the WTO can be enlisted as an active participant in addressing climate change. In 2007 the EU and the US put forward a joint proposal for a WTO Environmental Goods and Services Agreement [21]. The idea behind the proposal is to use the WTO as an institution to help remove existing tariffs and nontariff barriers to the diffusion of "green" goods and services. The World Bank has already analyzed the benefits of more open trade in these goods and services, and WTO members have generated a list of more than 150 environmental goods suitable for tariff-free trade.

Even more ambitiously, Brewer points to emerging developing country leadership in some green technology sectors and calls for strategies to promote South-South as well as North-South exchanges of environmental goods, services, and technologies [22]. To realize the potential benefits will require the removal of existing tariff and nontariff barriers on every climate-friendly product and technology, from compact fluorescent lights, photovoltaics, and wind turbines to hybrid automobiles and biofuels. Brewer perceives that the WTO is not inclined to expand its agenda in this direction (it has not acted on the EU-US proposal), and therefore enlists the aid of the World Bank and regional development banks to promote this agenda.

Finally, some commentators have raised the question whether the allocation of emission allowances by governments to industrial firms or power producers at no cost might constitute an actionable subsidy [12 pp 60-64] [17 pp 206-211]. Although such a determination by the WTO cannot be ruled out, it seems unlikely that the issue would generate a formal trade dispute given the widespread practice of granting emission allowances by most countries adopting a cap-and-trade strategy for emission reductions. Moreover, the economic calculation of whether a firm or industry sector received a benefit from a free allocation, as opposed to a system of auctioning allowances, is far from simple because of the opportunity cost to the firm of not selling allowances received and the differential profit opportunity and cost exposure across industries. Environmental performance considerations further complicate the analysis [23][24].

3.5 Measures addressing leakage and competitiveness

Leakage is a climate policy term identifying the potential for relocation of manufacturing with the result that GHG emissions "leak" out of industrial countries that are aggressively reducing such emissions into developing countries where GHG emissions are not regulated, thereby counteracting the climate benefits of emissions reductions in industrial countries. Political anxiety about industrial competitiveness in international markets goes hand-in-hand with the concern over leakage. The competitiveness concern is particularly acute with respect to energy-intensive industries, which will presumably face high costs for GHG emission control that could erode their ability to compete with producers in countries not regulating GHG emissions.

The climate policy literature in the United States has become obsessed with the policy implications of trade measures proposed in bills in the U.S. Congress addressing leakage and competitiveness. The highly respected Peterson Institute for International Economics has produced two studies on these questions in less than a year [12][25]. The Brookings Institution, a major public policy center in Washington, held a conference in June 2008 on climate change, trade, and competitiveness [26]. A research group called Climate Strategies produced a "working paper" in March 2009[27], and the Bipartisan Policy Center's National Commission on Energy Policy came out with report in April 2009 [28]. As of the end of June 2009, a major comprehensive climate change bill that addresses leakage has been passed by the US House of Representatives, H.R. 2998, also called the Waxman-Markey bill [29].

There are three broad questions with respect to leakage and competitiveness:

1) How significant is leakage likely to be from a climate mitigation perspective? That is, how many US industries are likely to relocate or lose production to businesses in nations not subject to climate mitigation obligations?

2) To the extent leakage is anticipated, does it arise because of shifting terms of competition? This is sometimes phrased as the question whether the energy-intensive industries are vulnerable to international trade competition.

3) If competitiveness is a factor, what strategies are available to reduce leakage? Specifically with reference to trade, are trade measures a necessary or appropriate part of a national climate strategy?

In addressing these three questions, the policy discussion in the United States, Europe, Australia and elsewhere has focused on so-called energy intensive industries—usually including iron and steel, chemicals, paper, and aluminum, and sometimes cement. Electricity generation, which is also energy intensive, is excluded from consideration because electricity is generally not traded internationally and the sector is therefore not subject to international competition.

To the first question—Is significant leakage likely?—most economics studies answer "No." To begin with, in the US economy the energy-intensive industries account for only 6% of total GHG emissions [30], so the climate impact of leakage from these industries would be modest at best. Moreover, most economic analyses find only a modest percentage of emission reductions in developed countries would be offset by emission increases in developing countries in the same industries. For example, an analysis of a bill in the US Congress performed by the US Environmental Protection Agency concluded that only about 8-11% of emission reductions in the US would be offset by emission increases in other countries [31]. Leakage rates in other studies fall in a similar range. Moreover, one study demonstrates convincingly that "the developing countries from which these [energy-intensive] goods are imported [into the US] are, in many industries, less carbon-intensive on average than the United States" [25 p 46]. From a global climate change mitigation perspective, then, industrial leakage from the United States is not a major concern.

Part of the answer to the second question—Will changes in international terms of competition drive the GHG leakage?— is embedded in the answer to the first question. In short, careful economic analysis suggests only modest emissions effects from changes in competitiveness. For example, the leading source of US imports of the products of energy-intensive industries is another Kyoto Annex 1 country, Canada [25 p 45]. Most of the political concern over competitiveness (and emissions leakage) focuses on China, but for all energy-intensive products combined, China accounts for only 11% of US imports [12] [25]. Looked at from another perspective, the US is the destination for only 1-3% of total Chinese production of such energy-intensive products as steel, aluminum, and cement [30]. From these perspectives, efforts by the United States to prevent emissions leakage to countries like China are unlikely to have any significant effect on total GHG emissions. From a business perspective, however, the competitiveness question is somewhat more significant. US imports from China of these energy-intensive products could increase by about 7% if the US undertook serious emission reduction commitments and China did not.

The third question, and the biggest question at the intersection of climate policy with international trade, is whether trade measures should be part of the strategy to try to limit leakage and encourage GHG emission reductions in developing countries. To simplify, there are two strategic options. One is to reduce the cost (price) effect of climate policy on energy-intensive domestic producers. The other is to use some trade mechanism to impose a similar cost on competing foreign producers of the same products.

The most common suggestion under the first option is to give energy-intensive industries free allocations of emission rights under the national cap-and-trade system, even if emissions sources in other industries need to pay for their emission rights through auction or on the private emission-rights market. This is the option most frequently discussed in Europe [27][32] and is part of Australia's proposed system [25]. It is also the core option in the current draft of the Waxman-Markey climate change bill in the United States [29], which would distribute emission allowance "rebates" to energy-intensive industries until the year 2025 and then phase out the rebates over the following ten years.

Is "free allocation" or "rebating" acceptable under international trade law? The answer is not clear, though it is less problematic than the alternative of trade measures against imported products. From the trade perspective, the potential problem of free allocation only for energy-intensive industries is that it might be deemed a specific subsidy under the SCM agreement. It appears to be a subsidy because it is the grant by the government of a financial benefit (avoidance of the cost of buying the emission rights under the government-imposed system), and it is specific because the benefit is available only to a few industrial sectors. But, as discussed above in 3.4, economists do not see matters quite so clearly. The free allocation is not really free, because the emission rights thus acquired have market value, and firms will manage that valuable asset much the same as if they had paid for it. And whether the allocation is specific or not can only be judged in comparison to the number and diversity of other sectors also getting free allocations.

The final trade question under the SCM agreement would be whether a specific subsidy causes "serious prejudice" to foreign competitors competing with the subsidized US firms in their own domestic market or in third-country markets. Since the motivation of a free allocation scheme is to reduce competitive pressure on domestic producers, "serious prejudice" is the intended result, but whether the allocation scheme has that effect in the real world would be a factual determination in each case.

The other strategic option is to try to shield domestic producers from "unfair" foreign competition by establishing a border measure intended to offset the cost impact on the firm of complying with the domestic emission reduction mandate. If domestic firms had to pay a "carbon tax," the trade mechanism could be a so-called "border tax adjustment." There is some sketchy and controversial trade history and law regarding border tax adjustments, but no countries have adopted carbon taxes as their major mechanism for controlling industrial GHG emissions, so discussion of border tax adjustments remains an abstract intellectual exercise. More practically, under the prevalent cap-and-trade approaches followed in the EU and proposed for the US, the key trade issue is one of "border adjustments" more generally [32].

The most-discussed border adjustment mechanism, which first appeared in a 2008 version of proposed US legislation known as the Boxer amendment to the Lieberman-Warner bill, is included in section 401 of H.R. 2998, the bill passed by the U.S. House on 25 June 2009. In proposed amendments that would add sections 765-68 to the U.S. Clean Air Act, H.R. 2998 would establish some GHG emission allowances in an "international reserve," and then would require importers of energy-intensive products to acquire such allowances in sufficient quantity to equal the emission reduction credits associated with a like product from a US firm subject to cap-and-trade emission reductions, at a price equal to the US market price for domestic emission reduction credits. Under H.R. 2998, this carbon-based border adjustment would begin operating in 2020, giving other countries time to adopt equivalent carbon reduction requirements and thereby avoid the border adjustments for their producers. The expressed preference for a multilateral approach and the allowance of time to negotiate it before applying border adjustments are clearly designed to conform to WTO interpretations of the GATT Article XX environmental exceptions.

Commentators have noted that this international reserve allowance mechanism raises numerous trade law issues[12][25][30]. How many allowances the product would need at the time of import would depend on a system for determining or estimating the GHG emissions associated with the product in comparison to the GHG emissions for the like US product. There is a substantial possibility that any GHG calculation methodology would be skewed by political pressure from US producers. H.R. 2998, for example, assumes that the comparison would be done on a country average basis, not for each producer individually. No matter what system could be devised for assigning or determining GHG emissions related to the product (and just about any system would have comparability and fairness problems because of variation among individual firms and among countries in production facilities and energy sources), the system is inherently designed to distinguish among products based on how they are produced, and to discriminate between products from different countries. The anti-discrimination provisions of GATT articles I and III prohibit discrimination based on country of origin, and trade law generally does not permit regulation of products based on their method of production.

The only way to save such a trade provision from violation of trade law would be through the exception offered for many environmental measures under GATT Article XX. But the chapeau of Article XX itself prohibits "arbitrary or unjustifiable discrimination" between imported products from two different countries, or between imported and domestic products. It is difficult to say whether such a system would avoid such discrimination; much would depend on the details of the legislation and the way in which the law would be administered [12][27]. At the very least, the border adjustment mechanism is highly controversial as a matter of trade policy, and might be found in violation of WTO obligations. It has also aroused strong political opposition from key developing countries such as India and China [33]. A proposal by France to add a similar border adjustment in the EU ETS has met resistance from many other European governments [34].

4. Cooperative and multilateral approaches

Section 3 has described the many trade law issues that arise because unilateral national measures, even when imposed as a way to help achieve international goals, tend to be influenced by the selfish economic interests of the particular nation. Climate policies that substitute multilateral action or international cooperation are much more likely to avoid preferential trade treatment and thus avoid problems with WTO law.

Hufbauer et al. [12 pp 103-110], after explaining the infeasibility of negotiating climate-friendly amendment of WTO law directly, set forth in some detail possible provisions of a separate plurilateral code under WTO auspices (similar to the Government Procurement Code) through which code signatories could agree among themselves on the permissibility of impermissibility of many different trade-related issues in international and national climate policy. Their proposals cover everything from border tax adjustments for carbon taxes to subsidies for measures related to adaptation to climate change. Hufbauer et al. are correct in principle that a negotiated agreement on climate mitigation measures under WTO auspices would almost certainly avoid trade disputes related to the agreed measures, and their suggested code elements provide an excellent agenda of issues for consideration. As a practical matter, however, their suggested approach seems, to this author, idealistic and unachievable. As of this writing in September 2009, it appears that the major emitting nations are having enormous difficulty in coming to agreement on the global climate mitigation goal and timetable and on their respective national mitigation commitments. So it seems totally unrealistic to imagine that they could also agree on a separate legal framework that would anticipate and approve in advance the application of trade restrictions as national climate policy tools.

A more plausible scenario would be for the post-Kyoto climate agreement itself, insofar as it perpetuates trade-based approaches to emission reduction commitments, to directly incorporate, as a matter of climate policy, the preferred or expected national implementation approaches. Writing explicit permission to use certain trade measures in a multilateral environmental agreement does not provide the trade law certainty of a WTO-based plurilateral code, but it would likely suffice to avoid most trade disputes and offers a more realistic negotiation goal.

Another international approach, suggested by Houser et al. [25], is for key countries to negotiate sectoral agreements covering key industrial sectors associated with high GHG emissions. Sectoral agreements would be affirmative bilateral or multilateral agreements in pursuit of the same emission control and competitiveness management goals that motivate some of the national legislation being considered in the US, the EU, and Australia.

5. Conclusion

Mitigating the effects of human activity on the climate before those effects reach catastrophic proportions is going to require an unprecedented level of international commitment and cooperation. International cooperation among the diverse countries of the world under the principle of common but differentiated responsibility demands in turn international exchanges of goods and services and technologies. In this way, the rules of international trade, as represented by the WTO agreements, become part of the international framework within which climate mitigation and adaptation will take place. With some forethought by climate agreement negotiators and national policy makers, the WTO agreements should not be an impediment to reaching the world's mitigation goals. On the contrary, the same principles that underlie WTO law serve as a salutary check on national legislators and policy makers, so that they keep the international goal in view and do not distort climate and economic policy to serve narrow conceptions of national self-interest. Indeed, WTO principles, and perhaps even the administrative machinery of the WTO itself, can be enlisted to promote international climate cooperation through diffusion of climate-friendly products and green technology.

References

[1] Deardorff A V 2005 How robust is comparative advantage? *Research in International Economics Discussion Paper 537* (Ann Arbor, MI: University of Michigan)
[2] Dornbusch R, Fischer S and Samuelson P A 1977 Comparative advantage, trade, and payments in a Ricardian model with a continuum of goods *Am. Econ. Review* **67** 823
[3] Krugman P R and Obstfeld M 2005 *International Economics: Theory and Policy* (London: Pearson)
[4] Stiglitz J P and Charlton A 2006 *Fair Trade for All: How Trade Can Promote Development* (Oxford, Oxford Univ. Press)
[5] General Agreement on Tariffs and Trade 1994 **33** *Intl. Leg. Mat'ls* 1125
[6] Agreement Establishing the World Trade Organization 1994 **33** *Intl. Leg.Mat'ls* 1125
[7] Voigt C 2009 *Sustainable Development as a Principle of International Law: Resolving Conflicts between Climate Measures and WTO Law* (Leiden: Martinus Nijhoff) pp 189-260
[8] Report of the WTO Appellate Body,*Japan-Taxes on Alcoholic Beverages* 1996 WT/DS8/AB/R, WT/DS10?AB/R, WT/DS11/AB/R
[9] Report of the GATT panel *United States-Section 337 of the Tariff Act of 1930* 1989 L/6439
[10] Report of the WTO Dispute Settlement Panel, *European Communities-Measures Affecting Asbestos and Asbestos Containing Products*, 2000 WT/DS135/R
[11] Report of the WTO Appellate Body, *United States-Import Prohibition of Certain Shrimp and Shrimp Products*, 1998 WT/DS58/AB/R
[12] Hufbauer G C, Charnovitz S and Kim J 2009 *Global Warming and the World Trading System* (Washington DC: Peterson Inst.)
[13] Pauwelyn J 2007 U.S. federal climate policy and competitiveness concerns: the limits and options of international trade law (Durham NC USA: Nicholas Inst. for Environmental Policy Solutions)
[14] Report of the WTO Appellate Body, *United States-Standards for Reformulated and Conventional Gasoline*, 1996 WT/DS2/AB/R
[15] Sampson G 2005 Effective multilateral agreements and why the WTO needs them *The WTO, Trade and the Environment* eds G Sampson and J Whalley (Cheltenham, UK: Edward Elgar)
[16] Directive 2003/87/EC, as amended by Directive 2009/29/EC
[17] A Petsonk 1999 The Kyoto protocol and the WTO: integrating greenhouse gas emissions allowance trading into the global marketplace *Duke Envtl. Law & Policy Forum* **10** 185
[18] Marshall F and Murphy D 2009 *Climate Change and International Investment Agreements: Opportunities or Obstacles* (Winnipeg: International Institute for Sustainable Development) (discussion draft)
[19] Wara M 2008 Measuring the Clean Development Mechanism's performance and potential *UCLA Law Review* **55** 1759
[20] Victor D G 2009 *Global Warming Policy after Kyoto: Rethinking Engagement with Developing Countries* (Stanford CA: Freeman Spogli Inst. for International Studies, working paper #82) pp 13-16
[21] Office of the United States Trade Representative Nov. 30, 2007 Press Release
[22] Brewer T L 2009 Climate change and the international trade and investment agenda *World Bank Conference: Development and the Economics of Climate Change*
[23] Weishaar S 2005 *CO2 Emission Allowance Allocation Mechanisms, Allocative Efficiency and*

the Environment: A Static and Dynamic Perspective (Maastricht: Maastricht University Metro Inst.)

[24] Dworsky M, Goulder L and Hafstead M 2009 *Impacts of Alternative Emission Allowance Allocation Methods under a Federal Cap-and-Trade Program* (Stanford CA: Stanford Center for International Development, SIEPR) (discussion paper 08-048)

[25] Houser T, Bradley R, Childs B, Werksman J and Heilmayr R 2008 *Leveling the Carbon Playing Field* (Washington DC: Peterson Inst. & World Resources Inst.)

[26] Brookings Institution 2008 *Climate Change, Trade and Competitiveness: Is a Collision Inevitable?* (papers available at http://www.brookings.edu/events/2008/0609_climate_trade.aspx)

[27] van Asselt H, Brewer T and Mehling M 2009 *Addressing Leakage and Competitiveness in US Climate Policy* (Cambridge UK: Climate Strategies)

[28] Yudken J and Bassi A 2009 *Climate Policy and Energy-Intensive Manufacturing: Impacts and Options* (Washington DC: National Commission on Energy Policy)

[29] American Clean Energy and Security Act of 2009, H.R. 2998 (passed by the U.S House of Representatives, June 26, 2009)

[30] Bordoff J E 2008 The threat to free trade posed by climate change policy *Geneva Trade and Development Forum*

[31] EPA 2008 Analysis of the Lieberman-Warner Climate Security Act of 2008

[32] Reinaud J 2009 *Trade, Competitiveness and Carbon Leakage: Challenges and Opportunities* (London: Chatham House)(Energy, Environment and Development programme paper 09/01

[33] Max A 2009 India against US trade barrier for climate policy *Associated Press* 14 Aug. 2009

[34] Marciano C and Franey J 2009 EU ministers shun French carbon tariff proposal *Agence France Presse* 24 July 2009

Elaborating a coherent and adequate financial structure for a post Kyoto framework

A M Bassi[1a,b] **and S Magnoni**[c]
[a] Millennium Institute, 2111 Wilson Blvd, Suite 700, Arlington, VA 22201, USA
[b] University of Bergen, Postboks 7800, 5020 Bergen, Norway
[c] Baltic Sea Solutions, Fabriksvej 2A, Holeby, 4960, Denmark.

E-mail: ab@millennium-institute.org

Abstract. The present paper reviews the international climate change financial framework and aims at providing insights on its future post-2012 development. This study offers an overview of the good attributes and distortions of the current regime, while investigating the work currently done by many countries and international organisation, in proposing unique and original financial schemes for a post-Kyoto agreement. The objective is to define potential strengths and shortcomings of the current (or projected) financial regime, and put this in relation with the creation of an improved new financing scheme, that could transfer sufficient resources from North to South in an efficient, transparent and participatory way. Indeed, international climate change negotiations are now working in this direction, and the regular submissions from Parties and civil society to the UNFCCC's AWG-LCA witness the desire of governments and organisations to achieve an innovative climate change agreement that could overcome existing weaknesses in the global financial structure, while providing nations with suitable tools to handle the adverse consequences of climatic modifications. The paper will additionally focus on the role of CDM and credit-based mechanisms in a new future financial framework, in consideration of needed improvements in the current international credit system and country visions and AWG-LCA submissions.

1. Introduction

With the approaching 15th UN Climate Change Conference representing the climax of climate diplomacy in 2009, the global debate on environmental protection has become increasingly active, and has focused on the necessity of an international agreement that is as much inclusive as possible. The critical aspect of it is that many are the contrasting interests at interplay, many are the parties that need to be involved, and many are the issues that have to be encompassed. From technology transfers to improved governance, to adaptation, to programmatic approaches, international climate politics have indeed a great deal of relevant elements to discuss in their negotiation talks.

[1] Corresponding author, 2111 Wilson Blvd, Suite 700, Arlington, VA 22201, USA.
Tel.: +1 703 351 5081.
E-mail address: ab@millennium-institute.org (A.M. Bassi).

However, one of the decisive matters that would need to be appropriately addressed and agreed upon is the post-Kyoto structure of climate change financing. As proved by the recent submissions to the AWG-LCA -Ad Hoc Working Group on Long-term Cooperative Action under the United Nations Framework Convention on Climate Change (UNFCCC), the role of investment and financial flows in addressing climate change is perceived as fundamental by the Parties, who are unanimously asking for improvements to the current financial system, and, in some cases, advancing concrete new structures (i.e.: submissions from Mexico, South Korea, Switzerland, etc.). Differently from previous UNFCCC conferences, developing countries have taken up a more proactive role in the global climate talks, and have reiterated their call for a robust and sustainable financial framework, that is able to efficiently fund actions for mitigation, adaptation and technology cooperation. In the particular case of developing countries, moreover, sufficient and secure financing is vital in order to effectively tackle the impact of global warming at national level, and develop activities that could lead to less GHG intensive, sustainable development paths.

Various forms of financial support have been developed with the aim of transferring monetary resources from industrialised nations to poorer countries, while helping leveraging the potentially significant benefits of environmentally friendly technologies for the developing world. Many of the existing funds employ market-based mechanisms to address environmental issues, including the use of compliance and voluntary emission offsets in the area of climate change. The existence of these instruments is delivering positive benefits to both funds' participants (donors) and recipients (beneficiary countries): while complying with their quantified emission limitations and (mandatory/voluntary) reduction commitments, the former are transferring critical amounts of money to the latter, with a forward-looking view of achieving sustainable development based on the use of renewable energy. The transfer of environmental technologies and know-how that should come out as a by-product of these financial instruments, the diversity in the assistance schemes (grants, concessional loans, etc.), the global coverage and attempted conciliation of national interests, as well as the linking of financial support to technical assistance and capacity-building programs have made the current financing system a critical and necessary element for developing nations to address climate-related issues.

Nevertheless, the current financial framework presents visible flaws and inefficiencies, which make it a defective tool in relation with the proportion and complexity of the climate change challenge. Specifically, the majority of existing funds present weaknesses in terms of short-term perspective against the need of longer term monetary commitments for sustainable projects/interventions, substandard governance with a lack of a balanced and participatory representation of the interests of all parties (incl. developing countries), insufficient monetary flows mobilized and existence of different funds/financial mechanisms, leading to fragmentation and inefficient duplications to the current financing system [1].

In view of these elements, there is common agreement at the global level on the urgency of creating a more inclusive and predictable mechanism that guarantees financial stability and sustenance, while allowing for direct and immediate actions to counteract climate-related challenges [2].

Based on a longer report prepared by co-author Silvia Magnoni for WWF [3], the present paper aims at reviewing the main current financial instruments, within and outside the UNFCCC framework, so as to underline strengths and weaknesses, and investigate potential improvements towards an enhanced financing system that could be 1) more inclusive of beneficiary countries' views; 2) more efficient in channeling (significant amounts of) money and technologies from North to South; 3) more inclined to the use of programmatic or wholesale approaches, rather than a project-by-project basis.

The study also analyses some of the recent proposals (AWG-LCA submissions) for improved funds and financial mechanisms, so as to reach a deeper understanding of possible post-Kyoto financial developments. Interesting submissions, in this respect, are those advanced by both countries and institutional entities, where concrete proposals for structured financial schemes are advanced. The analysis has been done upon review of the mechanisms according to the following criteria: purpose of the fund, coordination and governance; monetary amounts and allocation criteria; inclusion of programmatic and sectoral approaches; temporal perspective and performance. Finally, the paper intends to comment on the role of CDM and credit-based mechanisms in the future post-Kyoto financial framework, taking into consideration expectations for a comprehensive climate change agreement and the needed modifications that current market instruments would need to go through for enhancing their effectiveness (and cost-effectiveness) in implementing mitigation actions.

2. Current status of the climate financing regime

There is a growing number of international funding initiatives designed to help countries address the challenges of climate change, in relation with their own capacity to contribute. In this respect, developing countries are facing the hardest hindrances, as their ability and resources to adapt to climate change and its potential disruptive impacts on the environment is rather limited. The ample availability of funds and mechanisms that could help tackle greenhouse gas (GHG) mitigation and adaptation issues globally, while transferring money and knowledge from industrialised to developing countries, may at first sight indicate that a considerable effort is being taken to carry out effective actions, thanks to an international financing scheme that offers many opportunities to choose from [4]. However, the situation is far more complicated and far less effective.

The existing funds analyzed in this study[2] offer different solutions and approaches (i.e. carbon markets, carbon funds, climate financial initiatives) to similar problems and objectives. Overlapping can be easily found, and the financial system's fragmentation suggests that individual governments and multilateral institutions, in the rush of showing their immediate commitment towards climate change, have preferred to recur to different funds, rather than articulating single actions/initiatives into a global strategic and politically coherent framework. Even the CDM, which has a global coverage and should represent an effective, unite, UNFCCC-supported mechanism for financing climate-related projects, presents critical weaknesses in its design and implementation procedures [5].

The review of the current financing regime shows the presence of various initiatives, operating as fragmented programmes with some overlapping in terms of purposes and geographical coverage. Such fragmentation brings about poor synergies among the different funds despite they are all in some way related to the broader UNFCCC framework. The reviewed carbon programmes tend, in fact, to be linked to the global Clean Development Mechanism and/or to be operated under the guidance and authority of the COP, serving as the Meeting of the Parties (CMP). Even when the relationship with the UNFCCC is not straight, as in the case of the WB CIFs, the financing activities of the instrument still have to be guided by the principles of the UNFCCC. This link with the UNFCCC (more or less direct, according to the financial scheme) puts all Funds' operations (apart from some exceptions[3]) under a short-term perspective, with emissions reduction projects having to generate credits by 2012. This implies a close contingency upon the post-Kyoto arrangements and the overall post-2012 UNFCCC financial system.

But a very critical issue regards the scale of the financial amounts catalysed. By comparing what literature suggests as recommended investments for GHG stabilisation and adaptation with the flow of financing currently available at global level, considerable differences arise. The IPCC report 2007 suggests values of USD 108bn to USD 1350bn annual mitigation costs, amounting to 0.2% to 2.5% of current world GDP [6] -which is around USD 54 trillion- by 2030[4] [7]. Similarly, the Stern Review estimates annual costs of cutting total GHG to 550ppm level at an average of approximately 1% of

[2] Climate Development Mechanism, Global Environment Facility funds (including the new Adaptation Fund, Least Developed Country Fund and Special Climate Change Fund), World Bank's Carbon Finance and Climate Investment Funds, International Finance Corporation's Emission Reduction Purchase Facilities, UNDP's MDG Carbon Facility, Asia Pacific Carbon Fund, European Investment Bank- KfW Bankengruppe Facility, Multilateral Carbon Credit Fund, and Latin America Carbon Program.

[3] The WB Carbon Finance Funds, unlike most ERPAs that expire in 2012, can purchase carbon until 2015. Similarly, the MCCF offers the option to acquire post 2012 carbon credits under specific conditions only.

[4] For a median stabilisation level of 535-590 ppm CO_2eq.

current GDP[5] (about USD 540bn) [8]. As far as adaptation is concerned, recent estimates from the UNDP suggests USD 86-109bn/year by 2015 for adaptation activities in developing countries, while the UNFCCC envisages a cost of adaptation of USD 28-67bn/year by 2030[6]. This amount is expected to rise to USD 44-166bn/year, if global adaptation activities (not restricted to developing countries only) are considered [9]. However, reality presents far a different scene. The analysed funds and financial mechanisms currently at place have so far allocated amounts for GHG stabilisation and adaptation measures that are not comparable to the far bigger recommended figures, and most of them handle limited financial amounts to cover multiple years of operations. The WB values the global carbon market at USD 64bn in 2007 [10]. This is certainly a considerable figure, but extremely far from the massive flows required by the academic and scientific literature reported above. Further, that sum also includes domestic/regional emissions trading schemes (i.e.: EU ETS, New South Wales, Chicago Climate Exchange) that do not have a link with the global trading market[7], and consequently, do not contribute to technology transfers and "green projects" implementation in developing countries. In 2007 such schemes accounted for USD 50,4bn out of the USD 64bn –with a key role played by the EU ETS (USD 50bn), therefore only minor figures were left for project-based mechanisms involving developing countries.

If using the sectoral approach lens, the review of the current financing schemes reveals an increased tendency towards accessibility, with a gradual shift from project-based focus to broader sectoral and policy programmes. If this is not an option in smaller carbon funds yet, it has become a more tangent possibility for bigger carbon initiatives (World Bank -WB-, Global Environment Facility -GEF-, Clean Development Mechanism -CDM-). With the introduction of national plans, priority actions, sectoral programmes and the like, the financial system is slowly moving from an individual project target to a wider perspective that consider projects as single steps for the implementation of broader programmes and policies -at national, sub national and sectoral level.

There is therefore a lot of promise in developing more structured and well-functioning carbon programmes based on wholesale/programmatic approaches, although such development has to be coupled with an appropriate revision of the governance systems, so as to avoid that elements of the revised/new funds which are worthy of success will be unnecessarily doomed to failure if governance shortcomings are not speedily rectified [11].

Table 1. Existing climate-related funds and financial mechanisms and their main characteristics.

Purpose	Many common targets: sustainable development, climate change mitigation and adaptation, technology transfer. Funds overlap in the pursuance of similar objectives.
Geographical coverage	Mix of global (CDM) and regional (European Investment Bank -EIB-, Asian Development Bank -ADB-, European Bank for Reconstruction and Development -EBRD-) carbon funds. Although international, global funds have so far been dominated by large developing countries for credits registration. Need for further complementarities between global and geo-focused funds.
Financing	Small amounts, compared to what would be necessary for GHG stabilization. Financial fragmentation: small amounts in many financing schemes. Each fund has own specific allocation criteria.
Governance	Aims towards transparent, open and universal governance, but so far: a. Weak representation of developing countries b. Over-representation of donors' interests

[5] Under assumption of GDP growth at 2.5% rate.

[6] Different estimates are explained by different methodologies used in accounting for adaptation costs.

[7] In phase II of the EU ETS (2008-2012), linking with the KP international emission trading mechanism is operational.

Participating entities	Many actors in the carbon market: Multilateral Development Banks -MDBs-; Brokers/Consultants; Developers; CDM Designated Operational Entities -DOEs-; Private Banks; non governmental organizations -NGOs-; Insurance Companies; National Authorities.
Pricing	Wide range of prices in different carbon markets. Price drivers: investment risk profile, track record of project participants, overall project quality, host country support, sustainable development benefits, transaction costs allocation.

Source: Magnoni, 2009 [3]

3. Proposals for a new climate change financial mechanism

The global climate financial mechanism has expanded considerably in recent years and consists today of a relevant number of instruments, that are able to mobilise non-irrelevant monetary sums. However, it has been internationally recognised that the restructuring of the global environmental financial engineering is urgently needed, in a way that generates sufficient financial resources in global markets, unleashes the potential of environmental technology, and allows developing countries to act, both for mitigation and adaptation [12]. A common global call (i.e. AWG-LCA) was recently rising for an enhanced financial mechanism that could achieve more immediate impacts in the mitigation and adaptation areas. Additional goals include concrete transfers of resources and know-how "from North to South", while promoting programmatic funding modalities and increased volumes of disbursed amounts. This need is particularly felt by the poorer countries that are largely vulnerable to the effects of climate change, and sense that the current financing system –based on governance schemes that have not much included developing countries' participation- does not address their relevant concerns and pressing needs in an appropriate way [13].

In view of these considerations, and due to the fresh increased political understanding of climate change issues among governments, the global financing system has lately been under close scrutiny by both stakeholders and observers, and diverse proposals of improvements and changes of the UNFCCC and Kyoto Protocol structure have been recently advanced as a way to transform criticism into concrete action [14]. There have been a number of proposals submitted to the UNFCCC aimed at generating additional revenue for climate change through action in the carbon market, or carbon or international travel taxes or levies.

This section reviews the more interesting findings out of the analysis of the most relevant financial mechanisms that have been newly proposed by governments and other entities as submissions to the AWG-LCA[8].

What is remarkable in these proposals' submission process is the active role of developing countries in putting forward concrete suggestions for a new UNFCCC financial architecture. This is the first time in recent years that so many developing countries and their groupings have presented such tangible and systemic proposals on the Convention's financial mechanism. Despite the different approaches, all such proposals stem from the common basis of a strengthened, truly global financial system, with ample resources for targeting technology transfers and adaptation/mitigation activities in a genuine participatory way.

The review of newly proposed financial mechanisms reveal interesting insights and perspectives in UNFCCC negotiations, which are summarized in Table 2. Surely it is remarkable to notice the proactive contributions of developing countries in building a new climate-related financial framework. These countries have taken climate change as a serious political agenda, and, building on the experience gained from existing mechanisms, have voluntarily moved towards deeper engagement, representation and ownership of any post-2012 new financial mechanism. This can also explain the

[8] Analyzed proposals include Norway's auctioning system, Pakistan's CDM's levy, IATAL's levy on airline travel, IMERS' emission charge, Tuvalu's Burden Sharing mechanism, Mexico's World Climate Change Fund, Switzerland's global carbon dioxide tax, G77 and China Enhanced Financial Mechanism and the Registry of "nationally appropriate mitigation actions" (NAMAs).

high number of proposals for a new regime put under the UNFCCC framework, or somehow linked to it. Developing countries, as a matter of fact, have a preference for the UNFCCC (and relative accountability to the COP), which they see as an adequate ground for addressing climate change issues in an open, universal, efficient and equal milieu.

Further, it does not surprise that recent submissions call for an enhanced and more transparent governance structure, where all Parties are equitably represented, as well as for a scaling-up of funding, to suitably meet the present and future requirements for adaptation and mitigation. And it is indeed adaptation, together with technology transfer, to have an emphasized role in the new submissions, which highlight the urgency for concrete measures in the adaptation area, and the need for an efficient deployment of technologies as useful tools to achieve sustainable development.

Finally, the role of the private sector and the pursue of constant synergies with the carbon markets are also supported by recent proposals, where recurrent principles and objectives can be recorded, in confirmation of a shared perception of what is (not) needed in the new post-Kyoto architecture. From the polluter pays principle, to the equity concept, to the adequacy of funds, the tendency in many submissions focuses on certain commonly acknowledged priorities.[9]

All these elements generate remarkable insights of the future directions in the UNFCCC negotiations for a new financial architecture. The political attention towards climate change matters has consistently increased, and global and local interests interplay in international negotiations. The positive remark in this aspect is that countries have realized previous errors, and have learnt valuable lessons from prior failures.

Table 2. Main points in recent submissions to the AWG-LCA.

Developing countries' active role	A good number of the new proposals are advanced or supported by a developing country or group of developing countries (Mexico, South Africa, G77 and China).
UNFCCC guidance	Recent submissions prefer to build on previous experience, and use UNFCCC as the appropriate context for transparent climate change negotiations.
Governance	Most of the submitted proposals by Parties to the AWG-LCA call for equitable, appropriate and balanced representation of all parties in the governance, capital easy access and low management costs, demand-driven approach -with recipients involved in definition of needs, transparency, authority and guidance of COP.
Funding	Recent proposals focus on increased funding resources, in addition to existing ODA. Any funding pledged outside the UNFCCC shall not be regarded as being in fulfilment of commitments by developed countries under article 4.3 of the Convention.
Adaptation	Given the fact that financing for mitigation is more readily available and easier to access than for adaptation (e.g. the private sector is not much interested in building a seawall), there is a common sense towards concerted adaptation measures.
Technology transfer	The proposed schemes tend to emphasise the role of technology, whose diffusion can be pursued through funds for research and development, functional windows for technology or venture capital for emerging technologies.
Complementarity with carbon	The reviewed proposals complement existing international carbon markets. However, market-based mechanisms alone are not seen as the ultimate solution

[9] The "polluter pays" principle implies that all nations assume a fair share of their responsibilities for the climate change problem. On this basis, industrialised nations bear the highest share of contribution to the fund, with a percentage of 76 percent of total inflows. The burden sharing is therefore accompanied by differentiated responsibilities and results to be more equitable, while allocating an adequate amount of funds (that developing countries otherwise would not be able to guarantee).

markets	to climate-related issues, since prior experience shows some distortions and inefficiencies into such markets, which require corrections.
Common principles and purposes	a. Adequacy of funds, predictability and stability, to meet the needs of adaptation, mitigation, and technology transfer; b. Equity; c. New and additional funds above the current commitments on ODA and 0.7% target; d. Polluter pays principle.

Source: Magnoni, 2009 [3]

4. Analysis

Climate-related funds currently at place are undoubtedly offering different opportunities, under and outside the UNFCCC umbrella, to tackle environmental degradation and climate change adverse consequences. The analysis of the existing financial system, however, reveals widely acknowledged problems and distortions [1]. The latest submissions to AWG-LCA also indicate common perceptions in terms of more urgent needs and outline of the improved climate change agreement.

Many elements of existing funds need to be considered and revised. Firstly, given the ample impact that GHG emissions have at the global level, independently of their source of origin, it is fundamental not to restrict the climate change action to industrialized countries only, but to consent an active participation in the process also to developing countries, thereby widening the scale of global mitigation activities.

Secondly, the current fragmentation of financial sources does not guarantee an optimal scope of action, as the global dimension of climate change necessitates either (1) a participatory, financially-significant mechanism that has abundant resources at place to tackle the climate problem with proper, coordinated actions or (2) the interaction of different –but complimentary- financing schemes, each focused on a specific aspect of mitigation/adaptation (or geographical areas), which interact and work in synergy among one another, to avoid overlapping and duplication of efforts, and towards an integrated and symbiotic system of climate change financing. Nevertheless, the current picture of the global financial structure portraits neither the first nor the second case.

Thirdly, the actual governance problems in relation to balanced and equal representation is also a frequent problem in many financial mechanisms (CDM in primis) and if not whole-heartedly addressed it might create a recurrent path of (inefficient) climate change action based on the sole developed countries' perspective. Further criticism has additionally risen around the weak involvement of the private sector, whose resources can be abundantly mobilized, as well as about the project-based approach, whose individual focus is not accompanied by proper integration with policies and programs at national or sectoral level.

Some shared elements along the analyzed new proposals show the preferred post-2012 direction: new and additional funding, more involvement and representation of developing countries, open and transparent governance, increased resources for technology transfer and adaptation, polluter pays principle, enhanced policy-based approach. This is, in brief, what new proposals are seeking to achieve. LDCs have become more and more active in international negotiations –as the Mexican and G77 and China proposals prove- and have stressed their concerns for an enhanced and more universal financial regime. Ghana has submitted an idea for a technology crediting mechanism and, apart from Korea, other LDCs support the use of market instruments to finance NAMAs (i.e.: Paraguay, Indonesia, El Salvador). The Mexican and Swiss proposals are actually including non-industrialized nations in the category of donors, and the South Korean submission gives international credit to their domestic climate-related actions. This amplified representation of developing countries, and recognition of their mitigation efforts, can only be seen in positive terms, as it could put developing countries' needs and difficulties on the tray, and help rectify mistakes, distortions and inefficiencies in the existing financial architecture. In this respect, interesting talks can be expected at the upcoming COP15, when the proposed schemes will be further discussed. With the recent augmented involvement of developing countries in the international emissions reduction arena, and the broader engagement of such key nations as China and India, it might be that other critical countries (like the United States) would be stimulated in, for instance, adopting binding targets under the Kyoto Protocol.

5. Discussion

In view of the analysis above, improvements towards an enhanced financial regime might account for a higher reliance on UNFCCC support, adequate monetary flows, enhanced governance, programmatic action, and a longer-term view.

More in details, concerning UNFCCC support, it is clear that the UNFCCC is the most participatory environment for carrying out climate change negotiations. Any new financial scheme outside it would not be perceived by all countries as the outcome of open and transparent talks. Building up on existing structures can also be more efficient, as it can also avoid duplication of institutions and efforts.

Regarding monetary flows, the currently available funding is not sufficient for tackling the climate change adverse consequences at an adequate scale. Economic literature suggests far bigger amounts of monetary flows for mitigation and adaptation measures [6, 8]. The reviewed proposals work in this direction, and not only suggest more significant pledges from governments, but indicate market-based mechanisms and private sector involvement as possible fundraising areas, as well as advocate for contributions from developing countries (in measure of their capacity).

Governance is a critical element in structuring financial funding, and it is fundamental to build a new scheme that takes into account everyone's interests (both industrialized and non-industrialized nations). Good governance allows for an equitable and balanced representation of constituencies, consents easier access to funding from developing countries, accelerates procedures and ensures open and transparent operations. Without these elements, any new financial mechanism may just not make any durable impact in its actions in favor of global environment (and people).

Furthermore, recent negotiations under UNFCCC have focused on a gradual shift from project-based activities to programs of actions and cooperative sectoral approaches. The proposed financial schemes are all adopting a programmatic/sectoral perspective to overcome the limits of project-targeted actions, which are not able to reach the scale necessary for effective impacts (e.g. the Mexican and G77/China proposals seem much appropriate for a full implementation of programmatic initiatives). The CDM has also adopted Programmes of Activities, which can be defined as a discreet sum of individual projects. A critical element in the design of policy-based programs is to use apt methodologies that allow evaluating actions in an adequate way. Scale-up efforts require methodological approaches that emphasize the emission reduction trends and transformational impact rather than tracking each ton of emission reduction.

Finally, given the need for carrying out significant mitigation and adaptation actions, whose impact can create lasting, sustainable benefits, any new financial scheme should work under a long-term perspective, so as to allow for the implementation of relevant initiatives that can bring about transformational change. This is particularly important if efforts scaling-up is put at place through policy-based, programmatic approaches.

6. Conclusions

The expected future costs for mitigation and adaptation at global level are estimated in the scale of billions of dollars [2]. With this in mind, and in consideration of the urgency of action needed, there has been in the past years a rising political attention to climate change issues. COP15 may likely offer a good opportunity for governments to show their commitments and concrete intentions for an improved global financial structure. Even if what can be realistically achieved in Copenhagen is probably just a consensus over a basic framework for a post-Kyoto agreement, this would already be enough for moving forward –and giving signals to market-based financial mechanisms of future actions and positive enhancements.

As observed, new proposed mechanisms could significantly contribute to a revised financial architecture, where all countries and interests are equitably balanced. The key element behind a global financial system is that, as such, it has to be indeed global. Therefore, much attention today is on developing appropriate governance structures that can contribute to a truly open and universal financial mechanism, where increased monetary flows are efficiently and transparently moving from North to South. Similarly, various country -and industrial- agendas have to be taken into consideration when formulating a global framework on climate change, therefore hybrid systems –national and sectoral- may need to be considered (e.g. adoption of a more programmatic approach). Adequate funding, within the UNFCCC, and a longer term perspective are also identified as winning characteristics of a post Kyoto framework, to mitigate the risk and create the stability needed by developing countries, as well as by various industrial sectors to address present and future climate-related challenges.

References
[1] Porter G, Bird N, Kaur N, Peskett L 2008. "New Finance for Climate Change and the Environment", WWF/Heinrich Böll Stiftung.

[2] UNFCCC 2007a. "Dialogue on long-term cooperative action to address climate change by enhancing implementation of the Convention", Fourth Workshop, Vienna, 27–31 August 2007.

[3] Magnoni S 2009. "Review of the CDM and Other Existing and Proposed Financial Mechanisms to Transfer Funds from North to South for Mitigation and Adaptation Actions", WWF, Washington D.C. http://www.panda.org/what_we_do/how_we_work/policy/development_poverty/macro_economics/our_solutions/gfm/

[4] UNFCCC 2007b. "Investment and Financial Flows to Address Climate Change", Bonn, 2007.

[5] Environmental Defense Fund 2007. "CDM and the Post-2012 Framework", Discussion Paper, Vienna, August 27 – 31.

[6] Intergovernmental Panel on Climate Change 2007. "Climate Change 2007", Fourth Assessment Report, Geneva.

[7] International Energy Agency 2006. "The World Energy Outlook 2006", OECD/IEA, Paris.

[8] Stern N H 2007. "The economics of climate change: the Stern review" Great Britain Treasury, New York: Cambridge University Press.

[9] Agrawala S, Crick F, Jette-Nantel S, Tepes 2008. "Empirical Estimates of Adaptation Costs and Benefits: Key Insights and Limitations", OECD, Paris.

[10] Capoor K, Ambrosi P 2008. "State and Trends of the Carbon Markets 2008", The World Bank, Washington D.C.

[11] Ellis J, Baron R, Buchner B 2007. "SD-PAMs: What, Where, When and How?", OECD/IEA, Paris.

[12] McKinsey Global Institute 2008. "The carbon productivity challenge: Curbing climate change and sustaining economic growth", Washington D.C.

[13] Parry J-E, Hammill A, Drexhage J 2005. "Climate Change and Adaptation: Summary", International Institute for Sustainable Development, Winnipeg.

[14] Stern N H 2008. "Key Elements of a Global Deal on Climate Change", London School of Economics, London.

Websites consulted
Asian Development Bank, www.adb.org

Bretton Woods Project, http://www.brettonwoodsproject.org/

EIB-KfW Carbon Programme, http://www.kfw-foerderbank.de/EN_Home/Carbon_Fund/EIB-KfW_Carbon_Programme.jsp

European Union, www.europa.eu.int

Global Environment Facility, www.gefweb.org

IMERS, www.imers.org

International Finance Corporation, www.ifc.org

Multilateral Carbon Credit Fund, http://www.ebrd.com/country/sector/energyef/carbon/mccf/index.htm

Multilateral Fund for the Implementation of the Montreal Protocol, www.multilateralfund.org

OECD, www.oecd.org/dac/environment

Pew Center on Global Climate Change, http://www.pewclimate.org/

PLAC, http://www.caf.com/view/index.asp?ms=12

UNFCCC AWG-LCA submissions http://unfccc.int/meetings/ad_hoc_working_groups/lca/items/4578.php

UNFCCC, Bonn Climate Change Talks 2008, Bonn, 2-13 June 2008, http://unfccc.int/meetings/sb28/items/4328.php

United Nations Development Programme, www.undp.org

United Nations Environment Programme, www.unep.org

United Nations Framework Convention on Climate Change, www.unfccc.org

World Bank, www.worldbank.org

World Economic Forum, www.weforum.org

IOP Conf. Series: Earth and Environmental Science **8** (2009) 012004 doi:10.1088/1755-1315/8/1/012004

Climate change mitigation by carbon stock – the case of semi-arid West Africa

A M Lykke[1], A S Barfod[2], G Tinggaard Svendsen[3], M Greve[2] and J-C Svenning[2]

[1] Department of Terrestrial Ecology, National Environmental Research Institute, Aarhus University, Vejlsoevej 25, DK-8600 Silkeborg, Denmark

[2] The Ecoinformatics & Biodiversity Group, Department of Biological Sciences, Aarhus University, Ny Munkegade 114, DK-8000 Aarhus C, Denmark

[3] Department of Political Science, Aarhus University, Bartholins Allé 7, DK-8000 Aarhus C, Denmark

E-mail: aml@dmu.dk

Abstract. Semi-arid West Africa has not been integrated into the afforestation/reforestation (AR) carbon market. Most projects implemented under the Clean Development Mechanism (CDM) have focused on carbon emission reductions from industry and energy consumption, whereas only few (only one in West Africa) have been certified for AR carbon sequestration. A proposed mechanism, Reducing Emissions from Deforestation and Degradation (REDD) to be discussed under COP15 aims to reduce emissions by conserving already existing forests. REDD has high potential for carbon stocking at low costs, but focuses primarily on rain forest countries and excludes semi-arid West Africa from the preliminary setup. African savannas have potentials to store carbon in the present situation with degrading ecosystems and relatively low revenues from crops and cattle, especially if it is possible to combine carbon stocking with promotion of secondary crops such as food resources and traditional medicines harvested on a sustainable basis. Methods for modelling and mapping of potential carbon biomass are being developed, but are still in a preliminary state. Although economic benefits from the sale of carbon credits are likely to be limited, carbon stocking is an interesting option if additional benefits are considered such as improved food security and protection of biodiversity.

1. Introduction

The European Union (EU) faces a major challenge under the Kyoto Protocol due to its ambitious commitment to achieve an 8% reduction of its greenhouse gas (GHG) emissions in 2012 relative to 1990 and a 20% reduction in 2020. As deforestation constitutes about 17% of the total GHG emissions in the World [1], carbon (C) stocking in tropical forest ecosystems may be an attractive way to mitigate climate change by promoting afforestation/reforestation (AR) and slowing down deforestation and forest degradation [2]. Mechanisms to promote C stocking will be an integrated part of the upcoming COP15 agreement with a 2012-2020 commitment period.

Africa is currently experiencing climatic changes, which in a few decades are predicted to increase in severity, posing a severe threat to the economies and the livelihoods of millions of people that live in rural districts [3]. Arid, semi-arid and dry sub-humid areas, which cover 43% of Africa and maintain 40% of the population, are particularly vulnerable to climate change as many people here live by subsistence farming in an environment that is at the limit of being cultivatable, while the national governments at the same time have limited economic means for assuring food security. Impacts of climate change are therefore seen as a very substantial threat against development and economic growth. Hence, a key issue is to develop mitigation strategies to limit the magnitude of the impending climate change, especially if these also have additional positive effects, e.g., in terms of increasing the adaptive capacity of local communities and ecosystems. C stocking has such a potential in areas with degraded savannas, woodlands and unproductive agricultural soils, where changes in land use may be an efficient means to sequestrate GHG, thereby decreasing the effects of climate change [4] [5]. In a global analysis, Sub-Saharan Africa was found to have a high suitability for C stocking AR projects [6].

Tropical deforestation already causes large losses of C [7]. The Clean Development Mechanism (CDM) provides instruments for reversing this process by active C stocking through AR [8], but out of the almost 5000 projects approved until now, only one reforestation project has been approved in West Africa [9]. A new approach to lowering GHG emissions, called Reducing Emissions from Deforestation and Degradation (REDD), will be discussed during COP15 in Copenhagen in December 2009. REDD has a high potential for increasing C stocking at low costs [10], and will allow developing countries to protect forest ecosystems by selling Certified Emission Reductions (CERs). However, REDD is primarily focused on rain forest and presently does not target savanna and woodland ecosystems such as those found in West Africa [11].

The present article aims to discuss potentials and constraints of C stocking mechanisms, seen from a West African perspective. On the basis of climate mitigation mechanisms for GHG emission trading, the underlying economic models, C stocking models and vegetation ecology, concrete economic benefits from C stocking are discussed and possible additional benefits and constraints are highlighted as a basis for conclusions and recommendations in relation to carbon trading between Europe and West African countries.

2. Climate mitigation instruments

2.1. *The Kyoto Protocol*

The Kyoto agreement was signed by 149 countries in 1997 and represented two major breakthroughs in the efforts to reduce greenhouse gas emission at a global level. Firstly, a five-year commitment period (2008–2012) was defined in which 38 countries, listed in the Annex 1 to the protocol, committed themselves to meet specified emissions target. Secondly, the Kyoto agreement allowed countries to use three flexible mechanisms in their attempts to lower the overall costs when achieving emissions targets. The first one is emissions trading which implies that trade of credits can take place between the Annex I countries. The second mechanism is joint implementation (JI) which opens up for emissions-reduction projects to be shared between Annex I countries. The third mechanism, which is of major interest for developing countries, is the Clean Development Mechanism (CDM), which allows Annex I countries to buy Certified Emission Reductions (CERs) from projects in developing countries.

The Annex I countries that signed the Kyoto protocol are committed to an emission threshold for the greenhouse gases carbon dioxide (CO_2), methane (CH_4) and nitrous oxide (N_2O). The agreement also comprises three types of chlorofluorocarbon (CFC): hydrofluorocarbon (HFC), perfluorocarbon (PFC) and sulphurhexafluoride (SF_6). The Kyoto agreement establishes individual target levels on a so-called '8-7-6-5' basis, *i.e.* the EU must reduce emissions by 8%, the United States by 7%, Japan by 6% and the Annex I countries by 5.2% in average. Although among the original signatories to the Kyoto Protocol, the United States never ratified nor officially withdrew from the Protocol. Within the

EU, burden-sharing leads to big differences. For example, Portugal and Greece are allowed to increase their 1990 emissions by 27% and 25%, respectively, whereas Germany and Denmark both have committed to a 21% reduction. An evaluation of emission reductions by 2005 showed that the EU had reduced emissions only by 2.1%. Thus, in order to meet the target level by the end of 2012, a 6% reduction is still needed. Many of the EU countries, e.g. Austria, Denmark, The Netherlands, Germany, United Kingdom and France, made little progress meeting the agreed target levels [12].

2.2. Clean Development Mechanism

The Clean Development Mechanism (CDM) is an instrument under the Kyoto Protocol, which allows industrialized countries with a greenhouse gas reduction commitment (Annex I countries) to invest in projects that reduce emissions in developing countries in return for CERs. Each CER is equivalent to the reduction of one ton of CO_2 (CO_2e). Overall, the CDM may help to stimulate further environmental protection and economic growth worldwide at a lower cost. CDM may be of particular interest to the EU since it may help in achieving its ambitious GHG reduction targets.

CDM measures must be real, measurable and additional, so the assessment is vital to measure emission reductions and determine whether they would have happened without the CDM as well. Until now most of the projects implemented under the CDM have focused on CO_2 emission reduction. Carbon stocking through AR has been rare; globally only five afforestation and 37 reforestation projects have been approved, with nine in Africa. Only one reforestation project has been certified in West Africa [9].

2.3. Reducing Emissions from deforestation and degradation

The Reducing Emissions from Deforestation and Degradation (REDD) mechanism aims to maintain natural C sinks by financially compensating countries that are willing and able to reduce emissions from deforestation and degradation. The idea of REDD was first presented under COP11 in 2005. The Forest Dialogue is preparing a proposal on the structure of REDD. Decisions about what the REDD mechanism will finally look like need to be agreed upon under COP15 in December 2009 in order to be included in a post-2012 framework. REDD has the potential to simultaneously address climate change and rural poverty, while conserving biodiversity and sustaining ecosystems. Reaching consensus on REDD is considered of paramount importance for a global deal on climate change [11].

Not all developing countries will benefit equally under the proposed REDD mechanism. Developing countries with high forest cover will generally benefit more, whereas countries with little forest left and low deforestation rates will find it difficult to benefit. Except for Liberia, all West African countries belong to the latter group. Some of the countries most vulnerable to climate change, e.g. Niger, Mali, Burkina Faso, Togo and Benin, have therefore not been considered in the preliminary REDD documents. The impact of REDD for different countries will depend on the choices made for measuring baseline conditions, describing country circumstances, defining 'forest' and deciding on inclusion of non-forest landscapes.

The FAO definition of 'forest' presently used as a basis for REDD is rather broad: 'land with tree crown cover (or equivalent stocking level) of more than 10% and an area of more than 0.5 ha. Trees should be able to reach a minimum height of 5 m at maturity in situ'. According to this definition, forests cover 30% of the earth's land surface, whereas 'other wooded land' covers 10%. Most of the ecosystems generally defined as savanna and woodlands in West Africa are considered as forest by the definition of FAO, e.g., Senegal and Burkina Faso have approximately equal amounts of forest and other wooded land [13]. In spite of this, the focus in preliminary REDD documents has been on dense forest formations, mainly rain forest, and accordingly the benefits for West African countries will depend on a future focus on forests in the broadest sense.

Focus has been on deforestation and less on land degradation because of a lack of global datasets on degradation, although the latter is an important contributor to GHG emissions [14]. This puts countries in West Africa at a disadvantage because they are experiencing generally low deforestation rates whereas land degradation is widespread due to unsustainable management and human induced fires [15] [16].

3. Economic Incentives

3.1. Economic incentives in Europe

For a European country, an incentive to establish C stock forests under CDM or protect existing forests under REDD will be an interesting option when it is cheaper to reduce emissions in a developing country than in an Annex I country. Gains from CO_2 trade compared to a situation without trade are depicted in figure 1.

Assume that the CO_2 (CO_2e) price by free trade in the EU market is p^* (15 €) and that the ownership of one CER gives the right to emit 1 ton of CO_2e. In response to this price companies in the EU will reduce or increase their individual GHG emissions until all their individual marginal costs (MC), or the cost of one more unit of CO_2 reduction, are equal to the CO_2 price. The EU (g) generally faces significantly higher marginal reduction costs than a developing country (f) due to its more expensive technological options. The EU does not have the option of low-cost reforestation on a larger scale, because it already exploited the cheapest solutions. It is therefore forced to turn to relatively expensive technological alternatives such as improved energy efficiency, fuel switching, wind and solar energy. In contrast, a developing country has the option of low-cost C stocking by AR. Thus, the marginal cost reduction curve for the EU (MC_g) is situated above the marginal costs for a developing country (MC_f). Next, the developing country has not committed itself to any obligation, whereas EU is obligated to reduce its GHG emissions by 8% in average. In that case both countries will save costs by CER trading compared to the non-trade situation. For ease of interpretation let q^* correspond to one CER, q_f to two CERs and q_g to three CERs. A developing country will reduce q_f at the cost of area Odq_f. Since a developing country has no reduction obligation, it can sell these two CERs and receive a net gain corresponding to $Op^*d = Op^*dq_f - Odq_f$. Assume then that the EU has already achieved a one unit reduction (q^*) but that it has committed to reduce three CERs all in all (q_g). In other words, the EU will need two extra CERs to comply with the targets, and it will choose to buy these two CERs from a developing country, because it is cheaper to buy these two CERs at price $p^* = 15$ € than to reduce emissions by itself. The cost of buying two CERs from a developing country corresponds with the area q^*acq_g, i.e. 30 €. If the EU had to reduce on its own without the trade option, total cost would amount to area q^*abq_g, i.e. 60 €. Overall, the EU receives the net gain of area *abc*, i.e. 30 €, from its CDM investment in this hypothetical example. EU simply gets cheaper reductions than back home. In conclusion, both countries gain economically from this type of CO_2 trading.

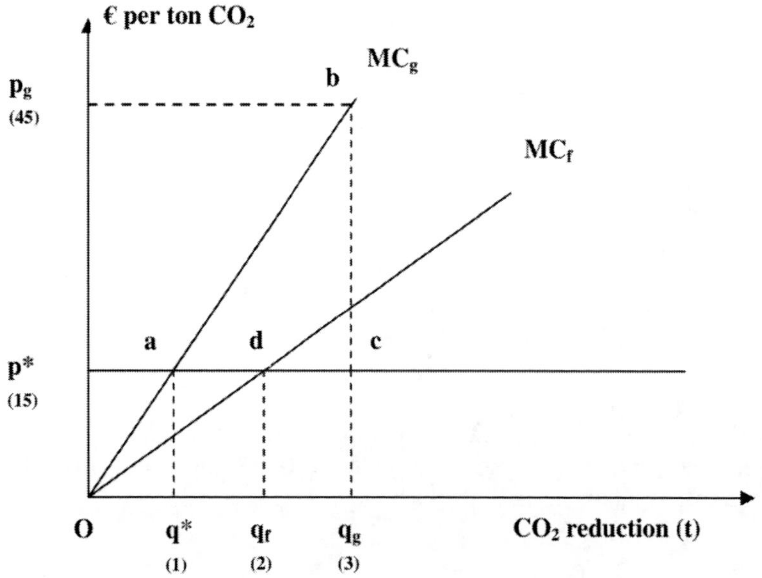

Figure 1. Model of marginal costs (MC) for CO_2 reduction in the EU (g) and in a developing country (f) for estimating gains from CO_2 trade compared to a situa-tion without trade. p*: CER price in the EU market. q*: one CER.

3.2 *Economic incentives in developing countries*

Incentives for people in developing countries to engage in C stocking projects will be determined by potential revenue, income security, time-frame and opportunity costs. In most of rural West Africa, opportunity costs from alternative occupations such as subsistence agriculture and livestock rearing are low, and in rural areas it is generally difficult to find employment outside the agricultural sector. Furthermore land prices are low, and viewed on a global scale the rate of C uptake is large compared to land prices [17]. The revenue from C stock forests can come from CER trading and from increased income from secondary crops, which can be increased considerably where economically important species are conserved. Secondary crops assures income stability through diversification and can provide important supplementary returns to household economies [18]. The low opportunity costs and difficulties in finding employment will encourage people to engage in emission reduction projects through CDM or REDD. Under CDM, funding received from CERs and secondary crops must exceed the cost of AR establishment and opportunity costs, whereas under REDD it needs only exceed opportunity costs and eventual costs for conservation. The major drawback for C stocking to be economically attractive is the long time-frame; about 20 years before a C stock has developed and 10-15 years before secondary crops can be harvested in large amounts, which may discourage many from getting involved, especially the poorest people who struggle for their bare existence.

4. Carbon biomass

Terrestrial ecosystems remove 3 billion tons of C year^{-1}, with the 4 billion ha of forests (30% of the global land area) as the major C sink and reservoir [19]. Much of West Africa is naturally covered by savanna and woodland, with forest covering wetter areas (figure 2). According to the FAO forest definition used under REDD, many areas usually referred to as woodland and savanna are included under forest [13]. The open savanna in the Sahelian zone has a productivity of less than 1 tC year^{-1}, whereas the Sudanean zone has a productivity up to 6 tC year^{-1} [20].

Figure 2. Sahelian and Sudanian vegetation types. Upper left: Sahelian savanna with moderate grazing. Upper right: Sudanian savanna. Lower left: Sudanian woodland. Lower right: Sudanian savanna with an area to the left protected from fire and human impact; the latter gives an indication of potential vegetation type under optimal management conditions. All photos are from the dry season.

4.1. Carbon biomass measurements

In order for C markets to function in practice, measuring and monitoring of C need to be cost-efficient and effective. One unresolved problem is the difficulty involved in determining baseline figures for vegetation biomass and C storage, and monitoring increases herein under the principle of additionality as implemented in the CDM regulations established by the Conference of the Parties serving as the Meeting of the Parties COP/MOP and the CDM Executive Board. Remote sensing approaches for forest C estimation are being developed, but are still inaccurate and need much improvement [21]. Allometric models are being developed worldwide to assess biomass, above-ground as well as below-ground, for a given vegetation type, with the aim of allowing non-destructive biomass estimation. These statistical models predict total individual biomass from dendrometric measures (stem diameter-at-breast-height and/or tree height). Most of the allometric models are developed in ecosystems outside Africa, and are expensive and time-consuming to develop for this region. Therefore, the models presently used as default estimators are not well adapted to the region, probably introducing uncertainty in the biomass estimates [22]. Thus, some technical problems need to be resolved. Once the woody biomass has been accurately estimated it is possible to estimate the amount of C stored in it and the corresponding amount of CO_2. The C fraction of dry woody biomass is approximately 0.5 of the absolute weight, while the corresponding CO_2 is calculated as C*3.67, as the mass of CO_2 (44.0 g mol^{-1}) is 3.67 times that of C (12.0 g mol^{-1}) [23].

4.2. Carbon modelling

To assess the scope for additional C stocking in West Africa, the potential amount of C biomass that the region could support under optimal management conditions was modeled. We used the "New IPCC Tier-1 Global Biomass Carbon Map for the Year 2000" data set, which quantifies biomass C stored in above- and below ground living vegetation for the year 2000 on a global level at a ~1-km spatial resolution [24], to represent the current vegetation biomass C stock across the West African region (figure 3). The realized C biomass is expected to be influenced by a number of factors, notably environmental factors such as climate and soil as well as anthropogenic disturbance, which may decrease the C biomass sequestered in the vegetation [25].

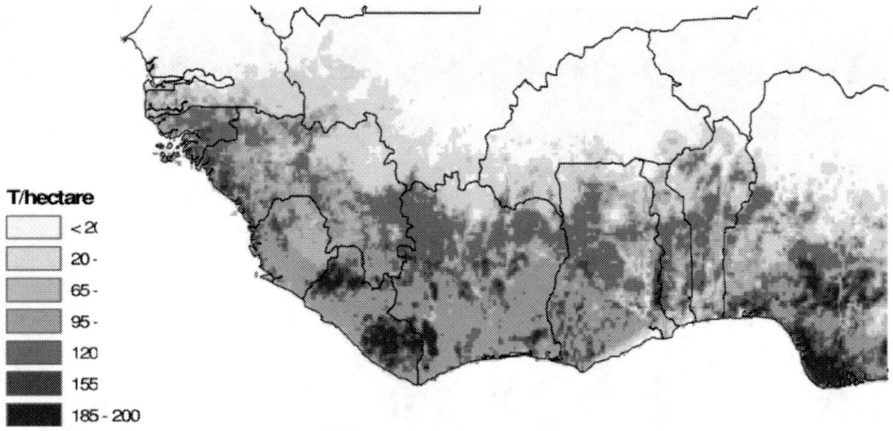

Figure 3. Map of carbon stored in living biomass above and below ground (from [24]).

Quantile regression was employed [26] to estimate potential C biomass across the West African region given the current climatic and soil conditions, but with optimal management, particularly in the absence of the intensive anthropogenic disturbance and over-use, which currently characterize extensive areas of the region [25]. In standard ordinary-least-squares linear regression, the mean of the response variable is estimated as a function of the predictors. However, quantile regression can be applied to estimate other parts of the response distribution, e.g., a given quantile of the response variable frequency distribution for a given combination of the predictor variables. Here, a 90%-quantile regression analysis was employed to conservatively estimate the maximum potential vegetation biomass C stock that an area could support given its climate (mean annual temperature, mean annual precipitation, precipitation seasonality) and soil (sub-soil pH) (figure 4).

Figure 4. Map of potential carbon stored in living biomass above and below ground under optimal conditions, based on 90% quantile regression. Letters indicate areas used as examples in table 1.

To compare present vegetation biomass C stock to maximum potential C stock, we subtracted the former from the latter (figure 5). Additionally, we selected seven sites in the Sudanian and Sahelian vegetation zones, which have been subject to different, known amounts of human impact, for further critical comparison of the current and potential vegetation biomass C stock (table 1).

Table 1. Estimates of current and potential carbon biomass and income possibilities through sale of CER at optimal management conditions in selected areas across West Africa.

Point in figure 4	Country	Vegetation zone	Current carbon ($tC\ ha^{-1}$)	Potential carbon ($tC\ ha^{-1}$)	Potential carbon increase ($tC\ ha^{-1}$)	Potential income from CER sale (€ ha^{-1})*
A	Senegal	Sudanian	4	111	107	5890
B	Senegal	Sahelian	4	55	51	2808
C	Senegal	Sudanian	51	121	70	3854
D	Burkina Faso	Sudanian	42	130	88	4844
E	Benin	Sudanian	44	98	54	2973
F	Niger	Sudanian	3	53	50	2753
G	Burkina Faso	Sahelian	4	44	40	2202

* Potential income is calculated based on a CO_2e market price of 15 € (potential C increase*3.67*15 €).

There is a general trend for the current C stock to decrease from the southern to the northern parts of West Africa (figure 3), as may be expected from the climatic gradients of the region (increasing aridity towards the north). The distribution of the C stock is patchy, however, particularly in the southern parts of the region, where farmlands with relatively low C stocks have replaced high C stock forest in many areas [27]. The quantile regression model (figure 4) suggests that a much higher potential for C storage in the vegetation exists across extensive areas of the region (figure 5 and table 1). The Sahelian zone sites all display low levels of current C biomass, though the quantile regression models suggest that an increase in vegetation C biomass is possible. In the Sudanian zone, the potential C storage increase is generally larger than in the Sahelian zone. It is apparent that the protected areas (national parks and classified forests) represented also display a potential for an increase in C storage, though not as much as the unprotected sites. Protected areas in this region are also impacted by human activities, especially fire, which may explain why they show a significant potential for C stock increases. Areas with high human population density show the greatest potential for C stock increases, which shows that human activities can significantly decrease the C stock through clearance of forests for agriculture, overgrazing, frequent burning and logging. Our results show that management aimed at reducing the negative anthropogenic impacts on vegetation cover could lead to substantial increases in vegetation biomass C stock (figure 5, table 1). Hence, even West Africa, which to a large extent is semi-arid, offers an important opportunity for climate change mitigation through CDM and REDD.

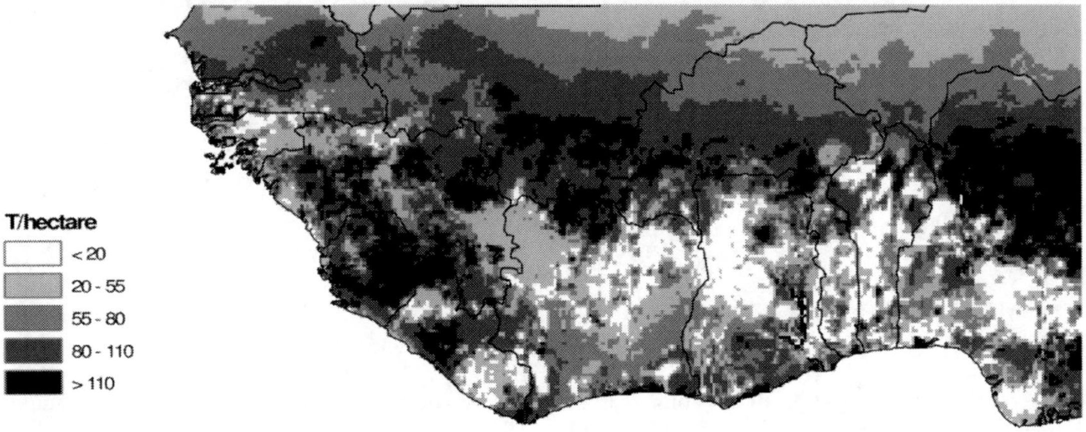

Figure 5. Carbon stocking potential, estimated as the difference between the potential (figure 4) and present vegetation biomass carbon stock (figure 3).

5. Carbon stock forests

5.1. Economic benefits

Economic benefits from CER investment through establishment of 1 ha of forest are potentially higher in the Sudanean zone, where the potential C stock can reach 155 tC ha^{-1}, compared to the Sahelian zone, where the potential C stock can reach 95 tC ha^{-1}, but mainly is <65 tC ha^{-1}. In the Sudanian zone potential increase from the present C stock typically ranges 55-80 tC ha^{-1} (or 202-294 t CO_2 ha^{-1}) corresponding to 3028-4404 € ha^{-1} after 20 years at the present CO_2e market price of about 15 €. In the Sahelian zone, the potential C increase typically lies between 20-55 tC ha^{-1}(or 73-202 t CO_2 ha^{-1}) corresponding to 1101-3028 € ha^{-1} after 20 years. A single tree with a biomass of 1 t would contain 0.5 tC corresponding to 1.8 t CO_2 and a value of 28 €.

Figure 6. Locust bean tree (*Parkia biglobosa*), a native species with a potential for combining carbon stocking, income generation from secondary crops, and conservation of biodiversity and ecosystems. An economist discusses the economic potentials of the tree with local people. Lower right corner: Seeds are fermented and used as a protein-rich food seasoning.

Opportunity costs from alternative sources such as pastoralism or agriculture of millet or sorghum correspond to 350-700 € ha^{-1} $year^{-1}$ depending on the zone and year, which would sum up to around four times the CER investment over 20 years. Secondary tree crops, however, can provide substantial additional income and food security, because they provide a variety of products, such as leaves used for spinach and fruits used in meals or raw, in addition to a number of commercialized non-food product. Few economic data are available, however, as secondary crops are marketed locally and therefore excluded from official statistics, but secondary crops derived from *Parkia biglobosa* (figure 6) can bring in the same income or often more than traditional agriculture 10 years after plantations are established or management of existing populations is initiated (own data). An example of economic benefits over a 20 year period is presented in figure 7. The CDM investment is relatively small compared to the expenses for establishment of plantations, as the costs of planting one ha can reach 5000 €, including plant nurseries, seedling transportation, fencing, hole staking, planting and supervision by experts. Hence the economic input from secondary crops and thereby the selection of tree species for AR is crucial for obtaining acceptable revenue from plantations.

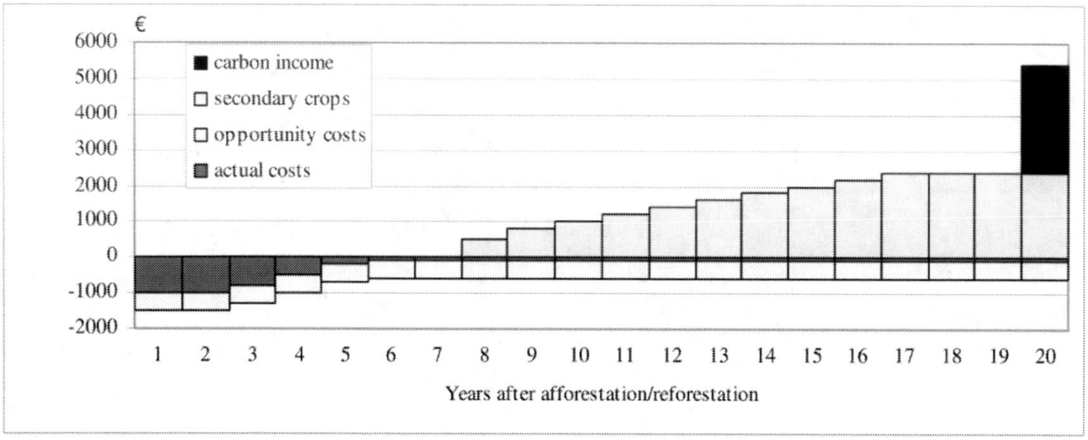

Figure 7. Example of costs and income generation over 20 years in an afforestation/reforestation (AR) project. The profitability of AR to a large extent depends of the revenue from secondary crops.

Economic benefits under REDD where people are compensated for avoided deforestation makes C stocking more attractive [28], as protecting forest is less costly and labour intensive than AR. In order to be attractive in a longer perspective, however, conserving a forest must provide an income from CER and secondary products at least on par with opportunity costs for the best alternative for landowners. Apart from this, there will be additional costs for monitoring and administration. If the market price would be 15 € for one CER, funding would range from 2753-8258 € ha^{-1} from CER sale for forest with 50-150 tC ha^{-1} (184-551 t CO_2 ha^{-1}). This is four times income from selling of 1 ha of forest for firewood, but less than the income from cultivation of crops over a 20 year period.
Although mechanisms for C stocking exists, the long-term economic benefit coupled with a high administrative load, makes it less attractive for local people to engage in C stocking projects [29]. Economic incentives and institutional capacity building are probably needed to make the projects realizable. Since C stocking have such a beneficial effect on sustainable economic development in rural areas developmental agencies should be willing to allocate money for this purpose. If Denmark for example were to achieve 1 percent point of its CO_2 reduction commitment through CDM, in an area with a potential C increase of e.g. 88 t ha^{-1}, Denmark would have to establish 22 km^2 of AR each year at a cost of 10.425.000 € per year (see calculation in table 2).

DK emissions in baseline year in CO_2 equivalents (CO_2e)	69.500.000 t CO_2
Kyoto commitment of 21% reduction in CO_2e	14.595.000 t CO_2
1 percentage point of commitment in CO_2e	695.000 t CO_2
1 percentage point of commitment in carbon (1 tC ~ 3.67 t CO2)	189.580 tC
Area needed at a carbon increase of 88 tC ha^{-1}	21.54 km^2
Annual costs for DK for reducing 1 percent point of commitment at 15 € t per CER	10.425.000 €

Table 2. Example on annual costs and areas employed if Denmark were to reduce 1 percent point of the Kyoto commitment through CDM in an area in West Africa with a potential increase of 88 tC ha^{-1}.

In a longer time perspective the income from secondary crops will far exceed CDM investment, if the tree planting and subsequent resource management is well organized. There are several challenges to be overcome such as selection of the most optimal species for planting, development of protocols for C measuring and legal interpretation of the regulations. If C stocking becomes a success it will increase the investments in the developing countries and eventually lead to a shift towards more sustainable land management and poverty alleviation.

5.2. Additional benefits

Carbon stocking by use of native tree species provides a number of additional benefits, which are not immediately measurable in an economic context. It may be an effective means to assure ecosystem resilience to future climatic change. It is the ability of the ecosystem to resist negative effects of climate change that determines its ability to render services to society. Reestablishment of the forest cover may contribute positively to local water circulation by causing biophysical changes in cloud formations and reflection of sunlight [19]. An increase in plant biomass will enable the ecosystem to sustain more wildlife, both in terms of individuals and in terms of species. Sound management may lead to more complex network of interactions with positive spin-off effects e.g. on pollination of potential fruit trees and pest control. Higher biomass will also lead to soil improvement because of increased litter decomposition and penetration of roots deep into the soil. Carbon stocking by means of native tree species could help to alleviate the destructive impacts of wildfires but again this depends very much on the management plans implemented. For the smallholders settled in the area replanting and enriching existing forest with economically important species not only provides new opportunities to generate cash income, but will also facilitate product diversification, thereby increasing food security and, by buffering against fluctuating world market prices on individual products, income stability.

5.3. Constraints and difficulties

A number of bureaucratic problems will have to be solved before establishment of C stock forest can become a reality, many of them arising from the approval process [30]. In West Africa well-functioning institutions for successful implementation of climate mitigation and adaptation mechanisms are lacking. Local communities are generally reluctant to support CDM and REDD initiatives unless they are convinced of immediate economic benefit and long term added benefits [31]. There are unfortunate examples of local farmers not receiving a fair share of the C revenue, and in some cases having been excluded from exploiting the forested lands [32]. The question is how to funnel the money gained from selling CERs back to the local communities in order to create the necessary incentives to make C stock forests economically and politically sustainable? This very much depends on land tenureship. One way of circumventing the problem would be to invest in cottage industries processing products extracted from the forest and to increase access to local markets by improving the infrastructure. Adding value to the product at the village level would have several positive spin-off effects such as creation of local jobs and lowering of transportation costs. A financial mechanism is needed for creating local incentives and assuring equitable distribution of benefits. A major problem associated with management of C stock forests is their vulnerability to disturbance that reemit C. This applies especially to fires that lead to substantial increases in C emissions [33]. Improved fire management imposes a major challenge to C stocking on the continental scale. Africa alone is responsible for 40% of the fire induced release of C into the atmosphere [34]. Another difficulty pertains to the estimation of biomass sequestered in the ecosystem since the upper 60 cm of the soil typically contains 39-69% of the total C stock [35]. The existing models for monitoring C stock need to be refined and tested on-ground by means of destructive sampling. Presently models are available that can monitor loss of C due to deforestation and selective logging with reasonable accuracy in rainforest [36], but they need to be adapted to dry forest and savanna environments where the relationship between aerial and underground biomass is probably much more variable. Real-time remote sensing of the amount of C sequestered could be an effective instrument in revealing deviations from agreed management plans and would undoubtedly solve several of the bureaucratic impediments mentioned earlier. Finally, it should be mentioned that the sequestration potential may likely change, in many cases decrease, as a consequence of climate change, because the precipitation in West Africa has been declining over the last decades and is likely to decline further as a consequence of climate change.

6. Conclusion

Most of the African continent is covered by savannas and woodlands that are often highly disturbed by anthropogenic factors and maintained in a degraded state. We have argued, that C stock forests in rural Africa have a potential and should be considered further. If planned correctly, C stock forests in these semi-arid ecosystems will not only help industrialized and rapidly growing developing countries to meet their ambitious goals for reduction of GHG emissions, but will also promote sustainable economic development in rural Africa by strengthening the natural resource basis. For local communities to engage in CDM and REDD project on a participatory basis, however, considerable technical support is needed due to complexity of the incentives. Significant transaction costs in terms of political and administrative barriers are most likely to occur when implementing the system. Future research should therefore try to address and solve implementation problems such as dealing with corruption problems and compensating local farmers in practice.

We urge the negotiators at the upcoming COP 15 meeting to consider semi-arid ecosystems in their discussions of the REDD mechanism. Hopefully a final definition of REDD that includes and focuses on forests in their broadest form will be adopted to allow for future C stocking projects in West Africa and other semi-arid regions worldwide.

7. Acknowledgements

Funding for the study was provided by SUN, tools for management and sustainable use of natural vegetation in West Africa (EU FP7 031685) and the Danish Natural Science Research Council (272-07-0242).

8. References

[1] IPCC 2007 Climate Change 2007 Synthesis Report. Contribution of Working Groups I, II and III to the Fourth Assessment Report of the Intergovernmental Panel on Climate Change. Core Writing Team, Pachauri R K and Reisinger A (eds) (Geneva: IPCC)

[2] Pascala S and Socolow R 2004 Stabilization wedges: Solving the climate problem for the next 50 years with current technologies *Science* **305** 968-972

[3] Boko M, Niang I, Nyong A, Vogel C, Githeko A, Medany I, Osman-Elasha B, Tabo R and Yanda P 2007 *Climate change 2007: Impacts, adaptation and vulnerability. Contribution of working group II to the fourth assessment report of the Intergovernmental Panel on Climate Change* Parry M L, Cansiani O F, Palutikof J P, van der Linden P J and Hanson C E (Cambridge: Cambridge University Press) 435-467

[4] Walker S, Pearson T, Munishi P and Petrova S 2008 Carbon market opportunities for the forestry sector of Africa *Winrock International and FAO* (Khartoum: Presented at the 16. session of African Forestry and Wildlife Commision)

[5] Grace J, José J S, Meir P and Miranda H S 2006 Productivity and carbon fluxes of tropical savannas *J Biogeogr* **33** 387-400

[6] Zomer R J, Trabucco A, Bossio D A and Verchot L V 2008 Climate Change mitigation: a spatial analysis of global land suitability for clean development mechanism afforestation and reforestation *Agric Ecosys Environ* **126** 67-80

[7] Cramer W, Bondeau A, Schaphoff S, Lucht W, Smith B and Sitch S 2004 Tropical forests and the global carbon cycle: impacts of atmospheric carbon dioxide, climate change and rate of deforestation *Philos Trans R Soc Lond Biol Sci* **359** 331-343

[8] UNEP 2003 *Clean Develpment Mechanism* (Roskilde: UNEP Risø Centre)

[9] UNEP Risø Centre 2009 CDM pipeline (www.cdmpipeline.org)

[10] McKinsey & Company 2009 Pathways to a low-carbon economy (www.mckinsey.com/clientservice/ccsi/pathways_low_carbon_economy.asp)

[11] Parker C, Mitchell A, Trivedi M and Mardas N 2008 *The little REDD book. A guide to governmental and non-governmental proposals for reducing emissions from deforestation and degradation* (Oxford: Global Canopy Programme)

[12] EEA 2007 EU greenhouse gas emissions decrease in 2005 (www.eea.europa.eu/pressroom/newsreleases/eu-greenhouse-gas-emissions-decrease-in-2005)

[13] FAO 2005 Global forest resources assessment (www.fao.org/forestry/fra2005/en/)

[14] Griscom B, Shoch D, Stanley B, Cortez R and Virgilio N in press Implications of REDD baseline methods for different country circumstances during an initial performance period

[15] Tiessen H, Feller C and Sampaio E V S B 2009 Carbon sequestration and turnover in semiarid savannas and dry forest *Clim Change* **40** 105-117

IOP Conf. Series: Earth and Environmental Science **8** (2009) 012004 doi:10.1088/1755-1315/8/1/012004

[16] Tappan G G, Sall M, Wood E C and Cushing M 2004 Ecoregions and land cover trends in Senegal *J Arid Environ* **59** 427-462

[17] Benitez P, McCallum I, Obersteiner M and Yamagata Y 2007 Global potential for carbon sequestration: Geographical distribution, country risk and policy implications *Ecol Econ* **60** 572-583

[18] Kronborg M, Lykke A M, Ilboudo J-B, Hien M and Balslev H in prep *Parkia biglobosa* as an economic resource for rural women in southwestern Burkina Faso *Ecol Econ*

[19] Canadell J G and Raupach M R 2008 Managing forests for climate change mitigation *Science* **320** 1456-1457

[20] Perez C, Roncoli C, Neely C and Steiner J L 2005 Can carbon sequestration markets benefit low-income producers in semi-arid Africa? Potentials and challenges *Agric Syst* **94** 2-10

[21] Patenaude G, Milne R and Dawson T P 2005 Synthesis of remote sensing approaches for forest carbon estimation: Reporting to the Kyoto Protocol *Environ Science Policy* **8** 161-178

[22] Mbow C 2009 Could carbon buy food? The stakes of mitigation versus adaptation to climate change in African countries *GLP News* **5** 20-23

[23] CarbonFix Standard 2009 CarbonFix Standard Draft ver. 3.0. Criteria and methodology (www.carbonfix.info/CarbonFix_Standard/Standard_Review.html?PHPSESSID=c3cmseiismb97c6ahvgqst4jt3)

[24] Ruesch A and Gibbs H K 2008 New IPCC Tier-1 Global Biomass Carbon Map for the year 2000 Carbon Dioxide information Analysis Center (http://cdiac.ornl.gov) Oak Ridge National Laboratory, Oak Ridge, Tennessee

[25] Sanderson E W, Jaiteh M, Levy M A, Redford K H, Wannebo A V and Woollmer G 2002 The human footprint and the last of the wild *BioScience* **52** 891-904

[26] Cade B S and Noon B R 2003 A gentle introduction to quantile regression for ecologists *Front Ecol Environ* **1** 421-420

[27] Mayaux P, Bertholome E, Fritz S and Belward A 2004 A new land-cover map of Africa for the year 2000 *J Biogeogr* **31** 861-877

[28] Coomes O T, Grimard F, Potvin C and Sima P 2008 The fate of the tropical forest: Carbon or cattle *Ecol Econ* **65** 207-212

[29] Hunt C 2008 Economy and ecology of emerging markets and credits for bio-sequestered carbon on private land in tropical Australia *Ecol Econ* **66** 309-318

[30] Basu P 2009 A green investment *Nature* **457** 144-146

[31] Pritchard D 2009 Reducing emissions from deforestation and forest degradation in developing countries (REDD). The link with wetlands. A background paper for *FIELD* 1-25 (www.eldis.org/go/topics/resource-guides/climate-change/forestry/reducing-emissions-from-deforestation-and-forest-degradation-redd&id=42802&type=Document)

[32] Jindal R, Swallow B and Kerr J 2008 Forestry-based carbon sequestration projects in Africa: Potential benefits and challenges *Nat Resour Forum* **32** 116-130

[33] Schultz M G, Heil A, Hoelzemann J J, Spessa A, Thonicke K, Goldammer J G, Held A C, Pereira J M C and Van Het Bolscher M 2009 Global wildland emissions from 1960 to 2000 *Global Biogeochem Cycles* **22** 1-17

[34] Williams C, Hanan N P, Neff J L, Scholes R, Berry J A, Denning A S and Baker F B 2007 Africa and the global carbon cycle *Carbon Balance Manag* **2** 1-13

[35] Wauters J B, Coudert S, Grallien E, Jonard M and Ponette Q 2008 Carbon stock in rubber tree plantations in Western Ghana and Mato Grosso *For Ecol Managet* **255** 2347-2361

[36] Tollefson J 2008 Save the trees *Nature* **452** 8-9

[37] IPCC 2006 Guidelines for National Greenhouse Gas Inventories, Prepared by the National Greenhouse Gas Inventories Programme. Eggleston H S, Buendia L, Miwa K, Ngara T and Tanabe K (eds) (Japan. IGES)

[38] R Development Core Team 2009 R: A language and environment for statistical computing (Foundation for Statistical Computing)

[39] Koenker R 2008 Quantreg: Quantile regression. R package vers. 4.26 (www.r-project.org)

[40] Burnham K P and Anderson D 1998 Model selection and mutimodel inference (New York: Springer)

[41] Spiess A N and Ritz C 2009 Qpcr: Modelling and analysis of real-time PCR data. R package (cran.r-project.org/web/packages/qpcR/index.html)

[42] Hijmans R J, Cameron S E, Parra J L, Jones P J and Jarvis A 2005 Very high resolution interpolated climate surfaces for global land areas *Int J Clim* **25** 1965-1978

[43] FAO, IIASA, ISRIC, ISSCAS and JRC 2009 *Harmonised World Soil Database* (Rome: FAO and IIASA)

IOP Conf. Series: Earth and Environmental Science **8** (2009) 012004 doi:10.1088/1755-1315/8/1/012004

Appendix 1. Description of the methodological approach for potential carbon modelling.

Environmental Layers

The Global Biomass C Map for the Year 2000 [24] quantifies above- and below-ground vegetation C biomass based on the relationship between C stocks and vegetation type and climate. This map is mainly based on the Global Land Cover Classification Map [27], and supplemented by three other datasets [24], which were used to classify different vegetation types. The IPCC GPG Tier-1 method [37] was used to estimate C biomass for different vegetation types under different climatic conditions, which were then mapped based on above-mentioned vegetation and climate maps. The 5-minute resolution version of the C biomass map was used here.

While subsoil pH is available for most of the African continent, some gaps do exist. Therefore, kriging was used to interpolate subsoil pH into the regions where values were unavailable. The kriging semivariogram was adjusted in the geostatistical analyst extension in ArcGIS 9.3. The interpolated areas were only used in the predictions; areas with missing data were excluded from the model-building process. All predictors were available at a resolution of ~1 km^2; therefore, they were resampled to the same resolution as the C biomass dataset (5 minutes) using the bilinear interpolation algorithm of the ArcGIS spatial analyst tool.

Models

Quantile regression models were constructed in R [38] using the package quantreg [39], whilst map predictions were constructed in ArcGIS 9.3. Models of the relationship between C biomass and the predictors were quantified across the entire African continent. We used Akaike weights to decide on the best model of C biomass in relation to the environmental predictors [40]. The Akaike weight of the best model was greater than 0.99 and included all predictors and their squares (i.e. all relationships between C biomass and the predictors were quadratic). Predictions made by this model were projected to the West African region only. Akaike weights were calculated using the R package qpcR [41]. Areas in which predictions of C biomass exceeded 200 $t*ha^{-1}$, which is the current maximum C biomass of the African continent, were reclassified to 200 $t*ha^{-1}$.

Beyond Kyoto: Addressing the Challenges of Climate Change IOP Publishing
IOP Conf. Series: Earth and Environmental Science **8** (2009) 012005 doi:10.1088/1755-1315/8/1/012005

REDD and PINC: A new policy framework to fund tropical forests as global 'eco-utilities'

M R Trivedi[1,1], A W Mitchell[1], N Mardas[1], C Parker[1], J E Watson[2] and A D Nobre[3]

[1] Global Canopy Programme, John Krebs Field Station, Wytham, Oxford, OX2 8QJ, UK
[2] University of Queensland, The Ecology Centre, Queensland 4072, Australia
[3] Instituto Nacional de Pesquisas da Amazonia, INPA, Escritório Regional do INPA, São José dos Campos, 12227-010, Brazil

E-mail: m.trivedi@globalcanopy.org

Abstract. Tropical forests are 'eco-utilities' providing critical ecosystem services that underpin food, energy, water and climate security at local to global scales. Currently, these services are unrecognised and unrewarded in international policy and financial frameworks, causing forests to be worth more dead than alive. Much attention is currently focused on REDD (Reducing Emissions from Deforestation and forest Degradation) and A/R (Afforestation and Reforestation) as mitigation options. In this article we propose an additional mechanism – PINC (Proactive Investment in Natural Capital) – that recognises and rewards the value of ecosystem services provided by standing tropical forests, especially from a climate change adaptation perspective. Using Amazonian forests as a case study we show that PINC could improve the wellbeing of rural and forest-dependent populations, enabling them to cope with the impacts associated with climate change and deforestation. By investing pro-actively in areas where deforestation pressures are currently low, the long-term costs of mitigation and adaptation will be reduced. We suggest a number of ways in which funds could be raised through emerging financial mechanisms to provide positive incentives to maintain standing forests. To develop PINC, a new research and capacity-building agenda is needed that explores the interdependence between communities, the forest eco-utility and the wider economy.

1. Introduction

"We are still learning the 'nature of value', as we broaden our concept of 'capital' to encompass human capital, social capital and natural capital. By recognizing and by seeking to grow or conserve these other 'capitals', we are working our way towards sustainability."
Sukhdev, P. in [1]

Tropical forests play a crucial role in the climate system [2, 3], which is being recognised by international policy. This is primarily because tropical deforestation contributes about a fifth of global carbon emissions [4] and atmospheric carbon levels will not stabilise if deforestation is allowed to

[1] To whom correspondence should be addressed

continue [5]. Tackling deforestation in some regions could be one of the quickest and most cost-effective ways to mitigate climate change and it has been shown that delaying action to stop deforestation will lead to greater costs later [6]. As a result, much attention is focused on tropical forests in the lead up to the Conference of the Parties (COP 15) to the UN Framework Convention on Climate Change (UNFCCC) in Copenhagen, December 2009 because of the 'urgent need to reduce emissions from deforestation and forest degradation' [7]. There is considerable scientific and political support for the COP to agree a successor to the Kyoto Protocol that provides incentives for developing countries to maintain their forest carbon stores and promotes reforestation on previously deforested land. The name for this proposed mechanism is 'reducing emissions from deforestation and forest degradation in developing countries' (REDD) [7].

The economic rationale for tropical deforestation is clear: the alternative uses for land and timber provide higher short term financial returns than standing forests [8]. However, the significance of tropical forests in harbouring biodiversity and providing globally significant ecosystem services (ES) is also clear [9]. One recent study put the annual monetary cost (in net present value terms) to the global economy of tropical deforestation, through the lost services, at $2-5 trillion [10] and yet sufficient resources have not been directed to halting this loss. Since carbon dioxide has a price on global financial markets, it could provide a proxy by which to maintain other values [11]. Recognising this, the UNFCCC makes reference to the 'co-benefits' of a REDD agreement while remaining focused on its mission to reduce carbon emissions. Unlike the Convention on Biological Diversity (CBD), the UNFCCC lacks a specific mandate to maintain or enhance ES provided by biodiversity. We argue, however, that the maintenance and enhancement of biodiversity and ES underpins the resilience of forest ecosystems and their capacity to retain stored carbon and should therefore be brought to the heart of any comprehensive climate change strategy. Climate policy should therefore take a broad perspective that recognises that biodiverse, natural tropical forests provide opportunities for both mitigation and adaptation to climate change. In this paper, we develop a framework that considers tropical forests as global 'eco-utilities' providing ES that are important for both adaptation and mitigation. We illustrate our discussion with the case of Amazonia, where research programmes such as the Large-scale Biosphere-Atmosphere (LBA) experiment have provided a solid scientific basis to our understanding of ES. We propose that an additional mechanism – Proactive Investment in Natural Capital (PINC) – is needed to complement REDD in providing incentives for the large areas of standing forest that are currently unthreatened but could become so as deforestation and climate change advance.

2. Humid tropical forests as 'eco-utilities'

Over half of the world's terrestrial biodiversity is found in tropical forests [12]. This biodiversity and the resultant ecosystem functions are the basis for a suite of ES that contribute to human well-being, particularly among the poor [13]. Like public utilities, these forests provide services that underpin water, health, livelihood, food and energy security for more than a billion people [9]. We propose that the world's three great humid tropical forest regions (Amazonia, Congo Basin and SE Asia) could be considered giant 'eco-utilities', generating ES upon which humanity relies at local to global scales.

2.1. The tropical forest eco-utility

At the local level, tropical forest canopies moderate air temperature through shade and evaporative cooling [14]. Canopy biodiversity could play a significant role in this process. Epiphytes and mosses trap rain and release it back as vapour [15] while plant-herbivore interactions in the canopy generate volatile organic compounds, which could catalyse cloud formation [16, 17]. Trunks and roots act like a sponge, controlling the flow of surface and ground water into river systems and helping to mitigate flooding [18]. Summing these local processes over vast expanses of forest influences regional to global climate and hydrology [19-21].

2.1.1. Carbon capture and sequestration – mitigating climate change. Atmospheric studies suggest that undisturbed tropical forests act as carbon sinks [22], absorbing CO_2 from the atmosphere and storing it at the rate of about 1 tonne of carbon per hectare per year [23]. Although the topic of debate, this could amount to a global sink of about 1.2 billion tonnes of carbon each year [4], worth £110 billion based on the social cost of carbon (£25/tCO_2 in 2007 [24]). Recent research suggests that the role these forests play in regulating global climate may be greater than previously thought [22] and will likely become even more important as alternative sinks become saturated [25] while forests can continue to act as sinks throughout a century of climate change [2].

2.1.2. Water pumping – moderating surface temperatures. Although tropical forests have been labelled 'the lungs of the world' a more appropriate analogy may be that they are its "beating heart": pumping water and heat from the tropics to the rest of the planet. Amazonian forests release around eight trillion tonnes of water vapour to the atmosphere each year [26]. This evaporation process uses enormous amounts of solar energy, equivalent to the daily power of 50,000 of the largest hydropower stations. As a result, humid tropical forests cool the planet's surface in addition to their role in removing greenhouse gases from the atmosphere [19, 27].

2.1.3. Rainfall recycling – supporting energy and food security. Up to 40% of precipitation in Amazonia is from local evapotranspiration [28]. Furthermore, it has been suggested that continuous expanses of forest such as Amazonia act as biotic pumps, drawing moist air from the ocean and recycling it large distances in-land [29]. If the biotic pump hypothesis is correct, deforestation could have immense impacts on regional and global climate. Even if deforestation has a limited impact on rainfall, this could still have a significant effect on agricultural production in the region [30]. The transpiration service supports many other functions and services of the forest and delivers moisture to southern Brazil. Moisture is also transported by low-level atmospheric jet streams, or 'flying rivers', thousands of kilometres south to the La Plata River Basin [31] where agriculture, hydropower and industry generate about $1 trillion for the five countries sharing the basin [32]. The amount of moisture transported from Amazonia and the extent to which the economy of the Plata Basin depends on this service is unknown.

We divide the Amazonian eco-utility into three parts for the purposes of discussion: (i) standing forests, (ii) frontier forests and (iii) deforested and degraded land. Standing forests support human populations that are not currently clearing large areas of forest. In contrast, frontier forests are under direct and immediate threat from human activities, as in the 'arc of deforestation'. In such areas, the eco-utility services are being replaced by other ecosystem services such as agricultural production and forestry operations. Finally, there are large areas that were once forest but have been converted to agriculture and are now degraded of their natural capital. Conservation and sustainable development activities should aim to support the standing forests, stabilise the frontier, and restore the degraded forests. Each requires different tactics, but we suggest that the overall goal should be long-term maintenance of the ES provided by the forest eco-utility.

3. Threats to the eco-utility from climate change and deforestation

3.1. The 2005 Amazon drought

The Amazon drought of 2005 provides an example of the kind of extreme event that could occur more frequently under climate change. The southwestern and western portions of Amazonia experienced one of their worst droughts in 60 years, compounded by extensive forest fires. The cause appears to have been warmer global temperatures, which led to raised sea surface temperatures in the northern tropical Atlantic Ocean, and ultimately lower rainfall in Amazonia [34]. The diminished rainfall resulted in exceptionally low water levels in the Amazon river, draining many floodplain lakes and streams and isolating hundreds of riverine villages and communities. The government called a state of emergency and mobilized the army to provide water and medical supplies to these communities and contend with the intense forest fires in Brazil's western state of Acre [35]. Many fires were clustered close to forest edges, indicating that human activities made the forest more fire prone [36]. The drought caused massive carbon loss from the forest, mainly through killing trees that could not tolerate the drier soil conditions. Amazonian forests normally absorb nearly 2 billion tonnes of CO_2 each year, but in 2005 the forest decreased in living biomass by more than 3 billion tonnes [37]. At the same time, however, canopy 'greenness' increased in intact forest areas during the peak of the drought [38], which suggests that Amazonian forests could be more resilient to drought than previously assumed.

3.2. Mitigation and adaptation

Mitigating climate change through rapid reduction of greenhouse gas emissions from all sources is essential, but since the atmosphere is committed to some degree of warming over the course of the next century, there is also a growing need for adaptation to climate change. Recent extreme events, such as the 2005 drought, provide an indication of the potential future impacts of climate change. Adaptation has been less reflected in international policy discussions [33] yet is crucial from the perspective of local populations, biodiversity conservation and the maintenance of ES, especially in developing nations which may be experiencing the impacts of climate change already.

Conservation and sustainable development actions need to acknowledge that the drivers of biodiversity and ES loss such as deforestation and climate change are expected to act in synergy [39]. As a result of these interacting pressures, half of the forest area of Amazonia might be cleared, logged or burned within 20 years [40], while the loss of 30-40% could push it into a drier climatic state [41], emitting carbon and accelerating climate change. Tropical forest loss might also promote warming by reducing evapotranspiration and cloud formation [19, 27]. Mitigation efforts that protect forests should therefore also be seen as being adaptive, by reducing the risk of fire [42] and maintaining cloud cover. This underscores the importance of developing an integrated mitigation and adaptation strategy for the entire Amazonian biome that considers the role of hydrological regulation as a supporting ecosystem service that underpins carbon storage, biodiversity and wellbeing in the forest in the face of climate change.

4. From REDD to PINC: a response framework based on ecosystem services

The proposed REDD mechanism offers incentives to countries to reduce their deforestation relative to a historical reference level calculated from their deforestation rate in a recent snapshot of time (e.g. 1990s). It is currently proposed that this mechanism would operate at the national level and link compensation to a country's success in reducing recent deforestation rates. However, there have been concerns that targeting only a subset of countries (those countries with high deforestation rates in the recent past) and connecting the incentive only to the annual deforested area (with no relation to total forest area) would compromise the effectiveness of the mechanism [43]. The drivers of deforestation are mobile and vary over time and geography [44]. So despite proposals to minimise leakage through national accounting, a REDD mechanism may inadvertently push deforestation to new areas.

Some countries such as Costa Rica and India have already had success with programs to restore their forests and feel they should receive compensation for these early conservation actions. Others, such as the Democratic Republic of Congo, have large areas assigned to logging concessions and are keen for REDD to support sustainable forest management [45]. Countries like Guyana and Suriname, with high forest cover and low deforestation rates (HFLD) [46], argue that they should be rewarded for good stewardship of their forests as, at this stage, they want to retain their standing forests.

We argue that a range of responses is required to suit the varying contexts in the three different parts of the forest eco-utility: standing, frontier and degraded land (table 1). The common principle underlying all responses is that of the 'beneficiary pays' [47], since all require some form of payment for ecosystem services (PES) mechanism. The precautionary and the 'polluter pays' principles also apply depending on the context.

Table 1. The range of services, responses and tools needed to maintain the forest eco-utility.

	Standing forest	**Frontier forest**	**Degraded land**
Climate change	Adaptation	←Adaptation - Mitigation→	Mitigation
Regulating ES:			
Carbon	Store and sink	Avoided emissions	Sequestration
Water	Rainfall recycling	←Rainfall - Watershed→	Watershed protection
Climate	Global cooling	←Global-Local→	Local cooling
Risk & uncertainty[a]	High	←High-Low→	Low
Response	Support	Stabilise	Restore
Principle	Precautionary & BP	Polluter pays & Beneficiary pays	
Payment for	Standing forest	Reduced deforestation	Reforestation
Cost	Adaptive management	Opportunity & management	Opportunity & management
Expense	Low	←Low-High→	High
Mechanism	PINC & REDD	REDD	A/R

[a] 'Risk and uncertainty' refers to the state of knowledge of the impacts of losing ecosystem services.
Abbreviations: BP Beneficiary Pays; PINC Proactive Investment in Natural Capital; REDD Reducing Emissions from Deforestation and Degradation; A/R Afforestation and Reforestation

4.1. Standing forests

Forest communities who maintain and/or depend upon standing forests may not be at risk from deforestation currently, but climate change could threaten their livelihoods and food security in the future. The precautionary principle should therefore be followed and they should be given support to prevent uncertain but potentially serious harm to their wellbeing. Standing forests provide important regulating ES such as rainfall recycling and climate moderation and are therefore especially important to the region from a climate change adaptation perspective. The fact that large areas of standing forests are remote and sparsely populated means that opportunity costs and management costs are relatively low [48] compared with frontier and degraded areas [43]. The forest could be maintained for as little as $10 per hectare per year [49]. Future zones of conflict between conservation and economic development and regions vulnerable to climate change are not necessarily coincident with current deforestation hotspots [50-52]. Therefore, the costs of conservation will rise over time in standing forests unless sustainable development and conservation are reconciled in the short term.

We propose that a new framework mechanism of Proactive Investment in Natural Capital (PINC) is adopted to promote adaptation in standing forests that may not benefit directly from REDD funds (figure 1). We use the word 'proactive' to reflect that PINC would be a mechanism to build the resilience to climate change of the stewards of the eco-utility rather than to compensate for the costs of transforming and enforcing the large-scale economic drivers of deforestation such as the agricultural sector, for which REDD funds would be needed. Short-term investment in climate adaptation strategies that include, for example, agroforestry, fire limitation, community forest management and sustainable livelihoods would also have long-term climate change mitigation benefits by decreasing the chances of future deforestation.

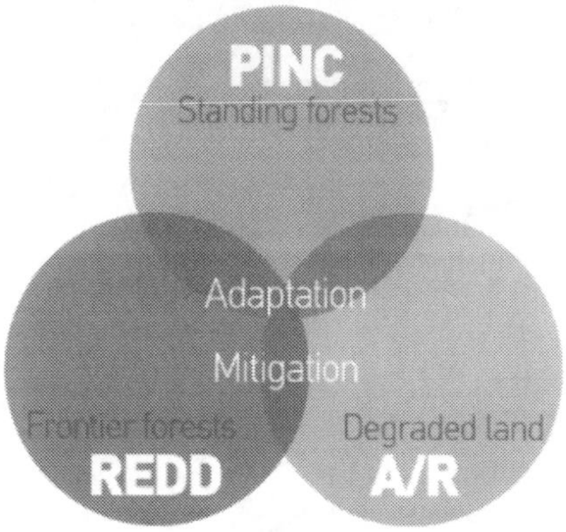

Figure 1. Three mechanisms for maintaining or restoring the tropical forest eco-utility. Mechanisms are represented by circles. Standing forests require an adaptation tool – PINC – while frontier forests and deforested land require REDD and A/R, respectively, in order to mitigate climate change. Mitigation activities such as fire prevention are also adaptive. Although PINC is aimed at climate adaptation, it would have long-term mitigation benefits by reducing the chances of future deforestation activities occurring.

4.2. Frontier forests

Deforestation in Amazonia is driven primarily by economic profitability (agricultural expansion and logging) and governance weaknesses (notably, lenient law enforcement), and only to a much lesser extent by deterministic poverty cycles [53, 54]: in the Brazilian Amazon, small-scale farmers are estimated to account for about one fifth of deforestation [8]. Actions to reduce deforestation in the frontier, e.g. preventing further agricultural expansion, will aim to avoid carbon emissions and in the process should help to secure biodiversity and ES. Since the majority of greenhouse gas emissions are from industrial nations, the polluter pays principle suggests that REDD will involve rich polluting nations compensating developing nations for (i) the opportunity costs of conservation; (ii) the costs of agricultural intensification and the certification of logging and agribusiness; and (iii) the management costs entailed in enforcement and monitoring at the frontier.

4.3. Degraded land

Forest clearing for cattle-ranching in Amazonia has left large areas of degraded pastureland, which are abandoned to be colonized by secondary vegetation and are now increasingly used for soybean cultivation [54] as demand for this commodity grows. Rehabilitation of the land is possible through agroforestry techniques [55], but the costs of reforestation can be significant. Afforestation and Reforestation (A/R) under the Kyoto Protocol's Clean Development Mechanism (CDM) can be used to generate funds to meet the costs of reforestation. There are currently very few A/R projects in developing countries due to the complicated and expensive process required to meet CDM criteria.

5. Paying for PINC

While ES are defined as 'the benefits that people obtain from ecosystems' [9], 'natural capital' is the capacity of the ecosystem to yield a flow of goods and services [56]. Payments for Ecosystem Services (PES) are usually made to cover the opportunity costs of the supply of a specific service rather than to pay for the actual value of the benefits. In the case of the eco-utility, the ES are provided by the natural capital of the Amazonian biome and PINC would aim to maintain this capital in the face of climate change, rather than pay for individual services, such as carbon. In order to keep the PINC system simple, the unit of measurement would be a hectare of intact forest, providing a bundle of services.

REDD and A/R are climate *mitigation* activities focused on paying for forest carbon storage. In contrast, PINC is a climate *adaptation* strategy and therefore needs to focus on the range of ecosystem services provided by forests that support resilience to climate change, specifically those related directly to biodiversity and water. These are key resources in the Andes-Amazon region and are likely to be significantly affected by climate change [57]. They are also crucial for regional and global populations, who may be willing and able to pay to maintain the resilience of the eco-utility through traditional sources (e.g. taxes and donations) or more novel ones (e.g. rainforest bonds, insurance premiums, micro-credit and patient capital). Below we outline some of the potential innovative financial mechanisms that could be used.

5.1. Rainforest bonds

The Prince's Rainforests Project's [58] 'Emergency Package for Tropical Forests' has proposed an international financing mechanism that hinges on the concept of Rainforest bonds. These are fixed income 'securities' that could be issued by governments in global private capital markets, with the potential to raise billions of dollars in upfront financing for rainforests from 2010; prior to the flow of funds in the UN's post-Kyoto system. Underwriting governments will have the time to generate revenues for repayment of the bonds from clean development investments, domestic carbon permit auctions or other schemes.

5.2. Insurance premiums

The costs of recent environmental disasters such as Hurricane Katrina (US$150 billion) [59] and the Indian Ocean Tsunami (total rebuilding cost: US$10.4 billion) could have been mitigated if there had been large-scale proactive investment in natural buffers such as wetlands and mangrove forests. Appetite is growing in the insurance sector for investment in natural capital. For instance, insurance firms and major shipping companies are financing a 25-year project to restore forest ecosystems along the 80-kilometre length of the Panama Canal, in order to reduce soil erosion and control the flow of freshwater into the canal, thus reducing insurance risks so that the shippers enjoy lower premiums [1]. The National Flood Insurance Program and the Federal Crop Insurance Corporation in the US have exposures of $1 trillion and $44 billion respectively, and these are expected to rise under climate change [60]. A levy on instruments such as weather derivatives or on the catastrophe element of insurance premiums could generate billions of dollars each year [58].

5.3. Microcredit and patient capital

Rural lending schemes have been very successful in helping to alleviate poverty and may help to build the resilience of poor smallholders to climate change. It may be possible in some standing forests, forest frontiers and degraded/deforested areas to use micro-credit to create social businesses that are self-sustaining but not based on profit maximisation [61]. Care is needed, however, in order to ensure that access to credit does not result in an increased ability to clear land [8]. Parallel to such grassroots initiatives that focus on the basic needs of the rural population, there is an urgent need for a scientific and technological revolution in Amazonia [62]. A decentralized network of new science and technology institutions throughout the Amazon could seed an innovative 'bio-industrial' model for the region [63]. Patient capital investment could help to develop the value chains of new forest products, which could add value to the biodiversity and ecosystem services of the forest [63].

Private patient capital investors are showing increasing confidence in the potential for a new market in ecosystem services. UK investment firm Canopy Capital (www.canopycapital.co.uk) bought the rights to market the ecosystem services provided by 350,000 ha of the Iwokrama rainforest reserve in Guyana in 2008, the first deal of this kind in the world. In the future, if regulatory frameworks can be created that recognise the value of standing forests, financial institutions such as pension funds and banks could find a place for such investments in their portfolios.

5.4. Eco-certification

Emerging certification schemes for agribusiness products such as beef, soy and biofuels are increasingly being asked to consider the 'eco-dependency' of product procurement on tropical forest ecosystem services (such as rain provision) and to mitigate the destructive forces that supply chains have on these forests, i.e. their 'forest footprint' [64]. Such schemes may be encouraged to include a PES element into the certification criteria.

6. Conclusion

Fundamentally, the 'eco-utility' framework values forests in equal measure for climate change mitigation and adaptation. Since the publication of the Millennium Ecosystem Assessment [9], the conservation and development movements have had the evidence-base upon which to join forces to improve wellbeing and create wealth through the improved management of biodiversity and ecosystem services. Climate change makes this effort imperative. Our framework for maintaining the tropical forest 'eco-utility' proposes that different mechanisms are needed to address climate change and deforestation: from REDD and A/R to PINC. REDD and A/R are part of an international carbon mitigation framework under the auspices of the UNFCCC. PINC would be an adaptation response, perhaps fostered by both the UNFCCC and the CBD, based on the recognition of the value of the ecosystem services provided by large areas of standing forests, including climate regulation, rainfall generation and carbon sequestration. The scientific understanding of these services is still under development and must be improved so that they can be recognised through policymaking. Hence, creating PINC will require a new research and development agenda that integrates climate change impacts, socio-economic assessment of ecosystem services, and mechanisms to raise funds and distribute benefits to rural communities.

7. References

[1] European Communities. The economics of ecosystems and biodiversity: an interim report; 2008. http://ec.europa.eu/environment/nature/biodiversity/economics/

[2] Gullison RE, Frumhoff PC, Canadell JG, Field CB, Nepstad DC, Hayhoe K, et al. Tropical forests and climate policy. *Science* 2007 **316** (5827) 985-6.

[3] Nabuurs GJ, et al. Forestry. In: Metz B, Davidson OR, Bosch PR, Dave R, Meyer LA, eds. *Climate Change 2007: Mitigation Contribution of Working Group III to the Fourth Assessment Report of the Intergovernmental Panel on Climate Change*. Cambridge, United Kingdom and New York, NY, USA: Cambridge University Press 2007.

[4] Denman KL, et al. Couplings Between Changes in the Climate System and Biogeochemistry. In: Solomon S, Qin D, Manning M, Chen Z, Marquis M, Averyt KB, et al., eds. *Climate Change 2007: The Physical Science Basis Contribution of Working Group I to the Fourth Assessment Report of the Intergovernmental Panel on Climate Change*. Cambridge, United Kingdom and New York, NY, USA: Cambridge University Press 2007.

[5] Socolow S, Pacala R. Stabilization wedges: solving the climate problem for the next 50 years with current technologies. *Science* 2004 **305** 968-72.

[6] Stern N. The Stern Review on the Economics of Climate Change. Cambridge: Cambridge University Press 2006.

[7] UNFCCC. Land use, land use change and forestry, Draft conclusions proposed by the Chair. *FCCC/KP/AWG/2008/L5, 11 June 2008, unfcccint/resource/docs/2008/awg5/eng/l05pdf* 2008.

[8] Chomitz KM, Buys P, De Luca G, Thomas TS, Wertz-Kanounnikoff S. *At Loggerheads? Agricultural Expansion, Poverty Reduction, and Environment in the Tropical Forests*. Washington, D.C.: World Bank 2006.

[9] Millennium Ecosystem Assessment. *Ecosystems and human well-being: biodiversity synthesis*. Washington, D.C.: World Resources Institute 2005.

[10] ten Brink P, et al. The Cost of Policy Inaction – in Monetary terms. In: Braat L, ten Brink P, eds. *The Cost of Policy Inaction: The case of not meeting the 2010 biodiversity target*. Wageningen / Brussels 2008.

[11] Miles L, Kapos V. Reducing Greenhouse Gas Emissions from Deforestation and Forest Degradation: Global Land-Use Implications. *Science* 2008 **320** (5882) 1454-5.

[12] Myers N. *The Sinking Ark: A New Look at the Problem of Disappearing Species*. New York and Oxford: Pergamon Press 1979.

[13] Fisher B, Christopher T. Poverty and biodiversity: Measuring the overlap of human poverty and the biodiversity hotspots. *Ecol. Econ.* 2007 **62** (1) 93-101.
[14] Pyke CR, Andelman SJ. Land use and land cover tools for climate adaptation. *Clim. Change.* 2007 **80** (3-4) 239-51.
[15] Wallace J, McJannet D. Modelling interception in coastal and montane rainforests in northern Queensland, Australia. *J. Hydrol.* 2008 **348** 480-95.
[16] Meir P, Cox P, Grace J. The influence of terrestrial ecosystems on climate. *Trends Ecol. Evol.* 2006 **21** (5) 254-60.
[17] Claeys M, et al. Formation of secondary organic aerosols through photooxidation of isoprene. *Science* 2004 **303** (5661) 1173-6.
[18] Laurance WF. Environmental science - Forests and floods. *Nature* 2007 **449** (7161) 409-10.
[19] Betts RA, Falloon PD, Goldewijk KK, Ramankutty N. Biogeophysical effects of land use on climate: Model simulations of radiative forcing and large-scale temperature change. *Agr. Forest Meteorol.* 2007 **142** (2-4) 216-33.
[20] Pielke RA, et al. An overview of regional land-use and land-cover impacts on rainfall. *Tellus B* 2007 **59** (3) 587-601.
[21] Bala G, et al. Combined climate and carbon-cycle effects of large-scale deforestation. *P. Natl. Acad. Sci. USA.* 2007 **104** (16) 6550-5.
[22] Stephens BB, Gurney KR, Tans PP, Sweeney C, Peters W, Bruhwiler L, et al. Weak northern and strong tropical land carbon uptake from vertical profiles of atmospheric CO2. *Science* 2007 **316** (5832) 1732-5.
[23] Baker TR, et al. Increasing biomass in Amazonian forest plots. *Philos. T. Roy. Soc. Lond. B.* 2004 **359** (1443) 353-65.
[24] DEFRA. *The Social Cost Of Carbon And The Shadow Price Of Carbon: What They Are, And How To Use Them In Economic Appraisal In The UK.* London: DEFRA; 2007. http://www.defra.gov.uk/ENVIRONMENT/climatechange/research/carboncost/
[25] Le Quéré C, et al. Saturation of the southern ocean CO_2 sink due to recent climate change. *Science.* 2007 **316** 1735-8.
[26] Intergovernmental Panel on Climate Change. *Climate change 2007: the physical science basis.* Contribution of working group I to the fourth assessment report of the IPCC. Cambridge, UK: Cambridge University Press 2007.
[27] Bonan GB. Forests and Climate Change: Forcings, Feedbacks, and the Climate Benefits of Forests. *Science* 2008 **320** (5882) 1444-9.
[28] Eltahir EAB, Bras RL. Precipitation recycling. *Rev Geophys.* 1996 **34** (3) 367-78.
[29] Makarieva AM, Gorshkov VG. Biotic pump of atmospheric moisture as driver of the hydrological cycle on land. *Hydrol. Earth Syst. Sci.* 2007 **11** (2) 1013-33.
[30] FAO. Forestry and Food Security. Rome: Food and Agriculture Organisation of the United Nations; 1989.
[31] Marengo JA, Soares WR, Saulo C, Nicolini M. Climatology of the low-level jet east of the Andes as derived from the NCEP-NCAR reanalyses: Characteristics and temporal variability. *J. Clim.* 2004 **17** (12) 2261-80.
[32] Vera C, et al. The South American low-level jet experiment. *B. Am. Meteorol. Soc.* 2006 **87** (1) 63-84.
[33] Pielke R, Prins G, Rayner S, Sarewitz D. Lifting the taboo on adaptation. *Nature.* 2007 **445** 597-8.
[34] Marengo JA, et al. The drought of Amazonia in 2005. *J. Clim.* 2008 **21** 495-516.
[35] Brown IF, et al. Monitoring Fires in Southwestern Amazonia Rain Forests. *EOS T. Am. Geophys. Un.* 2006 **87** (26) 253-64.
[36] Aragão LEOC, Malhi Y, Roman-Cuesta RM, Saatchi S, Anderson LO, Shimabukuro YE. Spatial patterns and fire response of recent Amazonian droughts. *Geophys. Res. Lett.* 2007 **34** (L07701):doi:10.1029/2006GL028946.

[37] Phillips OL, Aragao LEOC, Lewis SL, Fisher JB, Lloyd J, Lopez-Gonzalez G, et al. Drought Sensitivity of the Amazon Rainforest. Science. 2009 March 6, 2009;323(5919):1344-7.
[38] Saleska SR, Didan K, Huete AR, da Rocha HR. Amazon forests green-up during 2005 drought. *Science* 2007 **318** 612.
[39] Brook BW, Sodhi NS, Bradshaw CJA. Synergies among extinction drivers under global change. *Trends Ecol. Evol.* 2008 **23** (8) 453-60.
[40] Nepstad DC, Stickler CM, Soares B, Merry F. Interactions among Amazon land use, forests and climate: prospects for a near-term forest tipping point. *Philos. T. Roy. Soc. Lond. B.* 2008 **363** (1498) 1737-46.
[41] Oyama MD, Nobre CA. A new climate-vegetation equilibrium state for tropical South America. *Geophys. Res. Lett.* 2003 **30** (23) 2199.
[42] Golding N, Betts R. Fire risk in Amazonia due to climate change in the HadCM3 climate model: Potential interactions with deforestation. *Global Biogeochem. Cy.* 2008 **22** (GB4007):doi:10.1029/2007GB003166.
[43] Strassburg B, Turner RK, Fisher B, Schaeffer R, Lovett A. Reducing emissions from deforestation--the "Combined incentives" Mechanism and empirical simulations. *Global Environ. Chang.* 2009 **19** 265-78.
[44] Soares BS, et al. Modelling conservation in the Amazon basin. *Nature* 2006 **440** (7083) 520-3.
[45] UNFCCC. Views on issues related to further steps under the Convention related to reducing emissions from deforestation in developing countries: approaches to stimulate action. 28th Session, Bali, 3-11 December, 2007: Subsidiary Body for Scientific and Technological Advice; 2007. http://unfccc.int/resource/docs/2007/sbsta/eng/misc14.pdf
[46] da Fonseca GAB, Rodriguez CM, Midgley G, Busch J, Hannah L, Mittermeier RA. No forest left behind. *Plos Biol.* 2007 **5** (8) 1645-6.
[47] Pagiola S, Platais G. *Payments for Environmental Services: From Theory to Practice*. Washington, DC: World Bank; 2008.
[48] Mittermeier RA, Mittermeier CG, Brooks TM, Pilgrim JD, Konstant WR, da Fonseca GAB, et al. Wilderness and biodiversity conservation. *P Natl. Acad. Sci. USA*. 2003 **100** (18) 10309-13.
[49] Pimm SL, et al. Environment - Can we defy nature's end? *Science* 2001 **293** (5538) 2207-8.
[50] Cardillo M. Disappearing forests and biodiversity loss: which areas should we protect? *Int. For. Rev.* 2006 **8** (2) 251-5.
[51] Malhi Y, Roberts JT, Betts RA, Killeen TJ, Li WH, Nobre CA. Climate change, deforestation, and the fate of the Amazon. *Science* 2008 **319** (5860) 169-72.
[52] Lee TM, Jetz W. Future battlegrounds for conservation under global change. *P. Roy. Soc. B-Biol Sci.* 2008 **275** (1640) 1261-70.
[53] Butler RA, Laurance WF. New strategies for conserving tropical forests. *Trends Ecol. Evol.* 2008 **23** (9) 469-72.
[54] Wertz-Kanounnikoff S, Kongphan-Apirak M, Wunder S. *Reducing forest emissions in the Amazon Basin: A review of drivers of land-use change and how payments for environmental services (PES) schemes can affect them*: CIFOR; 2008.
[55] Fernandes ECM, Wandelli E, Perin R, Garcia S. Restoring productivity to degraded pasture lands in the Amazon through agroforestry practices. In: Uphoff NT, Ball AS, Fernandes E, eds. *Biological approaches to sustainable soil systems*: CRC Press 2006:305-23.
[56] Costanza R, et al. The value of the world's ecosystem services and natural capital. *Nature* 1997 **387** 287-60.
[57] ESPA-AA. *Challenges to Managing Ecosystems Sustainably for Poverty Alleviation: Securing Well-Being in the Andes/Amazon*. Belém, Brazil: Amazon Initiative Consortium; 2008. http://www.ecosystemsandpoverty.org/
[58] The Prince's Rainforests Project. *An Emergency Package for Tropical Forests*. London: The Prince's Charities Foundation; 2009.

[59] Burton ML, Hicks MJ. Hurricane Katrina: Preliminary Estimates of Commercial and Public Sector Damages: Marshall University: Center for Business and Economic Research; 2005.

[60] United States Government Accountability Office. *Climate Change: Financial Risks to Federal and Private Insurers in Coming Decades Are Potentially Significant*; 2007. Report No.: GAO-07-285.

[61] Yunus M, Weber K. *Creating a world without poverty: social business and the future of capitalism*. New York, NY: PublicAffairs 2007.

[62] Academia Brasileira de Ciências. *Amazônia - Desafio Brasileiro do século XXI*. Sao Paulo: Academia Brasileira de Ciências; 2008.

[63] Nobre C. A Scientific and Technological Revolution for the Amazon. 2008 http://www.forestsnow.org/burning.php?buid=7

[64] Mardas N, Mitchell AW, Crosbie L, Ripley S, Howard R, Elia C, Trivedi, M *Global Forest Footprints*. Oxford: Global Canopy Programme; 2009. http://forestdisclosure.com/

Acknowledgements

We thank the researchers of the Large-scale Biosphere-Atmosphere (LBA) experiment in Amazonia for the knowledge and dialogue that has provided the inspiration for this proposed policy framework. We also thank Paul Holland, Hylton Murray-Philipson, Wouter Veening, Stavros Papageorgiou, Richard Betts, Joanna Durbin, Yadvinder Malhi and Patrick Meir among a larger group of colleagues, who have provided insights, comments and discussion for this work. The UK Government's Ecosystem Services for Poverty Alleviation fund, the Prince's Rainforests Project and the Rufford Maurice Laing Foundation provided financial support.

IOP Conf. Series: Earth and Environmental Science 8 (2009) 012006 doi:10.1088/1755-1315/8/1/012006

SOYBEAN BIOMASS PRODUCED IN ARGENTINA: MYTHS AND REALITIES

S Semino[1], H Paul[2], J Tomei[3], L Joensen[4], M Monti[5] and E Jelsøe[1]

[1] Department of Environment, Social and Spatial Change, Roskilde University Universitetsvej 1, DK-4000, Roskilde, Denmark

[2] ECONEXUS, PO Box 1455, Oxford OX4 9BS, United Kingdom

[3] UCL Energy Institute, Central House, 14 Upper Woburn Place, London, WC1H 0HY, United Kingdom

[4] Grupo de Reflexión Rural, Rondeau, 812 Marcos Paz, 1727, Provincia de Buenos Aires, Argentina

[5] Direccion de Extension e Investigacion Agropecuaria, Ministerio de la Produccion, Provincia de Santa Fe, Pte Peron y Garay, 6100, Rufino, Provincia de Santa Fe, Argentina

Email: semino@ruc.dk and stella.semino@mail.dk

Abstract. Soybean biomass for biodiesel, produced in Argentina amongst other places, is considered by some to reduce greenhouse gas emissions and mitigate climate change when compared with fossil fuel alternatives. To ensure that the production of biofuels is 'sustainable', EU institutions and national governments are designing certification schemes for the sustainable production of biomass. In this paper, we question the validity of these proposed environmental standards, using the production of Argentine soybean as a case study. We highlight the negative environmental and social impacts of intensive soybean production, and conclude that certification schemes are unlikely to be able to address the detrimental impacts of increased biofuel production and trade.

Key Words: Soybean, Biodiesel, Argentina, Agrochemicals, Certification

1. Introduction

In the EU, biofuels are expected to make a contribution to reducing emissions from the transport sector, which accounts for around 33% of the EU's total CO_2 emissions [1]. Biofuels offer a potential solution to climate change as, in theory, they are carbon neutral; this means that although they release CO_2 when they are burnt, the carbon will have been removed from the atmosphere and stored in the plant during its growth. In 2003, the European Commission (EC) in its Directive 'on the promotion of the use of biofuels or other renewable fuels for transport' adopted indicative targets of 2% biofuels blending with fossil fuels by 2005 and 5.75% by 2010 [2]. In 2009, the Renewable Energy Directive was passed. It establishes a 10% binding minimum target for biofuels in transport to be achieved by each member state by 2020 [3].

However, concerns are increasing about the potential negative impacts that increased production and trade in biofuels may have; such impacts include direct and indirect land use change, loss of livelihoods and food shortages [e.g. 4 – 6]. The EC, and a number of private and public initiatives, are working on sustainability standards for biofuels, which aim to mitigate the potential deleterious impacts of increased production and trade in biofuels. Criteria on sustainable biofuels and bioliquids are part of the Directive (cf. Chapter 17 and Chapter 18) and supplementary standards are likely to be developed further from existing standardisation initiatives, including the Round Table on Responsible Soy (RTRS), the Better Sugarcane Initiative (BSI), the Roundtable on Sustainable Palm Oil (RSPO), Forest Stewardship Council (FSC), and the Roundtable on Sustainable Biofuels (RSB). The criteria established by the Directive are mandatory and supplementary standards will be established by the Commission as laid down in the Directive. The standard established by the Commission will be harmonized standards filling out the Directive. National law will have to follow the level of harmonization established by the Directive and the New Approach in respect to the use of standards. Regarding the aforementioned criteria and harmonised standards, Member States have the option of establishing national laws that are more environmentally friendly, and accepting voluntary standards e.g. inspired by the RTRS standards described below [7 – 8].

This article begins by presenting the role of Argentina as a major supplier of soy-based biodiesel. There follows an outline of the environmental and social impacts of soybean production in Argentina. The scenario described is the one within which the initiatives to certify soybean biodiesel are to be implemented. This leads us to examine whether certification schemes can address the already negative impacts of soy cultivation. It concludes with some reflections on the Argentinean experience and the scope for the re-establishment of sustainability offered by the proposed certifications. We scrutinise some of the criteria proposed, focusing on the RTRS standards for certification of biomass, as they are one of the most concrete proposals to date. [9]

2. Argentina supplies a new market

Argentina is well placed to respond to the expanding international markets for biofuels. Globally, the country is one of the top three producers and exporters of vegetable oils, and is the largest exporter of soybean oil in the world. Argentina has an economically efficient and export-focused agricultural sector, and soy is the most important cash crop currently cultivated in Argentina. Of the 45.5 million tonnes of soy produced in Argentina in 2007 [10], only 5% was consumed within Argentina; the remainder was exported, principally to Asian markets (as beans, flour and oil) and the EU (as cattle feed). Soy is the crop that both agribusiness and government are focusing on as the principal feedstock to supply global biofuel markets.

In 2006, more than 16 million hectares of soy were cultivated (see Figures 1 and 2). Over 98% of the soy cultivated in Argentina is genetically modified (GM), the production of which is associated with increased use of pesticides, particularly glyphosate, and conservation, or no-till (NT) agriculture.

IOP Conf. Series: Earth and Environmental Science **8** (2009) 012006 doi:10.1088/1755-1315/8/1/012006

Figure 1. Agricultural production of principal crops in Argentina (hectares), 1970 to 2006.

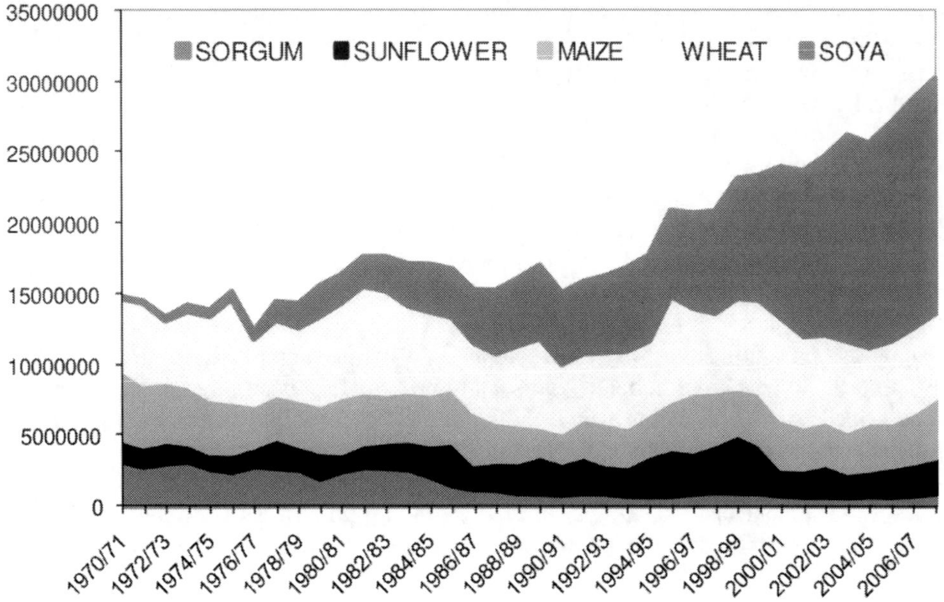

Source: [11]

Figure 2. Argentinean soy production hectares cultivated and yield per hectare, 1970 to 2006

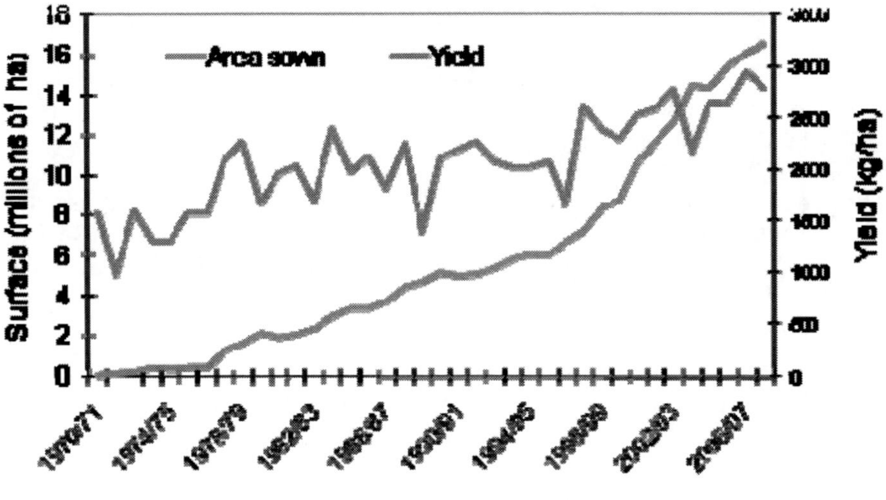

Source: [11]

Demand due to EU biofuels obligations is one of the main drivers for increased soy crushing plant capacity in the Argentinean soy region. In 2007, more than 20 biodiesel projects were launched. If these projects come on line, production could exceed 2 billion litres by 2010 [12]. Supplying this planned expansion will also require imports of soybeans from Paraguay, southern Brazil and Bolivia, most of which is likely to be GM. The expansion of soy crushing capacity is expected to deliver economic benefits to the country. However, it is likely that these benefits will only accrue to a small minority of dominant stakeholders, while the negative environmental and social impacts associated with the unplanned expansion of intensively produced GM soybean will be felt by all.

3. The environmental impacts of soybean production in Argentina

The intensive production of soybean in Argentina is associated with profound environmental impacts, including land use change, increased use of agrochemicals, and contamination of soil and water quality.

3.1. Land use change and deforestation

The RTRS standard indicator for deforestation reads "*Expansion for soy cultivation during field test period may not take place on land cleared of native habitat after May 2009. Exception: Producers who want or plan to clear native habitat after the cut-off date of May 2009 must produce scientific evidence from a comprehensive and professional third-party assessment of the area concerned that identifies the absence of: o all primary forest - other High Conservation Value Areas (HCVAs) - local peoples' lands.*" [9]

Over the past few decades, Argentina has undergone rapid land use changes, many of which are associated with the rapidly increasing cultivation of soybeans (see Figure 2). These include the loss of traditional agricultural systems, particularly livestock rearing, which are pushed to the margins, and land that is less suitable for production. Increased global demand for biofuels will lead to a further extension of the agricultural frontier in Argentina, as business tries to supply new markets with biodiesel produced from soy.

In Argentina, the clearance of native forests and scrub is carried out by machines, using fire, or through aerial applications of herbicides. Each of these methods of clearance will have far-reaching consequences for the communities and ecosystems that surround them. In the Chaco region, plans are already under consideration to incorporate in the medium term about 3,000,000 hectares of new lands for the production of biofuel feedstocks, including soybean [13]. However, since 2007, national legislation (the *Ley de Bosques* Law No. 2386) has forbidden the clearance of native forests. Yet, despite this legislation, it was estimated that in 2008 around 137,000 hectares of native forest in the north of Argentina has been cleared since the law came into force [14]. At the time of writing, this trend continues. [15]

Deforestation and loss of habitat lead to the direct loss of carbon both above and below ground, thus contributing to global climate change. When deforestation is followed by agriculture, further release of soil carbon occurs due to disturbance by tillage. Land use changes also lead to increased emissions of methane (CH_4), through wetland drainage and conversion to rice paddies, and nitrous oxide (N_2O) from agricultural activities, including the use of nitrogen fertilizers, irrigation, cultivation of nitrogen fixing plants, and combustion of biomass [16]. Deforestation has also led to public health impacts through the increased incidence of some diseases, including leishmaniasis and hanta virus [17 – 19].

Figure 3. Native forest clearance in Argentina

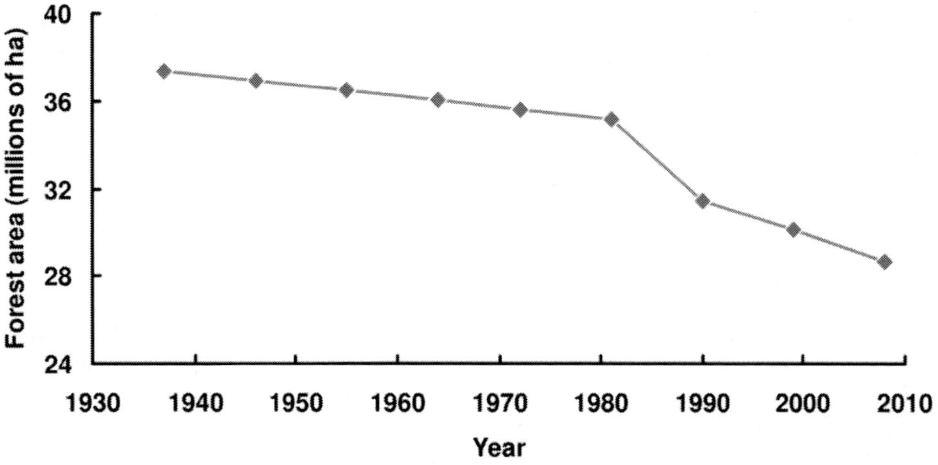

Source: UMSF 2007 [20]

3.2. Emissions from the cultivation of soy

The RTRS criterion reads: "*Efforts to reduce emissions of Greenhouse Gases (GHGs) are made.*" Emissions from soils cultivated with soybean have been established in several studies [21 – 23]. In Argentina, soybean cultivation is the main source of greenhouse gas emissions from agriculture [24]. A recent study by Panichelli *et al.* [22] used an economic allocation approach to analyse the environmental performance of soy-based biodiesel produced in Argentina for export. The authors found that the 'agricultural phase' of soy-based biodiesel accounted for more than 80% of global warming potential, primarily due to land use change. The study concluded that "the global warming potential of biodiesel production in Argentina... is higher than the fossil reference and consequently is not a good choice to mitigate global warming" (p.150). It is likely that the emissions from agriculture are poorly understood, as it is extremely difficult to measure nitrous oxide (N_2O) emissions as they are highly variable, both within and between seasons [23].

3.2.1. No-till agriculture

In Argentina, all GM soybean is cultivated using no till agriculture (NT). Although NT practices may vary, the dominant method is to sow or drill the seeds into the residues from the previous crops, and to eliminate weeds through the application of herbicides. According to the IPCC methodology, conversion from conventional tillage (CT) to NT systems leads to a 10% increase in the estimated sequestration of carbon in the soil. However, recent studies have cast doubt on these claims: Baker *et al.* concluded that the evidence for increased carbon sequestration in NT systems was not compelling [25]. Other authors have argued that the methodology for measuring carbon sequestration does not take into account increases in N_2O emissions [26]. Furthermore, there is little understanding of how tillage controls soil respiration in relation to N_2O emissions and denitrification, although recent research suggests greater denitrification potential in NT soils. Xuejun *et al.* [27] also registered higher CO_2 and N_2O fluxes in NT soil than in CT soil, irrespective of nitrogen source and soil moisture content.

3.2.3. Residues from soybean under no-tillage

To date in Argentina, research on what occurs during soy plant defoliation has been missing. In autumn, the leaves are lost on the field, just at the time of major rains and decompose rapidly, bringing the nitrates to the soils, which then form N_2O. In 2004, the Argentinean Second National Communication on greenhouse gas emissions (GHG) from agricultural activities acknowledged that "it is assumed that the agricultural residues are buried, however, conclusive information still does not exist about the possible emissions from [what at that time was] 14 million hectares under no-tillage" [24].

3.3. Pesticide Use

The RTRS pesticides criterion set reads: *"Negative environmental and health impacts of phytosanitary products are reduced by implementation of systematic, recognised Integrated Crop Management (ICM) techniques. All application of agrochemicals is documented and all handling, storage, collection and disposal of chemical waste and empty containers, is monitored to ensure compliance with good practice. Agrochemicals listed in the Stockholm and Rotterdam Conventions or banned by the Pesticide Action Network (PAN) Dirty Dozen are eliminated."*

The intense production of GM soy is dependent on the use of pesticides, which is widely associated with harmful impacts on biodiversity, water, soil, and human and animal health when pesticides are sprayed on to crops. These negative impacts include increased resistance to pesticides, which, in turn, leads to an increase in the quantity and types of pesticide used.

GM soy has been genetically modified to be tolerant to glyphosate, a wide spectrum herbicide that is used post-emergence. The growth in the cultivation of GM soy has been the key driver of the growth in agrochemical use in Argentina [28]. Research has also identified harmful effects of glyphosate and chlorpyrifos (an organophosphate insecticide) formulations on beneficial animals, such as earthworms, when applied at the nominal concentrations recommended for soy crops. These include reduced earthworm cocoon viability, earthworm avoidance of soils treated with glyphosate, and significantly increased DNA damage in earthworms exposed to chlorpyrifos treated soils [29].

3.4. Soil Compaction

The RTRS criterion on soil reads *"Soil quality is maintained or improved and erosion is avoided by good management practices."*

Soil compaction in the central Pampas region of Argentina is widespread and severe due to the lack of turning/ tilling of the soil, the continuous use of the land, and the use of heavy equipment during seeding and harvest, all of which are associated with NT agriculture [31].

Figure 3. Consumption of pesticides in Argentina, 1996 – 2008

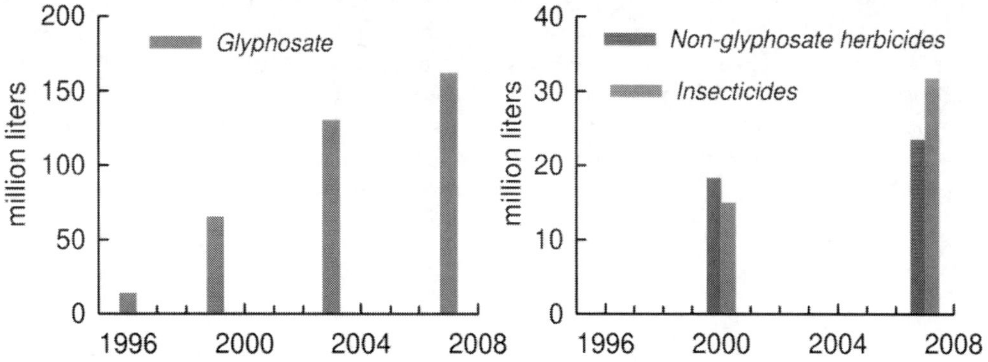

Source: CASAFE, 2007 [30]

After years of intensive production of soybeans, the soil is left with a reduced capacity to capture water as it drains superficially; increased runoff has led to river flooding and reduced the capacity of plantations to resist periods of draught. Soil scientists suspect that soil compaction is one of the main factors driving the increased use of fertilisers and herbicides on soybean under NT [32].

3.5. Soil Fertility and Demineralisation

In the Argentinean Pampas region, the agricultural heart of the country, continued increases in yields of soybean crops are followed by steep declines in soil nitrogen (N), phosphorous (P), potassium (K) and sulphur (S). Nitrogen deficiencies are especially high, despite the ability of soybeans to fix nitrogen biologically. These deficiencies are addressed by the use of synthetic fertilisers, whose production is energy intensive and whose use generates GHG emissions [33]. In 2007, a total of 3.69 million tonnes of fertilisers were sold and, of these, nitrogenised fertilisers accounted for 1.73 million tonnes, with phosphate fertilisers accounting for 1.58 million tonnes [30].

4. The social impacts of soybean production in Argentina

The RTRS criterion on social impact reads *"Channels are available for communication and dialogue with the local community on topics related to the activities of the soy farming operation and its impacts. In areas with traditional land users, conflicting land uses are avoided or resolved"*

Figure 4. Fertiliser use in Argentina (Thousands of Tons)

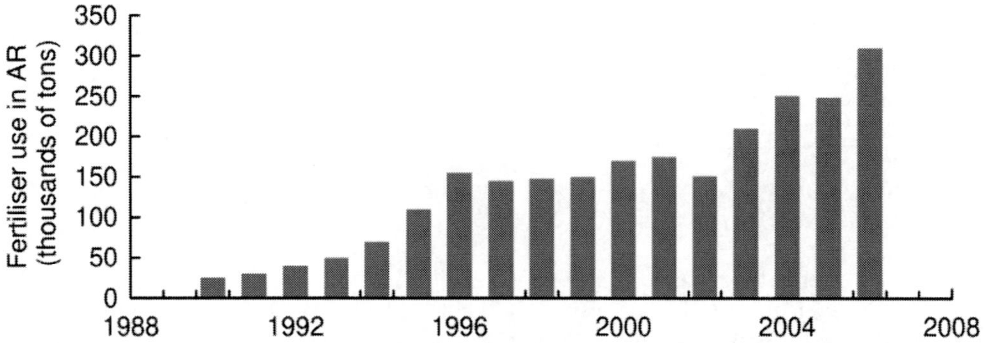

Source: Authors own elaboration from [30]

4.1. Human health impacts

The aerial and terrestrial spraying of herbicides (especially glyphosate) has led not just to contamination of ecosystems, but also to health problems amongst nearby human populations. Despite the existence of health and safety legislation on agrochemicals, personal accounts of local residents show that these are often ignored [34]. Evidence from local groups has pointed to a high incidence of cancers, skin and respiratory diseases in people living near crop-spraying areas [34], and a recent scientific study has confirmed birth malformations well above the normal rates in soya producing regions [35]. Research on human embryonic and placental cells has demonstrated that glyphosate can cause the disruption of sex hormones at non toxic levels, especially on fresh placental extracts. [36] Andrés Carrasco, a senior scientist at the government-funded research institute CONICET (Consejo Nacional de Investigaciones Científicas y Técnicas), who is currently researching the impacts of glyphosate, has described the situation thus: "from the eco-toxicological point of view, what happens in Argentina, with its nearly 18 million hectares [of soy cultivation], is almost a massive experiment" [37]. From the above, we conclude that there is an urgent need for well-controlled epidemiological studies to clarify the health risk of pesticides to human populations.

4.2. Rural exodus

The last Argentinean rural survey, carried out during 2002, when total soya cultivation was less than 11 million hectares, indicated a massive exodus from the countryside to the outskirts of large cities (see table 1).

Table 1. Argentinean agricultural census, 1988 and 2002

Agricultural census	**Year** 1988	2002
Total rural workers	1,321,215	775,296
Quantity of productive farms	421,221	317,816
Total surface of productive farms (in hectares)	177,437,398	171,331,163
Average surface productive farms (in hectares)	421	538
Number of cattle	47,075,156	46,964,059
Surface cultivated with oil seeds (in hectares)	4.328.847	10,835.300

Source: INDEC, 2002 [38]

The data show that this rural depopulation has gone hand-in-hand with the expansion of soy cultivation [38]. People migrate from rural to urban areas when farmers, agricultural workers, and their families, find they can no longer make a living from the land. There have also been reports of smaller farmers being driven off their land, often under threats and violence. Another recent phenomenon affecting rural livelihoods is the increase in land rented to external companies. These companies are managed by investment groups, or *pooles de siembra*, many of which have been linked to the agrochemical industry, which has an interest in increasing trade in biofuels. According to the Argentinean Farmers Union, the local production chain, and thus food security, is greatly affected by the *pooles*, whose focus is on the production of commodities [39]. Although recent empirical data on the depopulation of rural areas are lacking, there is anecdotal evidence of population increases in city slums due to deforestation and land use change. Thus, there is a clear need for more research into the impacts of increased soy cultivation on the movements and livelihoods of Argentina's rural populations.

4.3. Deforestation

In addition to the loss of traditional agricultural systems, the loss of native habitats also impacts on livelihoods. Indigenous communities, such as the Wichi, the last hunter-gatherer culture in Argentina, are being forced to leave their land and livelihoods, threatened by the advance of agriculture, led primarily by soy [40]. As stated previously, deforestation has also been linked to increases in several diseases including leishmaniasis and dengue fever, particularly in the north of Argentina. The *Ley de Bosques* has proven extremely difficult to enforce. In January 2009, a year after the creation of the *Ley de Bosques,* rural organisations from six Argentinean provinces reported that deforestation continues unabated [41]. As a result, indigenous and rural communities continue to suffer forced evictions and loss of livelihoods, while provincial governments continue to ignore their plight. The problem persists despite explicit requirements within the law for full participation of all affected stakeholders, including indigenous communities.

5. Discussion

Within Argentina, the direct and indirect environmental and social impacts of the production of GM soybean are severe and widespread. Even though the evidence is partly anecdotal, from the available knowledge certain conclusions can safely be reached.

The dominant agro-export model requires the intensive use of agrochemicals, and resistance to these is increasing, necessitating the use of other, more noxious, chemicals. This feedback mechanism impacts on ecosystems and communities, especially those living close to soy fields.

Where native habitats are converted to agriculture, soils are often prone to erosion. The soy monocrops create depletion of soil nutrients and the herbicides degrade the soil biota. The decline of soil quality and fertility jeopardises the long term ecological and economic sustainability of the region. which are important at local and global levels, particularly for the mitigation of climate change. In our opinion, these concerns cannot effectively be addressed by broad certification schemes, such as those proposed by RTRS and other initiatives.

With regard to the social impacts of soybean production, certification schemes alone are, in our opinion, unlikely to create public policies to protect the health, habitat and food security of populations living in areas of soybean production, or to provide the cultural and economic incentives for them to remain in these areas, rather than migrate to the cities.

A major impediment in making the study more specific is the absence of quantitative empirical data on the effects of agrochemicals on the environment, human health, and population shifts. Yet, such research is vital if one is to guarantee that biofuels production is restricted to situations where a genuine carbon (CO_2 equivalent) savings can be demonstrated. A meaningful contribution to local livelihoods should also be a prerequisite for such schemes. Going down this path without doing the research could lead to irreparable damage to fragile ecosystems and the communities that live in them.

A discussion of the institutional and economic conditions that underlie the present expansion of soy in Argentina is outside the scope of this paper. Nor was there space to review certification standards other than those of RTRS. In closing this discussion, we note that existing environmental legislation is widely ignored. There is no obvious reason why this situation should change in the foreseeable future.

6. Conclusions

In response to the question posed in the introduction of whether certification schemes can address the negative impacts of soy cultivation, we conclude that such schemes are unlikely to address the detrimental impacts of increased biofuel production and trade. While the use of pesticides continues to increase without consideration for the environment or the communities that live alongside soy and while environmental legislation is ignored by producers and not enforced by the authorities, the impacts on the environment and livelihoods will be profound. More research is needed to ascertain the impacts of intensive soy production on soil fertility, the provision of ecosystem benefits, emissions from production, eutrophication, to name but a few key areas in which research is lacking. Instead of mitigating the negative sustainability impacts of increased global trade in biofuels, certification schemes could exacerbate the problems associated with feedstock production. So, what is likely to be the effectiveness of measures such as the EU directive on biofuels and bioliquids? In our experience, national environmental legislation is largely ignored in biofuel producer countries. We see no reason why this should be different for a law imposed through the influence of a distant power. Will the EU play an active role in the evaluation of the biofuels coming to European ports, or will this task be left to a third party, who may themselves be dependent on funding from the industry they are supposed to be policing? Will the EU turn back a shipment because one of the documents is not in order, or will they prefer a questionable biofuel to a fossil fuel? Only time will tell, but in the meantime massive changes are happening on the ground that are likely to be irreversible.

Acknowledgement

N. Bonanos of the Risø National Laboratory for Sustainable Energy, Technical University of Denmark, is thanked for helpful comments on the manuscript.

References

[1] EurActiv, 2009. *Biofuels for transport.* Available from: http://www.euractiv.com/en/transport/biofuels-transport/article-152282.

[2] COM, 2003. Directive 2003/30/EC of the European Parliament and of the council of 8 May 2003 on the promotion of the use of biofuels or other renewable fuels for transport. L123/42 Official Journal of the European Union, 17 May 2003.

[3] Directive 2009/28/EC of 23 April 2009 on the promotion of the use of energy from renewable sources.

[4] Searchinger T, Heimlich R, Houghton RA, Dong F, Elobeid A, Fabiosa J, Tokgoz S, Hayes D and Yu TH, 2008. *Use of US Croplands for Biofuels Increases Greenhouse Gases Through Emissions from Land-Use Change.* Science **319** (5867): 1238.

[5] Doornbosch R and Steenblik R, 2007. *Biofuels: is the cure worse than the disease*? OECD, Roundtable on Sustainable Development. OECD, Paris.

[6] OFID, 2009. *Biofuels and food security.* OFID Pamphlet Series. Vienna, Austria.

[7] Schelgel S and Kaphengst T, 2007. *European Union Policy on Bioenergy and the Role of Sustainability Criteria and Certification Systems.* Journal of Agricultural and Food Industrial Organization, **5**: Article 7.

[8] RSB, 2008. *Version Zero of the RSB Principles and Criteria.* Roundtable on Sustainable Biofuels (RSB), Cen-Energy Center, Ecole Polythechnique Federale De Lausanne August 2008 http://cgse.epfl.ch/page70341.html.

[9] RTRS,2009 *Principles and Criteria for responsible soy: field testing version*/28 may 2009. Produced by The RTRS Principles, Criteria and Verification Working Group (DG) approved

and ratified by the General Assembly on May 28 2009. http://www.responsiblesoy.org/ files/356.pdf.

[10] FAOSTAT (2008). Food and Agricultural Organisation Statistics. www.faostat.fao.org.

[11] SAGPYA, 2008. *Superficies Cultivadas*. Secretaria de Agricultura Ganadería y Alimentos. Buenos Aires, Argentina http://www.sagpya.gov.ar.

[12] USDA, 2007. *Argentine Bio-Fuels Report 2007*. USDA Foreign Agricultural Service, GAIN Report. http://www.fas.usda.gov/gainfiles/200706/146291490.pdf.

[13] Pengue WA and Morello JH, 2007. *La comida... para el auto*? Eco-Portal, 28th April 2007. http://www.ecoportal.net/content/view/full/68740.

[14] UMSF, 2008. *Pérdida de Bosque Nativo en el Norte de Argentina Diciembre 2007 – Octubre 2008*. Unidad de Manejo del Sistema de Evaluación Forestal, Dirección De Bosques, Secretaría De Ambiente y Desarrollo Sustentable (SAyDS). Buenos Aires: Argentina. http://www.ambiente. gov. ar/archivo/web/UMSEF/File/deforestacion07-08_ley26331.pdf.

[15] C. Montenegro, UMSF, Personal communication.

[16] Ellis E and Pontius R, 2008. *Land-use and land-cover change*. In: Encyclopaedia of Earth. Eds. Cutler J. Cleveland (Washington, D.C.: Environmental Information Coalition, National Council for Science and the Environment) [First published in the Encyclopaedia of Earth July 17, 2008]. http://www.eoearth.org/article/Land-use_and_land-cover_change.

[17] Salomon OD, Orellano PW, Quintana MG, Perez S, Sosa Estani S, Acardi S, and Lamfri M, 2006. *Transmision De La Leishmaniasis Tegumentaria En La Argentina*. Medicina, Buenos Aires, **66**: 211-219.

[18] Seijo A, 2008. *Boletin de temas de salud de la Asociación de Médicos Municipales de la Ciudad de Buenos Aires*. Suplemento del Diario del Mundo Hospitalario, Año 15 Nº 138 Julio de 2008. http://www.medicos-municipales.org.ar/bts0708.htm.

[19] Salomon OD, personal communication.

[20] UMSF, 2007. *Informe sobre deforestación en Argentina 2007*. Unidad de Manejo del Sistema de Evaluación Forestal, Secretaría de Ambiente y Desarrollo Sustentable (SAyDS). Buenos Aires: Argentina. http://www.ambiente.gov.ar/archivos/web/UMSEF/File/deforestacin_argentina_v2.pdf.

[21] Dalgaard R, Schmidt J, Halberg N,Christensen P, Thrane M, and Pengue WA, 2008. *LCA of Soybean Meal*. International Journal of Life Cycle Assessment **13**: 240–254.

[22] Panichelli L, Dauriat A and Gnansounou E, 2008. *Life cycle assessment of soybean-based biodiesel in Argentina for export*. International Journal of Life Cycle Assessment **14**: 144 – 159.

[23] Sheehan J, Camobreco V, and Duffield J, 1998. *Life Cycle Inventory of Biodiesel and Petroleum Diesel for Use in an Urban Bus*. Colorado National Renewable Energy Laboratory, May 1998. http://www.nrel.gov/docs/legosti/fy98/24089.pdf.

[24] Taboada M, 2004. Inventario de gases efecto invernadero del sector agrícola Argentino. Facultad de agronomía de la Universidad de Buenos Aires, II taller sobre la segunda comunicación de cambio climático. 29 November 2004.

[25] Baker JM, Ochsner TE, Venterea RT and Griffis TJ, 2007. *Tillage and soil carbon sequestration – what do we really know*? Agriculture, Ecosystems and Environment **118**: 1-5.

[26] Mosier A, 1998. *A New Approach to Estimate Emissions of N2O Oxide from Agriculture and its Implications for the Global N2O Budget*. Department of Agriculture, Agricultural Research Service, USA, and C. Kroeze, Wageningen Agricultural University, Netherlands. Reprinted from

IGAC*tivities* Newsletter No. 12, March 1998. http://www.igac.noaa.gov/newsletter/highlights/1998/n2o.php.

[27] Xuejun X, Liu XJ, Mosier AR, Halvorson AD, Reule CA and Zhang F, 2007. *Denitirification and N2O emission in arable soils: Effect of tillage, N source and soil moisture*. Journal of Soil Biology and Biochemistry, **39**: 2362-2370.

[28] E-Campo, 2007. *Los Buenos Niveles De Humedad Aumentan El Riesgo: Atentos y en guardia ante la roya*. E-campo, 22nd February 2007. http://www.e-campo.com/?event=news.display&id=E96C178E-188B-7C0FF222D0F67F1925B9&.

[29] Casabé N , Piola L, Fuchs J, Oneto ML, Pamparato L, Basack S, Giménez R, Massaro R, Papa R and Kesten E, 2007. *Ecotoxicological Assessment of the Effects of Glyphosate and Chlorpyrifos in an Argentine Soya Field*, Journal Soil Sediments, **7**: 232 – 239.

[30] CASAFE, 2007. *Estadísticas del Mercado de Fertilizantes en Argentina 1980-2007*. Cámara de Sanitarios Agropecuarios y Fertilizantes, Estadísticas (CASAFE).

[31] Gerster G, Bacigaluppo S, De Battista, J and Cerana J, 2008. *Distribución de la Compactación en el Perfil del Suelo utilizando diferentes Neumáticos. Consecuencias sobre el Enraizamiento del Cultivo de Soja*. Instituto Nacional de Tecnologia Agropecuaria (INTA), 2008. Econoagro http://www.econoagro.com:80/verArticulo.php?contenidoID=646.

[32] Benbrook C and Baumüller H, 2003. *Argentina: trip report*. Available from: www.biotech-info.net/trip_report.pdf.

[33] Austin A, Piñeiro G, and González Polo M, 2006. *More is less: agricultural impacts on the nitrogen cycle in Argentina*. Biogeochemistry **79**: 45-60.

[34] GRR, 2009. *Pueblos Fumigados: Informe sobre la problemática del uso de plaguicidas en las principales provincias sojeras de la Argentina*. Buenos Aires, Argentina http://www.grr.org.ar/trabajos/Pueblos_Fumigados__GRR_.pdf.

[35] Benítez Leite S, Macchi ML, and Acosta M, 2009. *Malformaciones congénitas asociadas a agro-tóxicos*. Base de Investigaciones Sociales-Universidad Nacional de Asunción Séptima Región Sanitaria, Encarnación Ministerio de Salud Pública. BASEIS Documento de Trabajo 120-2009.

[36] Benachour N, Sipahutar H, Moslemi S, Gasnier C, Travert C and Séralini GE, 2007. *Time- and Dose-Dependent Effects of Roundup on Human Embryonic and Placental Cells*. Archives of Environmental Contamination and Toxicology, **53**: July 2007.

[37] Página 12, 2009a. *Lo que sucede en Argentina es casi un experimento masivo*. Interview with Andrés Carrasco, Professor of Embyrology at the University of Buenos Aires. Página 12, Domingo, 3 de Mayo de 2009. http://www.pagina12.com.ar/diario/elpais/1-124288-2009-05-03.html.

[38] INDEC, 2002. *Censo Nacional Agropecuario 2002 & 1988*. Instituto Nacional de Estadística y Censos (INDEC). http://www.indec.gov.ar/agropecuario/cna_principal.asp.

[39] Federación Agraria, 2007. *Preocupación en el sector agropecuario Argentina: los pooles de siembra expulsan a chacareros y empobrecen los pueblos*. Nueva Rioja, 10 Abril 2007 http://www.nuevarioja.com.ar/default.asp?contenido=nota&id_nota=33005&id_seccion=7&nombre_seccion=nacional%20/%20internacional.

[40] Zunino Mariela 2007 Argentina: lo que la soya se llevó...Desnutrición y hambre en el país de los alimentos- Ecoportal http://www.ecoportal.net/Contenido/Temas_Especiales/Pueblos_Indigenas/Argentina_lo_que_la_soya_se_llevo_Desnutricion_y_hambre_en_el_pais_de_los_alimentos

[41] Página 12 2009c. *A un año de la sanción de la norma que debía proteger los bosques nativos: hecha la ley, hecho el desmonte*. Página 12, Viernes, 2 de Enero de 2009. http://www.pagina12.com.ar/diario/sociedad/3-117597-2009-01-02.html

The role of PV solar electricity to power the 21st century's global prime energy demand*

W Hoffmann, Applied Materials
Vice President and CTO Energy and Environmental Solutions
President of the European Photovoltaic Industry Association (EPIA)
Member of the Scientific Board FhG-ISE and the Supervisory Board of ISFH

***Opinion article (not peer-reviewed)**

Primary Energy – Secondary Energy – End user's energy

When discussing the challenges and opportunities of the future prime energy demand it is useful to clarify the understanding of some basic terms. We differentiate in this context between primary energy (PE), secondary energy (SE) and end user's energy. Primary Energy is for instance gas, oil, uranium. These are materials that are more or less existent on earth. They are transferred - obviously with losses - into the so called secondary energy. SE is the result of transferring crude oil for example via a refinery into gas or fuel or via a power station where electricity or process heat is generated. This SE is again transformed into the so called *end user's energy*. As an example we take fuel that is filled into an otto engine for driving a car in order to get from A to B. Or we take electricity which powers a light bulb in order to produce the needed lumen. Here usually more then 90% of electricity is unused heat while less then 10% are the needed lumen. The global prime energy need as of 2000 was in the range of 120,000 TWh. Just to put these numbers in perspective: the average electricity consumption of a four person's household is about 5000 kWh, in the households in Germany 450 TWh of electricity is used. In 2000 out of these nearly 120,000 TWh 8% were from renewable energies, mainly hydropower. Today we see wind adding increasingly important to this secondary energy, and the future PV solar electricity is also expected to become an important part.

The rest of these TWh are produced by nuclear (7%) and 85% from fossil. During the process of transforming this PE into SE electricity nearly two thirds of the PE is wasted due to low efficiency and generates unused heat. Most of the global electricity power plants still work very low efficiently although we can find nowadays some power plants that work on an efficiency of close to 45%. To produce 18,000 TWh electricity it needs 50,000 TWh PE what is nearly the half of the the whole PE amount.

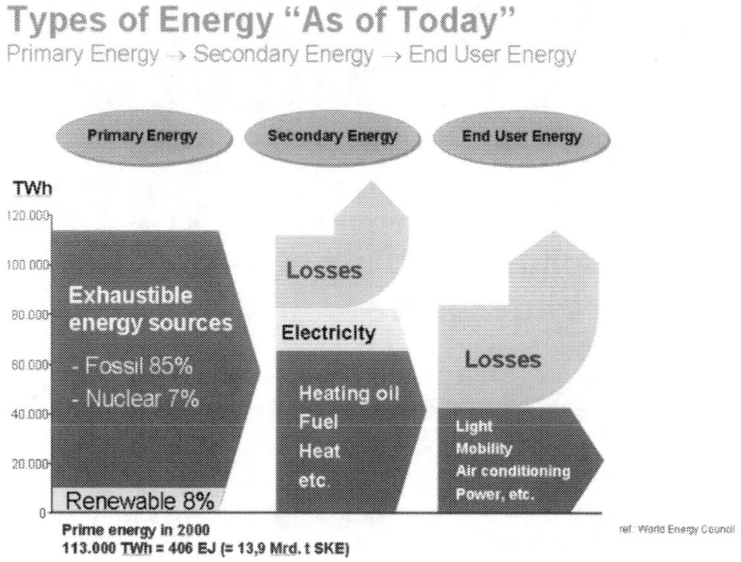

Figure 1: *Types of Energy "As of Today" (World Energy Council)*

To understand and to get a feeling for the situation as of today it has to be mentioned that one quarter the of world's population (what is nearly ~1,5 billion people) consumes three thirds of the primary energy (what means ~90.000 TWh). This shows a tremendous inequity: three quarters of the world's population can use only one quarter of the global PE. Today the living standard in the industrialized world means a consumption of nearly 60.000 TWh for 1 billion people. For the future this would mean that if there will be a world population of 10 billion people by the end of this century and all of them will have today's living standard, then there would be an energy need of 600.000 TWh - based on today's energy efficiency. This would mean that in 2100 instead of 600,000 TWh only something in the range of 200,000 and 300,000 TWh would be the world's prime energy demand. So this development shows obvious that the world's future energy demand can not be tackled with old solutions.

Examples for a tremendous increasing of energy efficiency with view on the end user's energy are as follows. The replacement of a conventional light bulb with a power of 100 W by a LED would then need only 10 W for the same lumen produced. The approach of the replacement of conventional light bulbs through energy efficient lamps or through (O)LEDs1 realizes the same quality of life with much less energy. Here, we at Applied Materials can make a big impact since 12% of the global electricity is used in lighting (~ 2100 TWh or ~6300 TWh prime energy input). The opportunities for (O)LED lighting becomes obvious in this perspective: if the ordinary light bulbs or fluorescence are replaced by (O)LEDs in a very conservative forecast the energy need can be at least halved. And if in addition the energy for those (O)LEDs is not longer generated on conventional ways but with renewables the future prime energy part is not longer 6300 TWh but only 1050 TWh. There are many more examples of emerging new technologies that may dramatically reduce the SE need for the same end user's energy. With the above example it becomes clear that it is possible to use much less energy without reducing the quality of our everyday lives. These new technologies are needed in order to reach a significant increase of energy efficiency, thereby reducing the energy needs while keeping the same quality of life.

[1] LED = Light Emitting Diode / OLED = Organic Light Emitting Diode

PV Market and regulatory framework

PV's environment today and in the middle-term until 2020 may be characterized as a transition period, since in the outlined time frame grid parity will be reached starting in southern regions and moving to northern countries. Grid parity is identified at that time when the generation cost of PV systems becomes equal to the retail price of private households. In most of these years market support programs, best known as feed-in tariffs, are still needed with decreasing reimbursement. The public awareness for the necessity of the usage of renewable energies is accelerated through the debates on how to face the climate and energy change. This is emphasized by the various support programs in many countries. The Intergovernmental Panel on Climate Change (IPCC), established to provide the decision-makers and other stakeholders with an objective source of information about climate change, has recently stated that today's rising CO2 emission is man-made and reason for real concerns for the globe. This is increasingly accepted on a global level as a matter of fact.

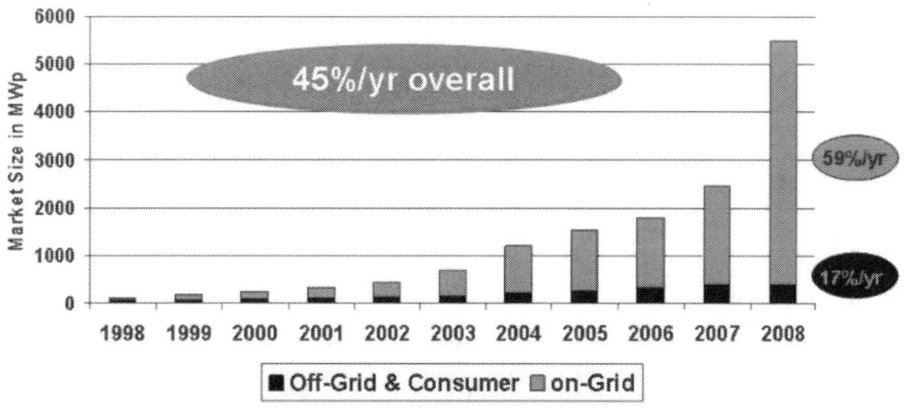

Figure 2: *World PV Application Segmentation (ref. European Photovoltaic Industries Association [EPIA], Navigant Consulting and Commerzbank)*

There has been an average growth of approximately 45% per year since 1998, reaching the impressive number of more than five GW new installations worldwide in 2008. This positive development is driven by the various feed-in-tariffs like in Spain, Germany and various other European countries as well as by support schemes like in Japan, the US and others. Moreover the established infrastructure in these countries supports this success. Nevertheless all support schemes and feed-in-tariffs need to be made with a sense of appropriateness. Otherwise a negative development like the straw fire in Spain during last year is observed. First, the feed-in-tariff was too high, which led to the creation of windfall-profits along the value chain. Second, the administrative hurdles were too complicated: one year lead time for the whole process of approval and installation is too long. Third, the Spanish example shows that feed-in-tariffs work best when the budget is distributed to defined population and not, as in the Spanish case, taken from the tax payer. In the latter case the responsible ministry cannot control the increasing flow of money out of his budget. The consequence was the introduction of a cap of maximum 500 MWp new installations per year. Those caps must be avoided. Finally, it has to be stated that feed-in-tariffs are better than investment subsidies since they share the vantage of a long-term horizon. In addition, reasonable degression rates over time such as in Germany with 8-10% each year for new systems is needed. This still implies a reasonable return on investment and avoids windfall-profits along the value chain.

The fact is that oil is at peak in the upcoming decade. In addition, political instability in the Middle East and conflicts of gas supply between Russia and Ukraine, influencing the whole Western hemisphere, may happen again in the future. It should furthermore be mentioned that uranium for fission reactors is also very finite and fast breeder reactors are very critically seen in the public due to their security risk. These are just a few examples that underline the increasing important security of supply issues during the last years.

Germany is the world's leading PV market of installed systems with more than 5 GWp of total and about 1.7 MWp of newly installed systems in 2008. Many steps had to be taken to reach this fruitful environment for PV solar electricity. First of all a competitive market environment had to be created: The costumers received the supporting money and therefore had the choice of product, creating a highly competitive environment.

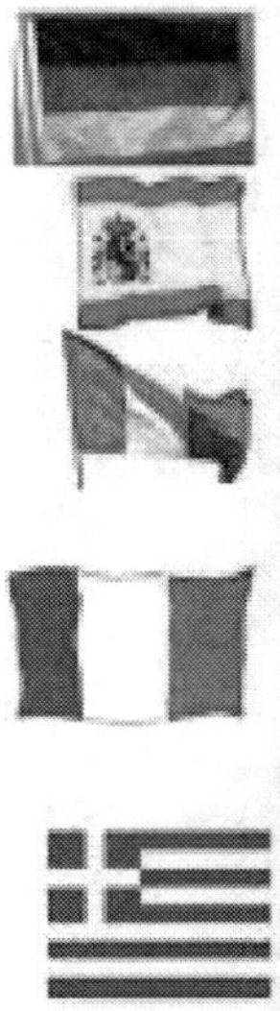

	Feed in Tariff 2009 [€ct/kWh]	Market size 2008 (est.)
Germany	31.9 – 43 (20 years)	1,500+ MWp
Spain	32 – 34 (25 years)	2500 + MWp
Italy	35 – 48 (20 years)	200 MWp
France	31.19 – 57.19 **50% tax reduction max 8,000/16,000 €** **grants in some regions**	150 MWp
Greece	40 – 50 **20%-40% grants for commercial plants**	15 MWp

Figure 3: *Support schemes of important European PV markets*

In most countries with a feed-in-tariff, according to various polls, more than 80% of the population is willing to accept higher electricity prices to promote renewable energies. Politicians just have to follow such a scheme.

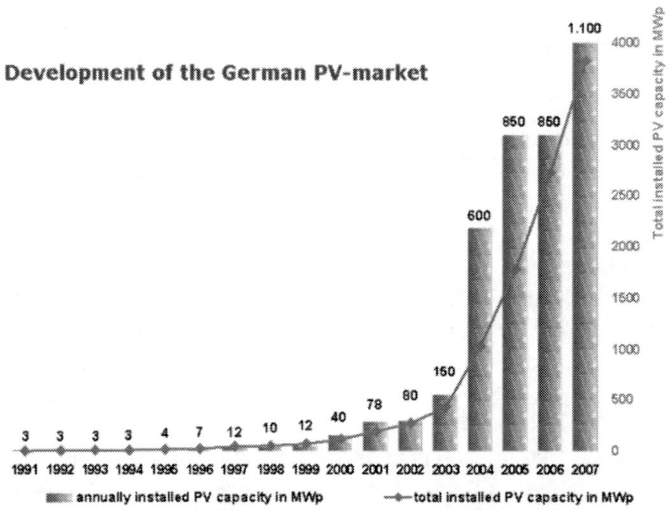

Figure 4: *Development of the German PV-market (BSW-Solar)*

Price Development and competitiveness of PV solar electricity

It is a matter of fact that electricity produced by PV systems worldwide are already today able to compete with their generating cost against peak power retail prices from utilities in a liberalized utility market, especially in areas with high solar irradiation where a PV system produces 1.500 to 1.800 kWh / kW. In places like Germany with a lower average solar irradiation producing 800 to 1.000 kWh / kW it will take approximately 7-8 years longer. There are three general facts that determine the point of competitiveness: (i) the price decrease of PV systems lead to an equal decrease in the actual generation cost. (ii) the development of a really liberalized electricity market is absolutely crucial. Finally (iii) there is the high correlation of the peak power price from utilities and the output of a PV system. The highest prices in electricity markets, mainly in summer afternoons, correspond directly to the highest output of a PV system. Electricity tariffs all around the world provide differentiated prices in liberalized markets during the daytime and the seasons. This entails that during summer and at daytime PV solar electricity is already today cost effective for the costumer in e.g. in California. However, even after reaching the so called "grid parity" support is needed for a few more years.

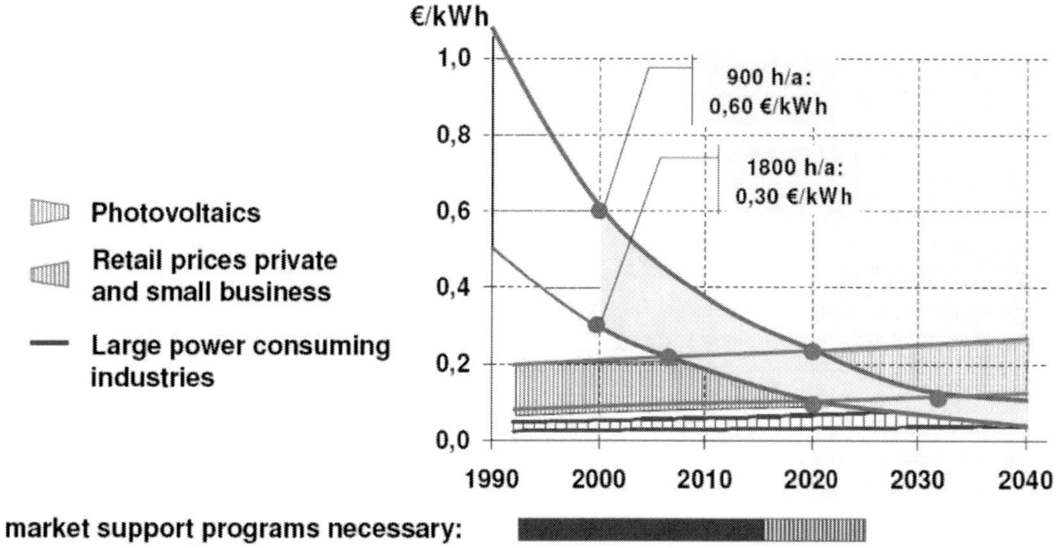

Figure 5: *Competitiveness between Electricity Generation Cost PV and Electricity Price (W.Hoffmann (1999), personal estimates)*

To make projections on the development of prices for PV solar modules, so called price experience curves can serve as a useful tool. Generally, a price experience curve shows in a double logarithmic plot the dependence of the component price versus the cumulative sold volume of said product. The general finding is that a straight line is observed where one can calculate from the slope the "experience factor" which serves as a measure how many % of price decrease are obtained by doubling the cumulative sold volume. For products where the smaller part of production cost is material-driven and the remaining, greater part, is "intelligence", one obtains experience factors at and above 30%. In contrast, if the dominating production part is material and energy cost, this value is at and below 10% and may even reverse the slope: that is a price increase by further doubling sold volumes (e.g. float glass or plastic foils). The price experience curve for PV solar modules shows an experience factor of about 20%, reflecting the fact that modules have some area related cost elements (glass, plastic) as well as high-tech content (wafer thickness, efficiency).

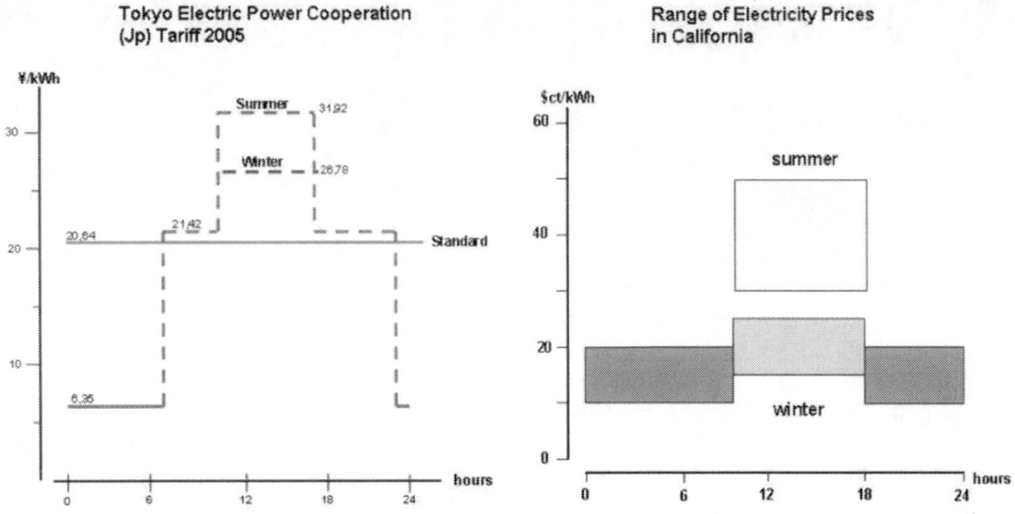

Figure 6: *Electricity retail prices for various countries (Japan: KEPCO office data; California: Alison Hyde of BSW*

The real prices deviated from the straight line as prices for poly silicon more than tripled due to the shortfall of poly silicon in recent years. Massive investments and a significant boost of production capacities in 2008/09 for poly silicon removed this bottleneck. As a consequence the module prices will return to the projected line very soon, where most has already happened until spring 2009. It is obvious that the price experience curve does not provide a direct answer as to the time when certain price levels will be reached. Together with the assumption of a growth figure the expected future cumulated volumes can be calculated and an estimated price figure can be obtained from the extrapolated price experience curve.

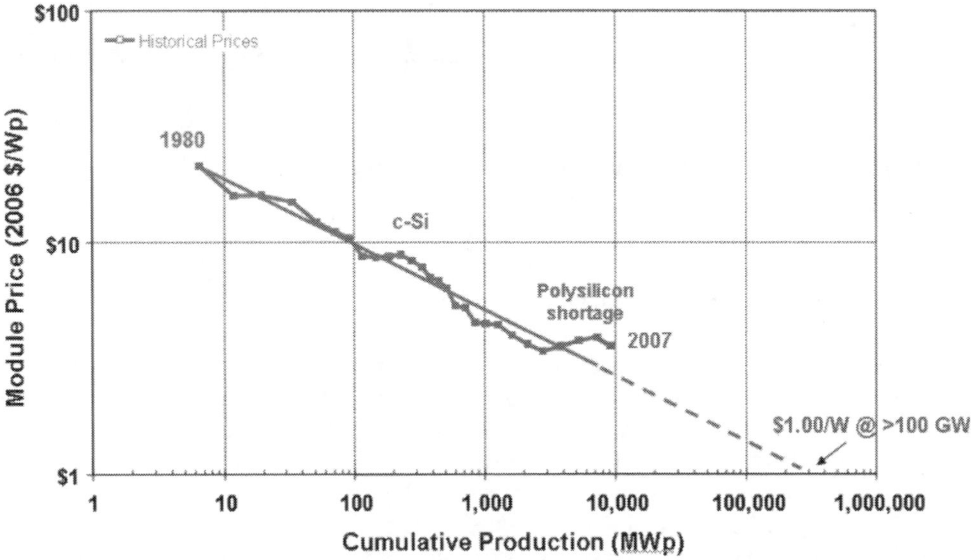

Figure 7: *Solar Experience Curve: Module Cost/Watt (Adapted from National Renewable Energy Laboratory)*

The fact that PV solar electricity will be mainstream for the future electricity production can be observed, when comparing it with the generation costs of different energy transforming processes like fossil (coal, gas) or nuclear. One finding of this comparison, examined by the European Association of Utilities (Eurelectric) in 2007, is an increase in all fossil technologies. While a kWh electricity, generated by fossil power stations, has by now reached a price level of 4 to 4.5 €ct, it will be at 6-7 €ct around 2030 and likely be at 6.5-9 €ct in 2050. The bandwidth for nuclear power generation is assumed to be in the range of 4 €ct in 2007 up to 5 €ct in 2050. The assumptions taken for the increase of fuel prices in this time frame are very modest and future will tell whether this will be realistic or not.

Unfortunately, in the above mentioned study PV solar electricity is not included. In order to take the price experience curve for solar modules serious, this should be changed in future studies. With a proven price experience factor of 20% and an assumed growth rate of ~ 30% this will lead to an estimated price decrease of about 8% per year. In other words this means that every 8 years generation costs are halved. In concrete numbers the result is that from today's level of 20 - 40 €ct the generation cost will be at about 10 - 20 €ct in 8 years (in 2017) and at about 5 – 10 €ct in 16 years from now (in 2025).

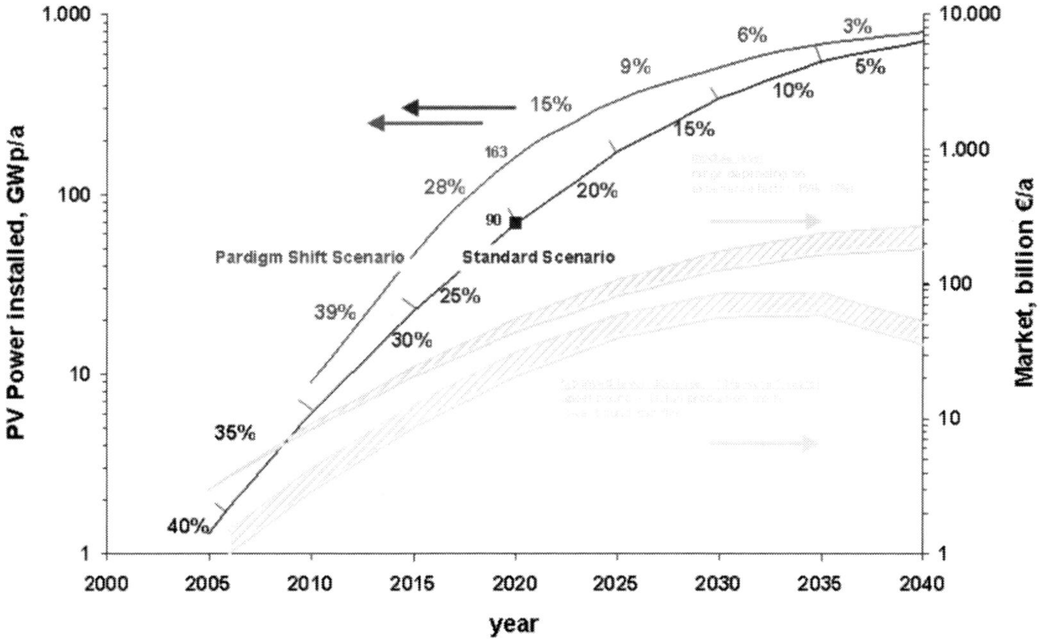

Figure 8: *Future growth of global PV solar electricity market in GWp a bn€ turnover (W.Hoffmann personal estimates, including SET for 2020 goal, see blue paradigm shift scenario curve)*

Future Market Development

PV solar electricity has the potential to be a dominant technology to solve our energy problems in the future. Due to different costumer needs there will be the necessity for different PV technologies in order to serve them. The technology needed depends on the specific application and function. At first there is the grid-connected segment: Here one can see roof-top applications and increasing building integrated Photovoltaics (BIPV) coming up as a contribution to cost-saving through the usage of functional parts of buildings. BIPV can be seen as a revolution in the building sector, in the urban planning and for architectural solutions. Thanks to feed-in-tariffs this segment is growing from a niche market to a mainstream segment.

All across Europe the deployment of PV will be a mainstream technology for future electricity. For instance the European Renewable Energy Council (EREC) sees for 2020 a 40% contribution of renewable energies for Europe's electricity demand. Furthermore, the European Photovoltaic Industry Association (EPIA) predicts in their study "SET for 2020" that 12% of Europe's electricity provided by PV in 2020 is feasible. In their study EPIA identified three possible deployment scenarios depending on changes in the system boundary conditions. The first scenario is called "Baseline Scenario". It does not require any changes in the current electricity system since the PV penetration of 4% PV can be integrated without any further efforts. The second, the "Accelerated Growth"-scenario, requires minor changes to the existing electricity system and moreover that the industry optimizes its supply chain and cooperates with utilities on (grid-) infrastructure in order to handle an installation of 6% of EU's electricity. The last scenario, dealing with the indicated 12%, is the so called paradigm shift scenario.

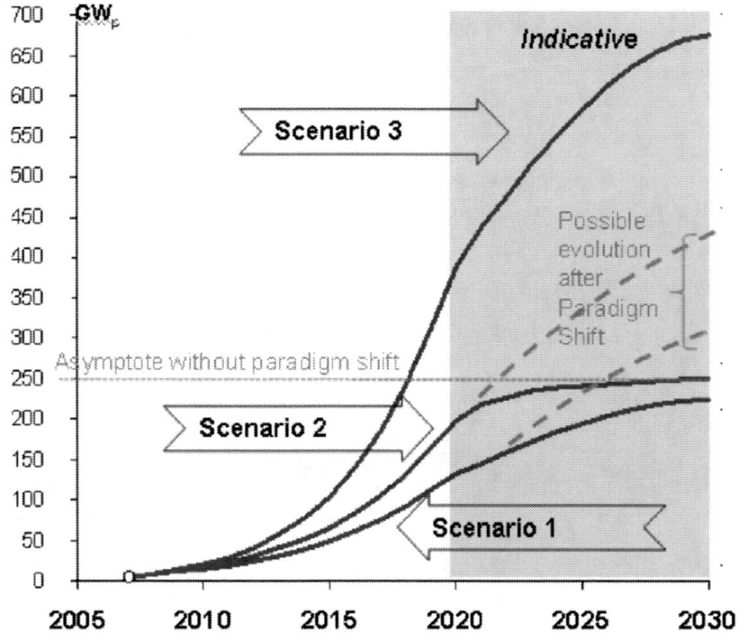

Table 9: *Three possible PV deployment scenarios depending on changes in the system boundary conditions that can take place by 2020 (EPIA, EU DG TREN "European Energy and Transport: trends to 2030, update 2007", Eurostat Data Portal, EU JRC Photovoltaic Geographical Information System, A.T. Kearney analysis)*

This paradigm shift scenario assumes important changes at the market, regulatory and utility levels. This becomes necessary as 12% electricity consumption in Europe by 2020 will amount to 460 TWh and would need about 390 GW cumulative PV installations. In comparison to an annual installation of 5 GWp last year this represents a challenging target. Although the technical requirements are ambitious, the real challenge is to reach significant changes in the existing electricity system (smart grids and meters) and at market and regulatory level, together with a strong collaboration with other players in the energy sector. Additionally, it will be key to change mind-sets and shape awareness of

people regarding their behavior in electricity consumption (demand side management). It is also important to mention that there will be a net present value benefit from PV deployment at EU level, which even increases when moving from scenario 1 to 3.
Rural electrification will grow considerably within the next decades and will provide billions of people with access to clean and affordable energy. From today's few regional on-grid markets a development towards a 50% share of PV solar electricity for rural electrification will be observed. The monthly costs today for battery driven radios and candles for lighting is basically around six to eight US$. With a 50 W PV solar home system these costs could be at around seven US$ which could be financed by micro credits.

The International Energy Agency (IEA) is projecting the global electricity consumption to increase from 16,000 TWh (in 2001) towards 36,000 TWh in 2040 due to the growth of the world's population. From these additional 20,000 TWh within 35 years most probably half will be needed for decentralized rural electrification. Assuming an average of 1.4 kWh/W this would need an installation of about 7,000 GW – fifteen times the volume of the paradigm shift scenario! And besides, every price increase of the conventional energy sources will speed up the contribution from renewable sources like PV.

Technology evolution

All existing PV technologies can be compared in terms of relative prices and module efficiencies. Today the crystalline Silicon (c-Si) technology has a market share of about 85% and Thin Film Technologies of 15%.

Although crystalline Silicon has today by far the biggest market share, Thin Film technologies and new concepts will catch up. Their relative market growth will most probably be higher than for c-Si. Due to the growth of the total market, c-Si will grow in absolute numbers as well. Both, crystalline and Thin Film technologies have experienced significant cost decreases within the last decades. For example, the thickness of c-Si wafers could be decreased from 450 μm to 180 μm while their efficiency could be increased from 12-14% up to 17-22%. This development will continue and a thickness of 100 μm and below with an efficiency of stable 19-24 % can realistically be achieved in the next decade.

With regard to Thin Film technologies there is one route aiming for lowest cost and price per m². This type of product is suited very well for BIPV applications. Here efficiency plays a secondary role, but homogeneous appearance and semi transparency are the main focus. Like in the architectural glass industry a substrate size increase from 0,6m² up to 5.7m² (SunFab) and a further development towards 10m² can be expected in the future. This development was driven by synergy effects from the LCD flat-panel display technology using very similar equipment for large area amorphous silicon deposition. In addition, there is a parallel development for low cost and price per Watt, like a-Si / μc-Si and II-VI compound semiconductor materials. Emphasis is placed here on efficiency and will develop from 8 -11 % module efficiency today towards 12-14 in the coming years with potential for further increase.

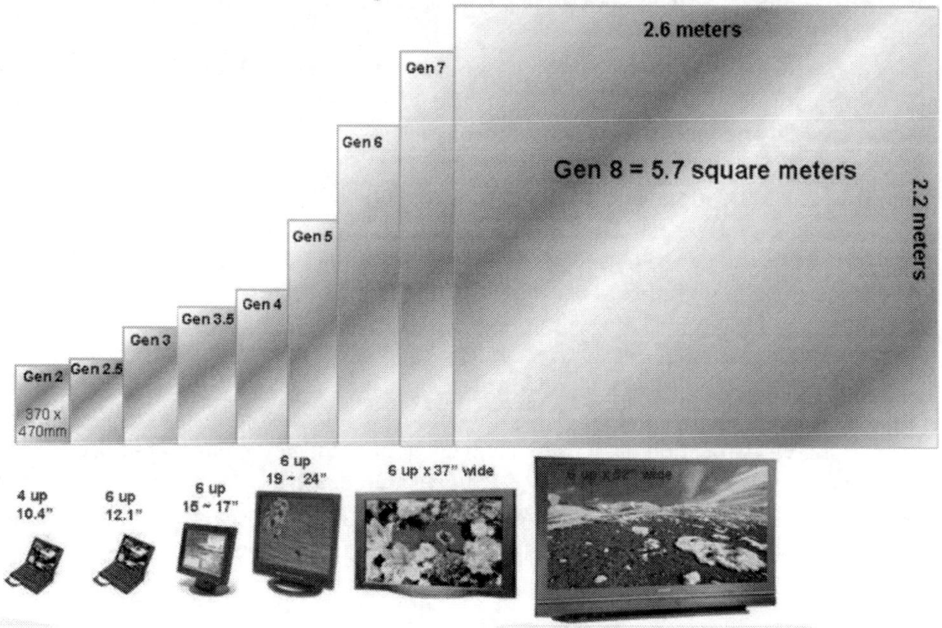

Table 11: *Substrate Size Expansion*

Overview Applied Materials

For over 40 years Applied Materials has been the world leader in nanomanufacturing technology, creating thin film solutions that help to make innovations like flat panel TVs, smart phones and powerful computers both possible and affordable. We are located in 18 countries across the globe, including Germany, Italy, France and the Netherlands, and we employ over 14,500 people worldwide. Over the last decade we have taken this proven track record and technological expertise and applied it to photovoltaics (PV) and energy efficiency technologies, such as lighting. At Applied Materials we believe that renewable energy technologies, in particular photovoltaics, will have a central role in meeting future energy demand. We believe that the future energy demand can be greatly reduced by intelligent use of energy efficient technologies while keeping the quality of our today's life at the same comfort level.

Acknowledgement:

the Author would like to thank Sebastian Goeke for his help to prepare this document.

IOP Conf. Series: Earth and Environmental Science **8** (2009) 012007 doi:10.1088/1755-1315/8/1/012007

References

Hoffmann, W and Waldmann, L (2008): PV Solar Electricity: From a Niche Market to One of the Most Important Mainstream Markets for Electricity. In: Petrova-Koch, V. et.al. (Editors): High-Efficient Low-Cost Photovoltaics. Recent Developments. Springer. Berlin / Heidelberg.

EPIA (2009): SET For 2020 Study

EPIA (2004): EPIA Roadmap

EREC (2008): Renewable Energy Technology Roadmap "20% by 2020" http://www.erec.org/fileadmin/erec_docs/Documents/Publications/Renewable_Energy_Technology_Roadmap.pdf

BEE (2009): Stromversorgung 2020 – Wege in eine moderne Energiewirtschaft. http://www.unendlich-viel-energie.de/de/wirtschaft/stromversorgung-2020.html>

IEA (2004): World Alternative Policy Scenario

Union of the Electricity Industry (Eurolectric) (2003): Joint VGB-EURELECTRIC Report on Efficiency in Electricity Generation. Brussels.

From occupying to inhabiting – a change in conceptualising comfort

Svenja D. Jaffari, Ben Matthews

SPIRE Center for Participatory Innovation Research, Mads Clausen Institute, University of Southern Denmark, Alsion 2, 6400 Sønderborg, DK

E-mail: svenja@mci.sdu.dk, matthews@mci.sdu.dk

Abstract. The concept of 'comfort' has been influential in shaping aspects of our built environment. For the construction industry, comfort is predominantly understood in terms of the balance between an ideal human physiological state and a finite number of measurable environmental parameters that can be controlled (temperature, humidity, air quality, daylighting, noise). It is such a notion of comfort that has informed the establishment of universally applied comfort standards and guidelines for the built environment. When buildings rigidly conform to these standards, they consume vast quantities of energy and are responsible for higher levels of GHG emissions. Recent researchers have challenged such instrumental definitions of comfort on moral and environmental grounds. In this paper, we address this issue from two different standpoints: one empirical, one related to the design of technology.

Empirically, we present an analysis of ethnographic field material that has examined how, in what circumstances, and at what times ordinary users employ energy-intensive indoor climate technologies in their daily lives. We argue that when comfort is viewed as an achievement, rather than as a reified and static ideal homeostasis between humans and their environmental conditions, it becomes easier to appreciate the extent to which comfort is, for ordinary people, personally idiosyncratic, culturally relative, socially influenced and highly dependent on temporality, sequence and activity. With respect to design, we introduce a set of provocative designed prototypes that embody alternative conceptions of 'comfort' than those to which the building industry typically subscribes.

Our discussion has critical implications for the types of technologies that result from a 'comfort standards' conception. Firstly, we show that comfort is not simply a homeostatic equilibrium—such a view is overly narrow, inflexible and ultimately an inaccurate conception of what comfort is for ordinary people; secondly, that it promotes technologies that treat environmental parameters as variables for our manipulation, rather than opportunities to reflect on, and adjust our own (temporal human) fit with our environment. Borrowing from Cole et al. [4], we argue for technologies that enable users to inhabit rather than occupy the built environment.

1. Introduction

A rather peculiar notion of 'comfort' has been responsible for the development of certain international building standards, in spite of widely variable global climate conditions. An optimal, productive and comfortable indoor environment is being held up as the product of stable, predictable and universal comfort conditions, in the form of noise, daylighting, air quality, humidity, and temperature established to correspond to a determinate human 'comfort zone.' Technological systems and products are designed to provide standardized comfort conditions in buildings; e.g. windows, radiators, thermostats, controls, and insulation, each of which can be responsible for (or alleviate) the consumption of large amounts of energy, especially in climates where outdoor conditions are highly variable, or far different from the standardized, regulated indoor conditions. Standardized comfort systems have been built into (and legislated for) the indoor environment. But it is the daily use of such systems that determine the energy costs associated with and facilitated by them. Thus, the relationship and interaction between built environment and user behaviour is something to pay attention to.

"The human dimension" [20] of energy use and comfort practice comes to be important, because one simply cannot theorize about the environmental problem-solution space without considering it to be a social and contextual achievement of an interaction between inhabitants and their built environment. This is not just a matter of collecting more statistics about use in the manner in which policy makers might be accustomed; such statistics also need supplementation by the building users' points of view and their everyday actions. In the majority of cases, the usual data about indoor environment and comfort does not include information about the user's behaviour and personal values [20], and the users' practical reasoning about their daily energy use has not yet been a focus of research. Furthermore, we share with other critical voices the assumption that actual energy-related comfort practice might be variable according to situations and activities in time and place. These variations are important to consider. To understand how practical, sustainable and non-sustainable courses of action are taken and which complex net of experiences, expectations, variable to situations and activity is to be found behind these, will doubtlessly have implications for the successful innovation (e.g. for producing a carbon-neutral society) of energy policy, building industry standards, and new energy technologies for everyday use.

In this paper, drawing on previous work, we want to develop the consideration of buildings from "an integrative and participatory [perspective] [...] that considers the relationships between inhabitants and building systems as interactive and multidirectional, not linear and predictable as is assumed in the conventional approach" ([4] p 329). Our development of this stance is first empirical, then technological. Although we begin with a general introduction to various conceptualisations of comfort, we do this in an effort to draw out from our empirical field work and consideration of design examples, some of the practical consequences of different conceptualisations of comfort to the design and use of energy-sensitive technologies. Our field study of ordinary individuals' practical and situated use (and non-use) of energy systems allows us to juxtapose the measurable environmental parameters with people's ordinary everyday activities and their (rather infrequent) interactions with systems. Our analysis suggests that comfort for everyday users must be understood in relation to much more than environmental conditions, namely the temporal and sequential aspects of a day's order of activities, people's practices of habit and routine, and the comparatively minor role that energy consumption plays in the ordinary consideration (and justification) of quotidian action. This leads us to critically consider the role of existing 'comfort' technologies (e.g. thermostats, heating appliances, control units etc.) and the conceptualisations of consumption that they embody. We contrast these with a set of provocative design exemplars that have recently been developed in order to challenge ordinary notions of energy use and climate control, and we propose these as a way forward towards making conspicuous and accountable the relationships between everyday action, local environmental conditions, and energy consumption.

2. Definitions of comfort

'Comfort' is an ordinary word that has a multiplicity of vernacular uses. It has also been applied as a technical or theoretical term in a number of different disciplines that each has a particular and selective interest in the notion. For example, in the study of human evolution, we can see a quite different perspective on comfort in that human beings have always been able to adapt to their immediate environment and create a shelter for themselves, which was comfortable enough to live in. In other words, "achieving thermal comfort pre-dates by thousands of years the development of the theory of heat exchange" ([9] p 5).

In architecture, heat and light have been essential elements in design for centuries, and are recognised as aspects which provide for not only the physical, but also the psychological, comfort of building inhabitants. Features such as fireplaces and front porches are understood to provide much more than heat, light, shade, or (purely) for people's physical needs, but also contribute to feelings of security, domesticity, practicality and aesthetics (e.g. [16] p 92). Incidentally, the (etymological) origin of the term comfort, from Latin, also connoted far more than physical conditions, meaning something that was responsible for "giving mental and physical strength, encouragement and consolation" ([19] p 43).

Yet these immaterial, psychological and/or circumstantial aspects of the notion of comfort have been lost in the way in which the term has been operationalised by the engineering disciplines. The idea of a 'comfort zone' emerged in the 1920s in engineering. A 'comfort zone' is said to describe a physical state, which is constituted of "comfort conditions [that are] those under which a person can maintain a normal balance between production and loss of heat at normal body temperature and without sweating" (Yaglou 1949 in [2]). And it remains the case that such engineering operationalisations of 'comfort', informed by a heat-balance model, have been profoundly influential within the building industry since the 1960s. Based on this definition, physical parameters, such as daylighting, noise, temperature, air quality etc., have been correlated to people's experienced levels of comfort; for each of these parameters, an optimum 'comfort' range for their measurable values has been calculated. This notion of comfort—one that is simply a human physiological homeostatic equilibrium with respect to measurable environmental conditions—has been employed in laboratory tests and surveys, during which subjects evaluate their comfort experience on various numerical scales of sensation (Bedford scale (1936), ASHRAE (American Society of Heating, Refrigerating and Air-conditioning Engineers) sensation scale (1966) in [2]). During these tests, it has been assumed that test subjects are able to reflect consciously on and immediately relate their current physical state of being, and that such reports 'are' comfort. From these quantitative methods, the 'comfort zone' has been formalized, and universal comfort conditions have been stipulated in international building standards and guidelines.

3. Comfort practice

It is not difficult to see how this particular engineering operationalisation of the notion of comfort has been influential in the design of environments, systems and technologies that relate to the provision (and commoditisation) of comfort. When comfort is conceived as a more or less stable and universal range of environmental conditions, then the problems to be remedied with technological solutions become related to the control of those comfort-relevant parameters. Thus far, these 'solutions' have tended to result in (a) information delivery systems, which monitor and make available to users (or building custodians) the moment-to-moment changes in sensor readings so that adjustments can be made; or (b) automated systems (e.g. 'smart buildings') which monitor and summarily adjust heating/cooling controls for the indoor environment in response to sensor readings.

It is important to bear in mind that people are not only recipients of information [4]. They do not simply take inputs and process them to an output, but they move in a complex network of social, cultural (both collective and individual), situated and personal circumstances in which their reasoning for their decisions is taken on a daily basis. One also finds that there can be a gap between what people say they do and what they actually do, i.e. what they value vs. how they act ([3] p 58). In that regard, the potential of systems based on pure information delivery to actually change users' behaviour is unlikely to succeed for the ways that it oversimplifies the bases on which people typically act. We suggest this in the hope of challenging those responsible (including ourselves) for shaping future technologies that facilitate interaction between building and inhabitant. We want to urge designers to try to grasp this complex network within which users are situated and for designers to act within it through the technology they develop, instead of treating the issue as either an informational one, or one of automated control. In this paper we hope to offer a third alternative for systems designers based on our own emerging understanding of what we will term users' comfort practice within their own environments.

3.1 Case: Indoor Climate and Quality of Life

It is the importance of this aspect—comfort practice—that we will illustrate in the following qualitative study. The relevance of people's everyday practices to their uses of and interactions with indoor climate technologies is something we can only demonstrate by actually being there as they live their daily lives. In this way, we can situate users' interactions with technologies and other comfort-related practices within the busy stream of their daily lives. We have conducted this study of users' daily routines in parallel with the collection of quantitative measurements of some relevant 'comfort' parameters, such as temperature, light, humidity and CO_2 levels.

This study is one component of a 3-year long project 'Indoor Climate & Quality of Life', which seeks to investigate what the notion of comfort means to people, how they actually experience it and what innovation potential rests in an appreciation of people's everyday practices and accounts of those practices. Within the project, four family visits were carried out during the winter months, where researchers accompanied the parents from their place of work, to their children's institution (daycare/kindergarten), to their home. Video, field notes, interviews and other participant/observation techniques (respondent diaries and card sorting activities) were conducted in each visit and each setting (work, childcare institution, home). Particular attention was paid to the types of building, heating/cooling installations, and most importantly people's activities and actions around comfort systems, as well as their justifications for these.

3.2 Examples from the field

When taking a look at people's everyday comfort practice, system designers might often recognize them to be 'irrational' in that people act and consume on widely varying, contrasting motives and habits. In many cases, people's way of doing and routines are also resistant to change, where non-environmental values such as aesthetics, beauty, maintaining social relations might outrun purely environmental ones; people also interact with each other and sustain certain common folk wisdom and hearsay about e.g. materials and systems which are thought to be environmentally-friendly but which are not. These values and social practices and understandings are recognized as creating a problematic interaction between people's comfort and energy conservation [13], [19]. On first impression of the material we had gathered in the field, we could recognize these and other motives and habits. Besides these, it also became rather quickly obvious that actual use practice is clearly non-static and not universal (in contrast to the standards of indoor comfort), but the more interesting because it is "the human dimension" [20].

For this reason, researchers within the Indoor Climate & Quality of Life project, took a closer look at the material and filtered out a family's activities during a day and juxtaposed them to the measurements of the two pervasive parameters humidity and temperature taken in parallel.

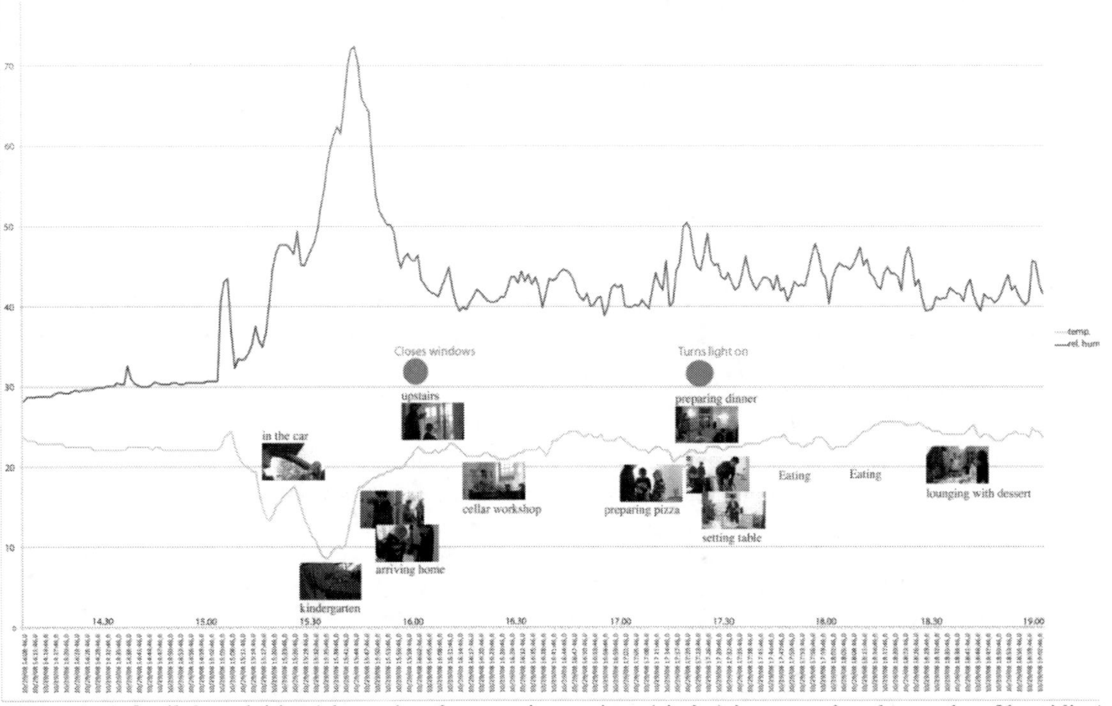

Figure 1: A family's activities (pictures) and system interactions (circles) juxtaposed to the graphs of humidity/ temperature from 14-19:00 o'clock.

The family above who the project team visited resided in a naturally ventilated, one-family, three-floor city house from the 1900s, which had old-fashioned double frame windows, which they were not allowed to change according to local heritage regulations. The researchers spent an ordinary day with the family, they followed the mother of two children aged 2 and 6 from the afternoon at work, picking up her children, coming home, preparing dinner, eating and relaxing until the evening. They tried to involve themselves in the family's daily routine, but also conversed about topics relating to comfort and environmental values when opportunities permitted. A portable data logger for temperature and humidity was worn by the mother while she was moving through her day. The conversations and activities were videotaped and interesting instances were noted down.

In the above graph [Figure 1], the measurements and activities are aligned with each other and direct interactions with indoor climate systems and products from afternoon to evening are highlighted. It becomes obvious that there are many activities happening during the family's day, which are directly or indirectly related to achieving comfort, which often is of energy-friendly manner. Only two direct 'system' interactions were observed during the afternoon/evening we shadowed the family: when arriving home with the children, the mother went upstairs and closed windows which were forgotten in the hectic morning routine; and when preparing dinner, the night set in and lights were turned on in the kitchen and living room. But other than these two actions, there were interesting 'others' happening, such as changing clothes (e.g. taking off the jacket when entering the house, but not when entering/leaving the car earlier), making a cup of tea (e.g. when coming home), being and playing in the kitchen (e.g. when the father comes home, the children start to 'build stuff' in the kitchen, with the kitchen functioning as the central place where the family gathers), turning the oven on, preparing dinner together (closely around the oven), lighting candles (e.g. when eating), eating a warm meal, and then drinking a cold 'smoothie' as dessert.

Some of these 'other-than-system' interactions are recognized as adaptive strategies that human beings have in their repertoire in order to achieve comfort e.g. by means of reorganizing space, using objects and people around them. In Humphreys [10], [11] actions such as stripping off or putting on clothing, curling up, eating a meal, visiting a friend, taking a blanket, cuddling, making a cup of tea, going to bed are seen as (energy-friendly) and powerful adaptive means by which people become comfortable, and reach an 'optimum' level of comfort on their own or with the help of others' (in case of cuddling) effort and engagement. These are among the actions they take towards becoming comfortable. Yet it is vital to recognise that there are many ordinary cases where 'comfort optimisation' is not in question. For example, in the above case, the mother stands outside in 9^{0}C having the same jacket on as in the car's 17^{0}C climate, while picking up her children. She talks to the kindergarten staff for ten or fifteen minutes. We might suggest that her concern for maintaining social relationships leaves her indifferent or tolerant towards the cold climate; yet it is also essential to recognise that the mother herself appreciates just how temporary her stay at kindergarten is likely to be.

We can see in this case that there may only be one or two direct system interactions while people move through their day, and these are often, especially in our hectic lives, part of a family's routine. The origins of these actions do not necessarily reside in a biological pursuit for homeostatic equilibrium. They may equally be based on a multitude of other considerations or contingent details of one's life. What we have learned to do out of habit (or family tradition), what we have learned about energy or consumption, what we can afford monetarily, what we have heard to be the case (e.g. folk wisdom and hearsay through the media or friends), our own sense of beauty and order, personal characteristics (e.g. forgetfulness or discipline), self-perception (e.g. inability to change), or time constraints. Each of these can vary according to situation, sequence, and time. The concept of comfort has to be expanded to include not only physical factors, but also social, psychological, behavioural and mundane, situational ones.

People are obviously able to adapt to their environments. But what becomes important as we seek to influence their behaviour towards more environmentally conscious action is to appreciate, and relate to, their moment-to-moment motivations for the actions they take. People need to be motivated in order for them to enact significant changes to their habitual patterns of behaviour if they are to adopt more energy-friendly actions [22]. This only seems possible when we can address their knowledge, and take into account their current values and ways of being. Otherwise engineers and policy makers who wish to direct people into certain behaviour towards their energy consumption must fall short, not just because people's routines and knowledge structures can be resistant to change, but because their overt actions and behaviours are clearly not the whole story. Furthermore, if building project teams design systems to overrule human action-taking they also, by so doing, take away human responsibility for and awareness of their immediate surrounds.

Let us consider another example of what people actually do on an everyday basis in order to keep comfortable. Below, the mother of two (the same family) was asked to keep a diary to collect her comfort actions, which is then used to reflect upon what she broadly counts as comfort. It is again juxtaposed to the available log data of humidity and temperature.

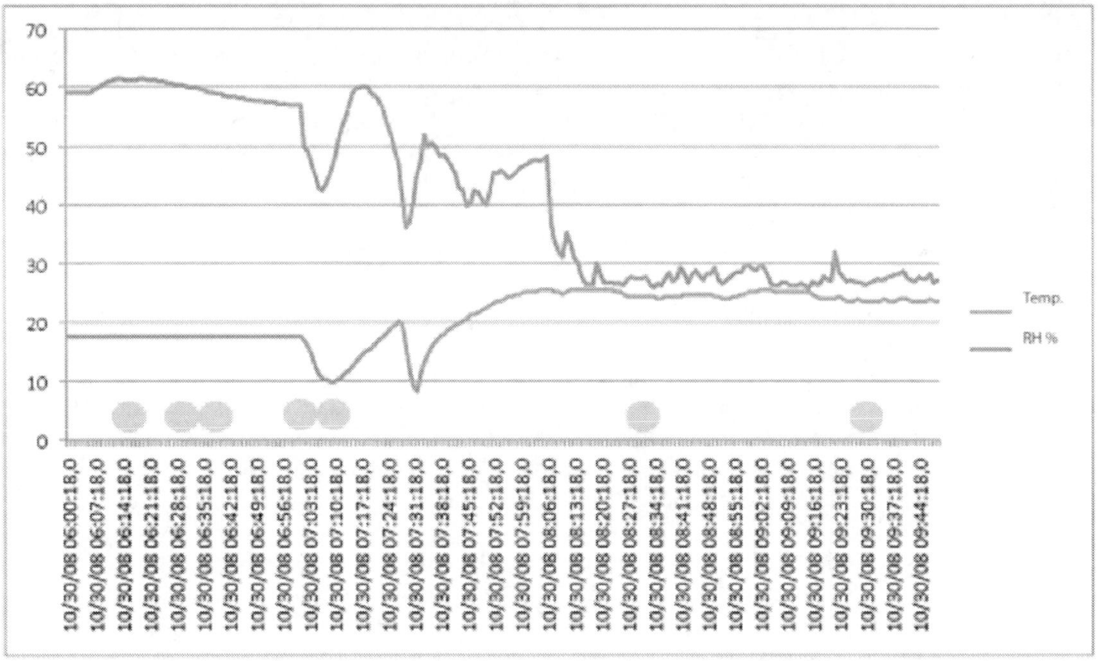

Figure 2: A mother's comfort actions (circles) juxtaposed to the graphs of humidity/temperature from 6-9:45 o'clock. Actions are listed below in Table 1.

Time	Place	Action
6:15	*Bed/blanket*	*"The alarm clock rings, I pull the blanket over my head."*
6:30	*Bedroom/windows*	*"I air out all bedrooms and shake all blankets."*
6:35	*Kitchen floor*	*"The floor heating is broken, I put on slippers."*
7:00	*Hallway/door*	*"I close the windbreak door before we leave the house to keep the warmth until we come back."*
7:05	*Car heating*	*"I fire it fully."*
8:30	*At work/kitchen*	*"I take a cup of warm coffee and warm my hands on the cup."*
9:30	*At work/kitchen*	*"I take a cup of warm coffee and warm my hands on the cup."*

Table 1: Anne's diary, October 30th, 2008

Above [Table 1], Anne accounts for how she keeps warm and how she interacts with the systems at her home and on the way to work. In just three hours, she employs many different adaptive strategies such as using a blanket to keep warm, enjoying the contrast between cold air (rather low at around 17^{0}C [Figure 2]) and warm bed, wearing slippers in order to not have to touch a cold floor, drinking a coffee to warm her hands, as well as her use of indoor climate controls, both routinely (windows) and selectively (car).

Later in the diary, she mentions that "fresh air and sunshine feels comfortable" while walking from one building to another one at work, and that when "it is cold outside, it feels really good to come inside and close the door behind us" when coming home with her children, with whom she later "cuddles under a blanket on the sofa and watches a children's movie." In the evening she "feels a bit sick", so she "turns on the radiators in the bathroom and takes an extra long warm bath to see if [she] gets better". These were the entries from when Anne, the mother of two, got up until she went to sleep, giving her own descriptions of common comfort actions a person takes during a winter's day.

We can notice that people are actively engaged in predicting, defining, changing their 'own comfort' and can be quite tolerant of a wider range of temperature etc. than might ordinarily be assumed by a 'comfort zone' model; people enjoy contrasts between warm and cold, they achieve a comfortable state by being together with their children, and interacting with other people and objects around them. It also shows the extent to which people are aware of and act in relation to their local circumstances, including how they feel right now ("feeling sick") or their environment.

A further observation, although fairly mundane, is actually highly relevant to our discussion, and relates to the immediate history and temporal sequence of actions. In both cases we have considered here, notice that deliberate actions to e.g. 'warm up' are not always (or even primarily) taken in response to the actor's current environmental conditions. In our first example [Figure 1], the windows at home are closed at a point in time where the indoor temperature is a (supposedly) 'comfortable' 22/23^{0}C. Whether the reason for this action is (a) in anticipation of the fact that, at 16:00 on a October day in Denmark the outdoor temperature is in rapid decline, or (b) that the windows should not have been left open in the hurried morning routine and would usually be closed, or (c) that open windows are responsible for high energy costs or high consumption or GHG emissions, or some other reason, it remains that it is not the current temperature of the house that is in need of remedy, or that can be cited as the cause for the action with the system (the windows). Similarly, in the second example [Figure 2; Table 1], the action (with the cup of coffee) takes place after Anne's ambient temperature has already been in the mid-20's for 30 minutes. This action (and her second cup of coffee one hour later) are listed (in her diary) as opportunities to warm her hands. Although the ambient temperature is (by fiat of engineering science) in the comfortable range, her experience of those conditions is not. In both of these cases (opening windows and holding cups of coffee), the actions taken have more to do with what the participants have previously been doing (travelling in a cold car, walking outside), with how they subjectively feel, and/or with what they ordinarily do (return to a house with closed windows for instance) than they are explainable in terms related to an attempt to remedy a current, local environmental condition.

The data presented above contributes to an empirical and analytical understanding of what people ordinarily count as comfort and how they practice comfort-related energy actions. This is a first step towards drawing implications of such an understanding for the design of technologies that are more sensitive to the multitude of ways in which comfort and comfort practices are woven into our daily lives.

4. Intervention in comfort practice through design

4.1 Creating awareness

When taking a look at how systems are engineered to provide comfort in buildings today, design choices are predominantly technology-driven, and paint a picture of user and activity in a frame of typification. Frequently, when designing a product and/or system the designer falls back on using "commonsense categories of social types" and attaching them to particular actions [5], but which were not derived from the particular context being designed for. When it comes to the situation of use, the 'typified' user is under no constraints to act 'typically', and often suddenly customizes, struggles or works around the (often rigid and inflexible) controls and systems provided for ([1] p 187). Most indoor climate technology, such as air-conditioning, mechanical ventilation, heating units, etc. tries to minimise people's interaction with the system, in many cases limiting them to a basic on/off switching of a button. Even in the latest zero-carbon housing concepts, it becomes obvious how developers quickly jump to conclusions about "a person [being] a passive receiver of sense-impressions" instead of being "in a dynamic equilibrium with the environment" ([10] p 1). The answer is frequently thought to be the automation of control of the indoor environment. Here, the allowance for adaptive actions to take place is limited, as temperature, humidity, lighting, levels of noise etc. are kept at an optimum level instead of giving choice and diversity, which takes no consideration of the possibilities afforded by the blanket, the cup of tea, changes of clothing and cuddling with family. For instance, the zero-carbon living design by Velfac in Aarhus, Denmark is giving the responsibility and control from the user to a computer.

From *The Guardian*, May 21st, 2009: *"Zero-carbon eco home is light years ahead"* (emphasis ours):

> "The interior climate is *controlled by a computer*, linked to a thermostat, *which opens and closes windows according to the temperature, season and time of day*. Chief engineer Amdi Worm assures me that there is a *manual override*. "If the occupants open a window, in an hour or so the window will automatically close a little," he says. [...]
>
> "If they really insist, they can choose to do it again, but I'm sure that *the house will tell them* that the way they are handling it is not energy efficient.""

The indoor climate in the above mentioned house is not only controlled by technology but also steps in and overrides the user's decisions, taking no account of the reasons they may have had for opening a window. Velfac's developers are also certain that by handing over the house to the inhabitants and having them live in it they will be told about energy-efficient living; here, energy-efficient living is understood as passive living where the control and decision-taking is left to the technology, and the family is not able to shape it according to individual preferences and understandings.

We are in doubt in how far this approach to reconciling comfort with energy conservation will succeed in the long run. We suspect that taking autonomy away from people is more likely to result in diminished (not improved) understandings of the correlations of their actions with sustainable living; that the reasons for the system's particular actions at particular points are not likely to be transparent to those who must live with their consequences; and that the loss of human autonomy and control is as likely to generate creative ways of 'working around' the system rather than straightforward, energy-efficient compliance with it. In contrast, we regard it as essential to engage people in the change towards energy efficient comfort practice, making them aware of their current practices and challenging some of their existing habits with the aid of technology design. Mazé and Redström ([17] p 57) argue that design has "a profound power to influence consumer and societal values" and can be successfully employed "as a vehicle for critical reflection". Clearly, design can tell someone about which thermostat setting, time of ventilation etc. is the "right one". But more than this, it can help to reflect on and to integrate new knowledge into the technical, social, material systems making up human practice [19], ([17 p 58); in order to do so, design has to be more than just the specification of form, however.

There is a potential in the creation of interactive, engaging design artefacts, which facilitate awareness of and responsibility for a building's energy-friendly operation. One must be aware of the danger that the automatic provision of a 'comfortable' indoor climate may serve to accustom people to an unsustainable and unnecessarily narrow range of environmental conditions, and create expectations of this type of environment in diverse private spaces including home, office, cafés, and libraries. Indoor climate technologies based on economic and social reasoning (or typification) form an idea or conceptualisation of the user of a building that then shapes the experience and a certain 'normality' of it ([8] p 411; "the ratchet" in [18]; [19] p 25). This critical voice about social consequences of employing a certain technology in a certain way and the need for moving in a different direction of engagement is further elaborated in Verbeek [21] (our emphasis):

> "A hearth was, in Borgmann's terms a 'focal thing': it was the focus of people's attention. Heating their houses gave people an *engaged way of interacting with their world* [such as the effort of gathering wood, chopping it etc.]. A central heating system takes away this engagement. Rather than engagement, it *produces consumption of the commodity 'heat'*."

In the above, Verbeek ([21] p 71), referring to the philosopher of technology Albert Borgmann (1984, 1999), raises the issue of people running the risk of becoming (or having become) passive consumers/occupants instead of active, dynamic inhabitants of a building. Here, the ability to measure and fully control leads to the inability to experience and be stimulated by the rich environment.

In order to avoid this road to be taken, we may consider alternative ways of how to employ technology in an engaging and responsive way to encourage self-sufficiency, awareness and responsibility. With the advent of the climate change crisis, the human need for adaptation must become a cost-effective (in terms of peoples' instead of a system's development and training) and morally acceptable way of future action. That means we may invest in technology that revitalizes people's effort (in the sense of 'attempt' or 'try', maybe 'slight exertion'), to provide for connectedness with their world, more specifically in this case, with the source of consumption, e.g. heat, light, air quality. In these forms of technology, people might use different senses, such as tactile, visual, auditory, to experience how their levels of warmth, coolness, air quality, noise gradually change; to recognize how their actions influence their environment and take responsibility for either changing themselves to it or adapting it to their everyday, situational needs and values. Through these engaging artefacts, we can take advantage of and change current habits towards reduced consumption of energy in that we both show and allow people to connect with the effects of consumption on their immediate, local (and ultimately global) environments.

4.2 Designing forms of engagement

In the following, we present a set of three rather different examples of how technology might serve such interests. Each of the examples we present are best considered as examples of what has come to be known as "critical design" (e.g. [6]). They are not intended to be commercial products (nor do we imagine there being a commercial market for them); on the contrary they are intended to provoke in users or viewers some kind of critical consideration of the topics they treat.

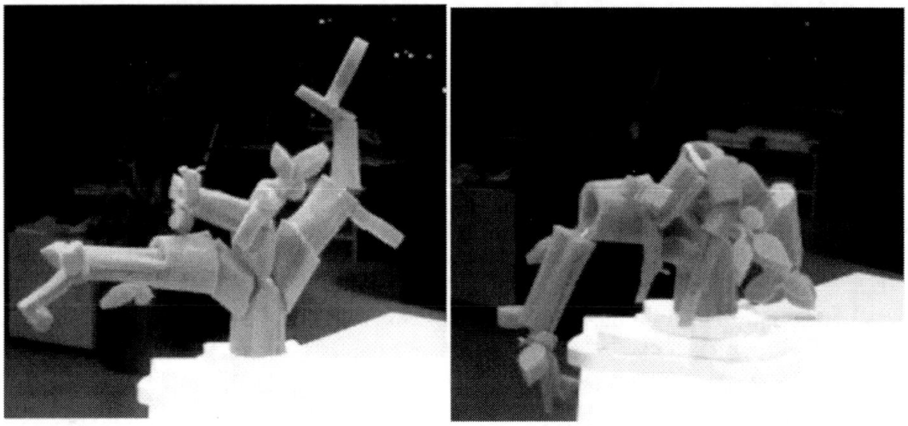

Figure 3: The CO_2 reactive tree, in upright and wilting position. Design by Liam Wu Zheng and Robb Mitchell.

The above design [Figure 3] has been developed as a part of the Indoor Climate and Quality of Life project. The concept is intended to be an engaging artifact that could raise awareness of users' patterns of comfort behaviour and motivate users to actively change these. The conceptual design of this interactive 'tree' reacts to levels of CO_2 in the room. Higher levels of CO_2 cause it to alter its position from upright to a wilting stance; lower levels mean it rises back upright. The branches can be twisted by the user to assume an individualized shape. The idea is to have an artifact that does not present information, but does serve as some kind of representation of the environment in real-time, with the idea that users can, after living with it over time, learn how their actions and presence in the environment they share with the 'tree' affects its behaviour (and by implication, affects the quality of the air that they share with the 'tree'). This flexible, reactive tree is one expressive and interactive idea of how technology might involve people towards realizations of, and changes to, their own practice, while possibly offering them the possibilities for new (local) understandings about the consequences of consumption. We associate this with 'critical design' in a number of respects.

For one, this product does not exist to serve any consumer's pre-existing 'need'; it is provocative and playful, but not functional in any strict sense. Furthermore, it is intentionally ironic. It takes the form of a tree, yet the representation embodied in the behaviour of the product inverts the relationship between plantlife and CO_2, as living trees breathe CO_2 and thrive in high atmospheric concentrations of it. The interactive 'tree' is an example of a form taken from the plant world that simulates our own (human/animal) relationship to carbon dioxide, rather than a tree's. Furthermore the very idea of a plastic or polystyrene tree that serves to depict some aspect of human beings' environmental footprint is clearly ironic. Although the idea of the artifact is to enhance public and individual attention to our affects on our local environment, it, in its very artifice (including its construction from non-natural materials), simultaneously embodies yet another instance of the natural world being simulated by, and replaced with, the artificial. It is only the symbol of the tree that is 'green' in this case. In these respects, the artifact challenges us to reflect on own practices of consumption, in part by presenting itself as another instance of that very practice.

Two projects conducted at the Interactive Institute in Kista, Sweden, are close in spirit to the interactive tree we have discussed. In the STATIC project [12], researchers and designers have worked with the possibility for intervention in energy use practice through making energy visible and interactive.

Figure 4: Disappearing Pattern Tiles. Design by Sofia Lagerkvist, Charlotte von der Lancken, Anna Lindgren, Katja Sävström.

The designers created technologically-enhanced ordinary household devices. The interactive tiles [Figure 4] are one such example. They have integrated technology into the existing fabric of a house's décor, but in a way in which interactive features allow the tiles to respond to conditions in the room. Building inhabitants can draw information from them by interpreting their behaviour as the patterns on the tiles change, fade in response to heat levels in the bathroom. A long hot shower will begin to strip the bathroom wall of its tile patterns. Embodied in a concept such as this is a subtle appreciation of some mundane aspects of people's ordinary motivations for action. These tiles do not simply make one form of energy consumption visible; they make it visible in a way that tends to matter to a house's inhabitants: by diminishing the aesthetics of the high-consumption bathroom. Its designers implicitly appreciate that minimising energy consumption is, in many cases, not yet incentive enough to alter people's comfortable patterns of behaviour. But 'de-aestheticising' the bathroom wall is more likely to be something people will notice and respond to in behavioural change. By making the aesthetics of the tiles responsive to heat, inhabitants can interact (through their energy behaviour) with something that they ordinarily value and pay attention to more than their consumption of energy for the sake of comfort.

Another project from the Interactive Institute's STATIC portfolio is the Power Aware Cord. In this case, the designers made energy use directly visible on the objects that consume it.

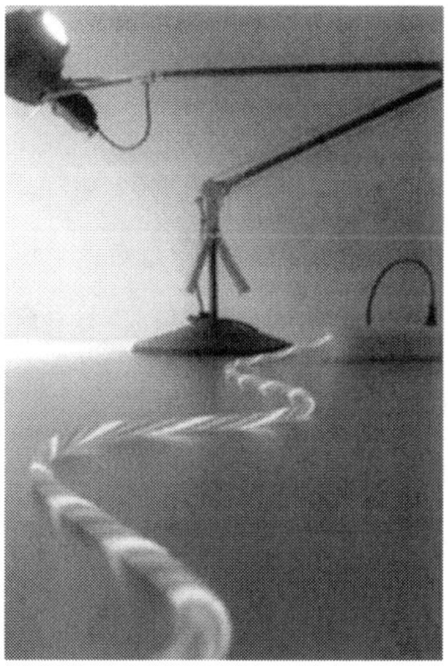

Figure 5: Power-Aware Cord. Design by Anton Gustaffson and Magnus Gyllenswärd.

The Power Aware Cord [Figure 5] is the cable of a power strip that, like the interactive bathroom tiles, makes visible energy consumption. The intensity, flow and frequency of pulses of light on the cord are proportional to the volume of current flowing through it. This is a straightforward mapping of energy use to a display of that use, attempting to influence people's energy use habits by making them publicly visible. The visibility of consumption might provoke discussion about each other's comfort patterns, in this case about the use of electrical lighting, and taking responsibility to rearrange e.g. one's seating or routine to make use of daylight instead. The 'critical' aspect of this idea also employs irony in the sense that the cord consumes (additional) energy in order to make the energy consumption of other appliances visible; once again we see the idea of a product that is itself an example of the behaviour and practices (e.g. of the use of energy-consuming technologies) that it seeks to make conspicuous.

5. Conclusion

In this paper we have sought to make two contributions related to the reconceptualisation of comfort. The first involved an empirical examination of people's actual comfort-related practices that are situated in the mundane routine of daily life. In the study we presented, we juxtaposed people's actions and activities to measurements of some environmental parameters that are, in an engineering orientation, taken to be definitive of 'comfort conditions'. This enabled us to demonstrate that people's actions and interactions with comfort technology are simply not understandable solely in terms of responses to their current environmental conditions, nor to a biological imperative to maintain a physiological equilibrium with respect to their environment (as per a 'comfort zone' conception of comfort). If we seek to understand the bases of people's actions with energy-consuming comfort technologies, we must first situate those actions within the human, social, environmental, temporal, circumstantial contexts in which such actions are taken; contextual details that are systematically stripped from operationalised definitions of comfort.

Additionally, we have sketched out some of the ways in which a naturalistic understanding of comfort and an appreciation of the situated and contextually-sensitive nature of comfort actions might inform and complement the design of different kinds of comfort technology. In particular, rather than developing technologies that provide information and control of comfort parameters, we might see the emergence of products and systems which, through their interactive behaviour, confront us with our own patterns of behaviour and our own affects on our environment. In these ways, the 'problem' to be addressed by technology design is not how to automatically optimise energy conservation on behalf of disempowered users, nor is it to treat people as information processing systems that make decisions based on new informational inputs regarding their rates of consumption. Rather, we have suggested that it is possible to design technology that implicitly acknowledges that agency and situatedness are inescapable aspects of our human condition, and that technology can play a role that enhances our awareness of ourselves and our own patterns of behaviour, and occasionally pricks our conscience.

6. Acknowledgements

The Indoor Climate and Quality of Life project at the SPIRE Center for Participatory Innovation, University of Southern Denmark, has been partly financed by the national program of the Danish enterprise and construction authority ('Erhvervs- og Byggestyrelsen'); it is a 3-year project (09/2008-08/2011) in collaboration with Danish industrial partners and two research institutions (SPIRE and Denmark's Technical University). We gratefully acknowledge our project team, particularly Monika Frontczak and Jesper Pedersen, who were participating in the presented user study and material analysis, Jared Donovan, Liam Wu Zheng and Robb Mitchell for the provocative design, and most importantly the families for sharing their everyday comfort with us.

7. References

[1] Ackerman M S 2000 The intellectual challenge of CSCW: the gap between social requirements and technical feasibility *Human Computer Interaction. A Journal of Theoretical, Empirical, and Methodological issues of User Science and of System Design* **15:2** 179-203

[2] Auliciems A 1981 Towards a psycho-physiological model of thermal perception *International Journal of Biometeorology* **25:2** 109-22

[3] Cole R J 2003 *Buildings, Culture, and Environment: Informing Local and Global Practices* (Canada: Wiley-Blackwell) chapter 5

[4] Cole R J, Robinson J, Brown Z and O'Shea M 2008 Re-contexualizing the notion of comfort *Building Research & Information* **36:4** 323-36

[5] Crabtree A, Hemmings T and Rodden T 2002 Pattern-based support for interactive design in domestic settings *DIS* 265-75

[6] Dunne A 1999 *Hertzian Tales: Electronic Products, Aesthetic Experience and Critical Design* (London: RCD: CRD Research Publications)

[7] Healy S 2008 Air-Conditioning and the 'homogenization' of people and built environment *Building Research & Information* **36:4** 312-22

[8] Heiskanen E and Pantzar M 1997 Toward sustainable consumption: two new perspectives *Journal of Consumer Policy* **20** 409-42

[9] Humphreys M A 1995 Thermal comfort temperatures and the habits of hobbits ed F Nicol and M A Humphreys (London, UK: Chapman & Hall) 3-13

[10] Humphreys M A 1995 What causes thermal discomfort? *Workplace Comfort Forum* (London:

Royal Institute of British Architects)

[11] Humphreys M A 1997 *An adaptive approach to thermal comfort criteria* ed D Clements-Croome (London: E & FN Spon) 129-37

[12] Interactive Institute 2004-05 STATIC, can be found on http://www.tii.se/static/

[13] Kempton W, Feuermann D and McGarity A E 1992 "I always turn it on super": user decisions about when and how to operate room air conditioners *Energy and Buildings* **18** 177-91

[14] Kjærsgaard M, Pedersen J and Djajadiningrat T 2003 A lost cause: the ever-improving developer's map *CHI* **April** 642-3

[15] Kolcaba K 2003 *Comfort Theory and Practice: a Vision for Holistic Health Care and Research* (New York: Springer)

[16] Lupton E and Murphy J 1994 Case study house: comfort and convenience *Assemblage* **24** (MIT Press) 86-93

[17] Mazé R and Redström J 2008 Switch! Energy ecologies in everyday life *International Journal of Design* **2:3** 57-72

[18] Shove E 2002 Converging conventions of comfort, cleanliness and convenience *Lancaster University On-Line Papers, Department of Sociology*

[19] Shove E 2003 *Comfort, Cleanliness and Convenience: the Social Organization of Normality (*Oxford: Berg)

[20] Stern P C, Aronson E, Darley J M, Kempton W, Hill D H, Hirst E and Wilbanks T J 1987 Answering behavioral questions about energy efficiency in buildings *Energy* **12:5** 339-53

[21] Verbeek P-P 2002 Devices of engagement: on Borgmann's philosophy of information and technology *Techné* **6:1** 69-92

[22] Winett R A, Hatcher J W, Fort T R, Leckliter I N, Love S Q, Riley A W and Fishback J F 1982 The effects of videotape modeling and daily feeback on residential electricity conservation, home temperature and humidity, perceived comfort, and clothing worn: winter and summer *Journal of Applied Behaviour Analysis* **15:3** 381-402

IOP Conf. Series: Earth and Environmental Science 8 (2009) 012009 doi:10.1088/1755-1315/8/1/012009

Media communication strategies for climate-friendly lifestyles – Addressing middle and lower class consumers for social-cultural change via Entertainment-Education

S Lubjuhn[1], N Pratt[2]

[1] Institute for Communication Studies, University of Duisburg-Essen, 45117 Essen, Germany
[2] UNEP/Wuppertal Institute Collaborating Centre on Sustainable Consumption and Production, 42107 Wuppertal, Germany

E-mail: sarah.lubjuhn@uni-due.de, nadine.pratt@scp-centre.org

Abstract. This paper argues that Entertainment-Education (E-E) is a striking communication strategy for reaching middle and lower socio-economic classes with climate-friendly lifestyle messages. On the international level (e.g. in the US and the Netherlands) E-E approaches are being theoretically grounded, whereas in Germany they are not yet. Therefore further theoretical discussion and mapping of E-E approaches is central for future research. As a first step towards providing further theoretical foundations for E-E in the field of sustainability, the authors suggest a threefold mapping of E-E approaches. The threefold mapping of E-E approaches for communicating climate-friendly lifestyles to middle and lower class consumers is based on recent results from academic research and practical developments on the media market. The commonalities among the three is that they all promote pro-sustainability messages in an affective-orientated rather than cognitive-orientated, factual manner. Differences can be found in: the sender of the sustainability message, the targeted consumer groups and the media approach in use. Based on this, the paper draws the conclusion that two new paths for further research activities in the field of Entertainment-Education can be proposed: (1) *Improving* the existing approaches in practice by using theoretical foundation from the E-E field. This comprises at its core (A) to do formative, process and summative effect research on the messages and (B) to use E-E theory from the field of social psychology, sociology and communication science for further improvement and (2) *Generating* new E-E theories by analyzing the existing practical approaches in the media to communicate climate change.

1. Introduction

Consumers, their lifestyles and consumption patterns, are a key factor in promoting a socio-cultural change towards sustainable development [1]. In particular, the mass media can be considered as an important medium for fostering climate-friendly consumer lifestyles [2]. For example, a survey by a German publisher [3] reveals that 73 percent of the German people indicate that they use mass media – television, radio, print, internet – as their most important source of prosocial (environmental and health) information. An American study emphasizes that around two-thirds of entertainment television viewers learn from the depicted messages content on prosocial issues [4]. Additionally, one-third of them 'take action' after watching these messages. Thus, viewers decide to put a lid on a cooking pot in order to save energy or they decide to become organ donors.

Consumer patterns differ significantly from each other and changes in awareness, attitudes and behaviors towards more climate-friendly lifestyles are complex issues. Interestingly, some scholars have shown that these changes function like logical progressions. For example "if we do this to people, they will behave this way" [5]. Singhal [6] argues that this may be a problematic prevailing mindset, because social systems and thoughts and actions of human beings could not be predicted, controlled, and measured in the same way as a machine, a clock, or the trajectory of an introductional ballistic missile. In other words, changing consumer patterns towards more climate-friendly lifestyles is not a cause-effect issue. Frequently, non-linear and contradictory developments arise, before changes in knowledge, attitudes or even behavior can occur. For scientists and practitioners it is challenging (1) to agree on what to do to initiate sustainable change, (2) to be certain about what the results will be, and (3) to determine clear factors, which have lead to changes from a retrospective viewpoint [7][1].

2. Structure of the paper

In order to analyze media communication strategies for pro-sustainability messages, which are especially suited for people, who like to be informed in an affective manner, i.e. people, who regularly come from middle and lower socio-economic milieus, this paper has the following structure: (1) First, affective-oriented media strategies are discussed, - as opposed to cognitive strategies. (2) In the second part, a literature overview regarding communicating environmental friendly lifestyles is provided. (3) The next paragraph introduces the E-E approach and discusses the application of E-E for sustainability messages. It is also argued, why a mapping of E-E approaches in this field is needed. (4) In the fourth and main part, the paper presents the threefold mapping of E-E approaches for sustainability, suggested by the authors: The Governmental Entertainment Approach, the NGO/Association Entertainment Approach and the Business Entertainment Approach. What the three approaches have in common is that they all promote pro-sustainability messages in an affective-orientated rather than cognitive-orientated, factual manner. What they differ in is the sender of the sustainability message, the targeted consumer groups and the media approach in use. Each approach is discussed in detail and case examples as well as a critical discussion are provided. (5) In the fifth part, a critical reflection on the mapping approach is presented on a broader level. (6) In the last part, the paper discusses recommendations for further research.

3. Cognitive vs. affective-orientated media strategies

The issue of climate change is communicated to consumers by different agents such as governmental organizations, NGO's and businesses. These agents address consumers in their different roles of their every day life.

[1] An example for a complex change issues is Rosa Parks' refusal to give up her seat in a segregated bus. Other African American people did this several times before Rosa Park, but her refusal led to cascading events and was one important step towards the civil right movement of African Americans.

For example, governmental organizations may see consumers in their role as a 'citizen' or 'voter'. NGO's may primarily look at consumers in their role as 'climate change activists', and businesses may initially see consumers in their role as 'clients for their climate-friendly products/services'. According to these perspectives the agents want to reach consumers in their 'role' and therefore apply different media strategies. In other words, a governmental organization sends other climate change messages and hence uses different media strategies compared with NGO's or businesses.

As will be outlined later, for all three agents, it can be differentiated between 'cognitive-oriented' and 'affective-oriented' strategies, with the latter being especially well suited for addressing consumers from middle and lower socio-economic classes.

Many communication campaigns have been implemented during the last years, which mainly focus on 'educating' consumers by providing cognitive information, i.e. appealing to reason rather than to emotions. This 'cognitive-orientated' strategy can be considered as the 'classical' strategy, for example "if we communicate facts and data to people, they will be convinced". The 'cognitive-orientated' strategy particularly addresses consumers from higher socio-economic classes, who are affine to cognitive information and interested in facts, figures and sophisticated elaborations [8]. Prominent examples for these cognitive-oriented media strategies are (1) the communication campaign of the German Federal Ministry for the Environment, Nature Conservation and Nuclear Safety *Climate protection pays off* (www.bmu.de/english/climate_protection_initiative/aktuell/42001) and (2) the German information campaign "*Encouraging sustainability*" implemented by several German associations as well as the Wuppertal Institute for Climate, Environment, Energy (www.mut-zur-nachhaltigkeit.de).

Studies from the field of communication science [9] clarify that 'cognitive-orientated' media strategies are not useful for addressing people from middle and lower socio-economic classes who are usually more interested in affective media strategies and consume most of the time in an affective manner. Thus, they recommended applying 'affective-orientated' media strategies for these socio-economic groups. For reaching them successfully with messages, it is suggested to investigate their emotional involvement in entertainment media formats. Affective communication is especially sufficient for describing the process during and after the audience has been exposed to entertainment media [10]. Beside the process of exposure, the relations between the narrative characters, affective dispositions and the experience of entertainment in a media format can be specified by affective communication research. For example, the affective disposition theory [11] understands empathy in the way that the audience first perceives access and then morally judges regarding the shown characters in entertainment television. In consequence, the audience members agree or disagree with the characters behavior and they develop positive or negative effects, which lead to affections like fear or hope according to the outcome of the story in which their beloved protagonist and the disliked/hated antagonist is portrayed. A prototypical entertainment experience like suspense or hope can be considered as a direct consequence of affective dispositions.

The advantages of applying affective communication research [12] such as comprehending the process of exposure, the relations between the narrative characters, the affective dispositions of audience members and the experience of entertainment in a media format may help to contribute to develop strategies that fit for specific media entertainment formats, campaigns and programs, which address middle- and lower socio-economic classes.

In recent years, scholars proposed using the E-E communication approach (see section 5), which communicates ecological messages in an affective manner [13] and which especially addresses socio-economic classes with climate change and other social issues.

The borders between affective and cognitive-oriented strategies however are subtle. There can be various mixed types. In this context, it may be central to take into consideration (1) what (which message) is communicated to the audience and (2) which strategies are used to communicate the message. Both the messages and the strategy can be cognitive or affective-orientated. For example, affective messages might be framed with a cognitive strategy and cognitive messages can be communicated in an affective manner. Especially the latter case is of interest for E-E approaches for lower and middle class consumers. In specific cases it might be useful for E-E approaches to also include 'cognitive-oriented' elements for middle and lower social classes, who may respond to very practical advice especially when it is close to their everyday concerns, e.g. how to save money or how to repair things in their home. If these messages are communicated in an affective format, such as do-it-yourself reality shows, environmental improvements could also be achieved.

4. Literature Overview: Communicating Environmental-Friendly Lifestyles

The issue of communication on environmental-friendly lifestyles has not been widely addressed in communication science. However, a number of sub-disciplines such as environmental psychology, environmental sociology and consumer studies have also started dealing with the topic. From the standpoint of environmental psychology, tackling with communication on environmental-friendly lifestyles has a longer tradition. Since the 1970s, theories have dealt with environmental attitudes, values and behavior patterns. A very prominent model is the one of Fietkau & Kessel [14], which takes intrinsic environmental values, environmental knowledge as well as extrinsic behavior motivations into consideration to explain why people finally act environmental-friendly. In environmental psychology two schools can be found: On one hand there are models focusing on reflected behavior [15] and on the other hand, there are models which focus on behavior based on norms [16]. The first school involves general cost and benefit expectations to explain environmental friendly behavior of individuals, whereas the latter focuses more specifically on personal norms (i.e. the feeling of duty for living an environmental friendly lifestyle), social norms (i.e. anticipated pressure evoked by 'significant other' persons) and attribution of environmental responsibility. In recent years, theories have been established that try bringing both schools together [17]. Those models additionally take components of routines of action into consideration, which have been identified to have a significant influence on environmental-friendly lifestyle behavior [18]. For relating these results to E-E, especially those models are striking that combine cost/benefit expectations, personal/social norms as well as routines of action of individuals.

In contrast to environmental psychology, environmental sociology does not investigate the relation of the individual to the environment, but the relation of human beings resp. of society towards their environment [19]. Environmental sociology deals with social factors that can cause environmental problems, the societal impacts of such problems, and approaches to address the problems. For this paper especially sociological classifications in different socio-economic classes and the environmental behavior of these classes are relevant. The institute, socio-vision, is known for its milieu-studies based on sociological data, personal interviews on norms, believes etc. and the analysis of home visits conducted by different social classes (e.g. analysis of furniture, fitments, designs etc.) [20]. Based on this analysis ten different socio-economic classes are identified. The main groups in the middle and lower socio-economic class identified are (for Germany): Consumer-materialists, traditionals, civic middle-class, hedonists and experimentalists. Representatives of these groups are not knowledable, interested nor in favor of environmental issues, e.g. they were found reluctant to spend more money for sustainable products and many agreed to the question of environmental problems were exaggerated. Reaching these groups, which make up a significant part of society, is crucial for sustainable development. As the authors argue E-E (see section 5) can be a useful means for reaching members of these groups.

In consumer or sustainable consumption research, the topic of environmental-friendly lifestyle messages is discussed with another focus. The main bulk of studies from the consumer field argues that beside specific trail depending consumer behavior, e.g. strategic consumption decisions as well as budget or availability limits, consumers can decide between more or less environmental and social friendly behavior alternatives [21]. However, consumers meet behavior barriers or restrictions, which exacerbate environment-friendly behavior systematically. This particularly comprises costs perceived as too high, paradoxical information signals, time pressure and a structural overload. Thus many studies from the consumer field argue that only a few, very engaged consumers will behave 'right', i.e.

environmentally friendly. The 'wrong' structures when acting environmentally friendly often cause burden rather than relief [22]. In recent years, consumer research established a heuristic of options (engaging factors) and restrictions (non-engaging factors) in sustainable consumption behavior [23], i.e. it has been claimed that a successful communication must emanate from options and restrictions for formulating appropriate messages. In this context, the following factors have positive influence on environmental-friendly lifestyle behavior [24]: perceived variety of behavior opportunities, knowledge of behavior consequences, positive consumption experiences with ecological or social products and services, positive appreciation of the consumption decision through social networks, positive behavior consequences through technical or communicational feedback-mechanisms (rewarding-effect) and target group communication through unconventional, affective strategies and campaigns. This target group communication may also take into consideration: informal setting, social networks and communities. These factors, and particularly the latter, are relevant for communicating the environment through an E-E approach.

The existing approaches reveal that there are many sub-disciplines starting to work on the topic of environmental communication. As outlined above, this article focuses on the E-E approach (see section 5), since it focuses on mainstream consumers from middle and lower socio-economic classes. Within this field, a mapping of different media approaches for environmental topics is missing and as the literature overview above indicated, there is no mapping, which E-E could draw from immediately. This is why this article aims at suggesting a mapping of media strategies within the E-E field (see section 6), in order to successfully apply E-E approaches.

5. The Entertainment-Education strategy for Sustainability Messages

In the past decade, the Entertainment-Education strategy has shown promise in addressing environmental and other social problems like climate change, environmental degradation, unchecked population growth, or gender inequality [25]. Entertainment-Education is the process of purposefully designing and implementing a media message to both entertain and educate in order to achieve desired individual, community, institutional and societal changes among the intended media user populations [26]. Entertainment-Education uses affective, emotional messages to show consumers how they can save energy and resources, live a healthier and an environmental-friendly life, and to show consumers how to address such issues in their communities (family, circle of friends, youth clubs etc.).

The idea of combining entertaining and educating elements is not new and goes as far back in human history as the art of storytelling. For thousands of years, drama, music, dance and other forms of folk media have been used around the globe for recreation, reformation and instructional purposes. The conscious combination of entertainment and education through mass media such as in television, radio, songs and comic books began in the 1950s and 1960s [27]. The first intentional endeavors came from the media side with the writer, producer and director Miguel Sabido. He was inspired by the audience success of a Peruvian television soap opera *Simplemente María* and produced a first series of seven E-E telenovelas from 1975 to 1982 [28]. Some years later, Sabido developed his methodology for E-E soap opera productions particularly based on the social cognitive theory. This theory outlines self-directed mechanisms of behavior change through social modeling and perceived self and collective efficacy [29]. The change in people's ability to carry out a specific behavior is primarily based on their vicarious experience modeled by positive, negative and transitional narrative characters portrayed in E-E radio and television soap operas.

Recent E-E research is particularly grounded in previously established theories of social psychology. This means that the focus lies on various social and psychological processes during and after media consumers encounter E-E narratives. This also suggests that the level of analysis is restricted to individuals. Scholars have recently discussed the danger of this individualistic, cognitive-proceeding centered and 'mechanistic' approach for social change [30]. They have made suggestions to employ a more holistic approach, e.g. taking the cultural background of the media users and the platforms and features on which the messages are sent into consideration [31].

On the international level, the theoretical foundation of E-E is particularly used by governmental organizations and universities/research institutes (see section 7) when designing and implementing E-E projects. The theoretical foundation at its core comprises (1) to do formative, process and summative evaluation on the messages and (2) to use an E-E theory (e.g. the Elaboration-Likelihood-Model, see [32]) for successfully reaching media users with messages. For NGO's, associations, as well as businesses, the theoretical foundation is so far rather untypical.

In Germany, NGO's and businesses do not make use of a theoretical foundation for their media strategies. In comparison to the international level, governmental organizations also do not do so. Nevertheless, many practical approaches exist, which may be related to the E-E strategy. These approaches can be characterized by a 'strategic practice', i.e. the messages are targeted at specific consumer groups.

According to these gaps, this paper proposes two new fields of application of E-E: (1) *Improving* the existing approaches by using a theoretical foundation from the E-E field via formative, process and summative effect research on the messages as well as using an E-E theory from the field of social psychology, sociology and communication science and (2) *generating* new E-E theories by analyzing the existing approaches.

6. Mapping of E-E approaches

In order to promote the application of E-E approaches for communicating climate-friendly topics to middle and lower socio-economic classes this paper suggests a threefold mapping. Due to the limited evaluations on the effects of E-E measures, further theoretical discussion and mapping of E-E approaches is central for future research. As this is a theoretical paper, it cannot provide detailed effect evaluations of E-E measures on a broad basis. But as a first step towards providing further theoretical foundations for E-E in the field of sustainability, the authors suggest the threefold mapping. This mapping could be used as a theoretical basis for practical case studies that evaluate the effects of E-E measures in the sustainability area. In this paper the authors can only provide indicative measures for evaluation the effects of E-E measures.

Within the field of Entertainment Education, approaches have been developed and implemented to diffuse affective-orientated and pro-sustainability messages via entertainment TV (see the formats *Echt Elly* in the Netherlands, *Wa$ted* in New Zealand or the format *Welt der Wunder* in Germany). However, there are still approaches outstanding, which clarify:

- Which groups of agents try to reach middle and lower class consumers with pro-sustainability messages? And
- Which characteristics their approaches have (targeted consumer groups and the media strategy in use)?

To answer these questions, this paper suggests a threefold mapping of approaches in the field of E-E for promoting pro-sustainability messages to middle and lower-class consumers, particularly climate-friendly messages:

1. The **Governmental Entertainment Approach** (see section 7) focused on sustainability issues with a high acceptance in society (e.g. promoting to save resources and energy or eating organic food),
2. The **NGO/Association Entertainment Approach** (see section 8) focused on marginal pro-sustainability issues from advocacy groups with a lower acceptance in society (e.g. promoting solar energy as 'the best' sustainable energy),
3. The **Business Entertainment Approach** (see section 9) focused on pro-sustainability messages from individual, for-profit businesses, with commercial intentions for pushing pro-sustainability messages (e.g. promoting to save energy and therefore to use services of a specific energy company).

Figure 1 shows a mapping of the different approaches. The main characteristic of all three approaches is that they communicate sustainability messages in an affective rather than cognitive and factual manner. This also allows reaching consumers with a lower information level according to sustainability issues, in particular from middle and lower socio-economic classes. Anyhow, the three approaches differ in (1) the sender of the messages, (2) the targeted consumer groups and (3) the media strategies in use.

Figure 1: Mapping of Entertainment-Education Approaches for Communicating Climate-Friendly Lifestyles

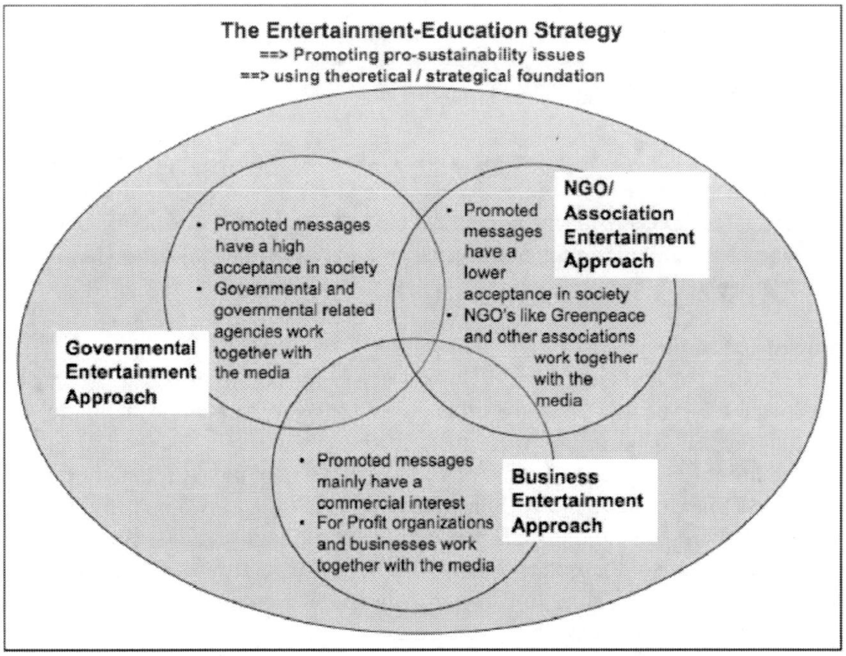

With this threefold mapping, this paper provides a sender-orientated focus to describe different approaches in the field of E-E. There are also other emphases possible. One example could be to look on the E-E field from a geographic/cultural perspective. In the international context (particularly in the US and the Netherlands), theoretical foundations are often used in the field of E-E related to the governmental approach. Because the theoretical foundation of E-E approaches is less diffused in German governmental organizations, NGO's/associations and businesses, they can learn from the experience made in other countries.

In the following sections the three sender-orientated entertainment approaches are discussed with respect to (a) description/definition (b) targeted consumer groups, (c) media approach, (d) case example and (e) a critical discussion of each approach (see section 7-9). Afterwards a critical reflection on the approaches is provided on a broader level (see section 10) and finally, recommendations regarding future research in this field are given (see section 11).

7. Governmental Entertainment Approach

7.1. Description/Definition

The governmental entertainment approach has mainly been implemented by governmental (related and in their behalf working) organizations like the German Ministry for Food, Agriculture and Consumer Protection, the German Federal Ministry for the Environment, Nature Conservation and Nuclear Safety, the German Federal Center of Health Education as well as universities and research institutes. In this approach, governmental organizations collaborate with entertainment media people (like producers, scriptwriters and journalists) to promote their sustainable messages. Regarding climate change issues, this approach primarily focuses on messages, which have a high acceptance in society and which can be transported in a short and easy manner, e.g. saving resources and energy or buying regional products (see section 7.4.). On the international level, e.g. in the US and the Netherlands, governmental organizations use effect studies to evaluate E-E messages and their influence on audience members [33]. In Germany, this has not been the case yet. Lubjuhn/Bouman [34] suggest implementing effect research as a strategic measurement for German future collaboration activities. Doing effect research in E-E projects may be important to prove whether the E-E messages reach the audience members in front of the TV sets – and generally – whether it makes sense to support the governmental entertainment approach in general.

The following section will briefly depict how E-E effect research could maybe look like in Germany. One option could be to evaluate a specific message with a pre-/post design to investigate, whether the message influences audience members in their knowledge, attitudes and behaviors [35].

As an example for illustrating future effect research may serve an E-E message of Hans Beimer in the soap opera *Lindenstraße* (see section 7.4). In one episode Hans Beimer emphasized that he buys regional products, because frozen ones consume four times more energy. Impact research may ask regular viewers of *Lindenstraße* before watching the episode, to analyze their buying habits in general. After watching the episode, impact research could investigate, whether the message (e.g. in combination with interpersonal communication) leads to changes in knowledge, attitudes and even behavior towards energy consumption and buying products and whether the viewers explicitly remember the message of Hans Beimer. As a control group, non-regular viewers of *Lindenstraße* could be investigated with the same research design.

7.2. Targeted Consumer Groups

The governmental entertainment approach reaches consumers who are interested in seeking information in an affective manner. These consumers mainly come from middle and lower socio-economic classes [36]. The following paragraph briefly depicts some characteristics regarding consumers who are affine for being engaged and persuaded by this approach.

7.2.1. Knowledge on Climate Change. Consumers who are receptive for these messages have less factual knowledge about climate change issues [37]. Liedtke et al. [38] differentiate between environmentally conscious and environmentally friendly behavior. Environmental conscious behavior arises when consumers reflect on the basis of facts, figures and sophisticated elaborations that it is necessary to protect the environment. For environmental-friendly behavior, consumers must not necessarily be interested in saving the environment. For instance when they abstain from a plane flight not for reasons of environmental protection but of saving money. E-E affine consumers mainly behave in order to environmental-friendly manners [39]. Interestingly, these consumer groups use less resources (e.g. not driving a car) than people from higher socio-economic classes because of money saving purposes [40].

7.2.2. Climate change aspects of interest. The messages provided by this approach reach consumers who are interested in narrative environmental messages, (1) which help them to save money, (2) which are easy to integrate in their every day life, (3) which are easy to remember and thus (4) which can be easily discussed with friends, family and community members (participative function). Energy saving messages (see section 7.4.) may be one of those that are very simple to adapt for consumers, because they receive a direct advantage from it.

7.2.3. Intention to change lifestyles. Environmental-friendly behaving consumers do not feel responsible for their environment and they have not internalized norms and values in order to promote sustainability issues [41]. If they change their lifestyles towards more climate-friendly ones, they do so unconsciously. Most of the time

- parasocial interaction with narrative characters in E-E series (for instance "I like the character and the clothes he wears and I can imagine also buying an organic T-shirt") as well as
- interpersonal communication with friends, family and neighbors about the E-E storylines

lead to processes that open up possibilities for changes in knowledge, attitudes and behavior towards more climate-friendly consumption patterns [42]. Consumers who are receptive for environmental-friendly messages from this approach have a high consumption rate of entertainment media in general. Additionally, they have a distinct reading culture and like to be informed in an entertaining and sensational way. Therefore, they use a wide variety of entertainment media (television, radio, internet, pod casts) to be entertained and to seek information from which they can learn and which they consider to be helpful in their every day life [43].

7.3. Media Approach

Governmental organizations use a mass media mix to reach the above mentioned consumer groups with their pro-sustainability messages. They do so, because they promote generally accepted topics, which can be easily diffused through different channels in the mass media (like television, radio and internet). Therefore, they collaborate with media entertainment people, supporting them with information and suggestions according to topics in the field of sustainable development. Governmental organizations try to prepare this information in a way that the entertainment media makers can use it for further ideas and storylines in soap operas, telenovelas, photo stories etc. For example, a governmental agency can suggest to a television producer of a telenovela to write a storyline about a narrative character who wants to make an energy consultation in his apartment for finding out where he can save resources, energy and money. Therefore, the governmental agency offers the TV producer the necessary background information (e.g. what is the typical course of a consultation and which main messages are given etc.) to write an accurate and authentic storyline.

7.4. Case Example

The first section of the paper depicts the campaign *Climate protection pays off* by the German Federal Ministry for the Environment, Nature Conservation and Nuclear Safety as an example for a 'cognitive-oriented' media strategy for reaching higher socio-economic consumers. Interestingly, the German Ministry extended its climate communication strategy by an 'affective-orientated' strategy, the E-E strategy. Therefore, an entertainment television series was used to communicate climate friendly messages to consumers from middle and lower socio-economic classes. In 2008, the German Federal Ministry for the Environment collaborated with the oldest German soap opera *Lindenstraße*[2] and the Department of Health and Environment of the city of Munich. *Lindenstraße* is broadcasted by the public broadcasting station ARD, and it is famous for dealing with social and environmental issues. Since 1988, the *Lindenstraße* broadcasted 13 episodes about environmental topics [45], and some of their characters are very sensitized to social and environmental topics.

The idea had been to initiate the environment day *Looking for climate, offer protection* on August 17th 2008 in downtown Munich and to broadcast a *Lindenstraße* episode seven days earlier, which dealt with the design and implementation of this environmental event. In the episode, the character Helga Beimer planned the environment day in her *Lindenstraße* neighborhood. She loves the nature, and in former episodes she has made several attempts to advocate for environmental issues. According to the above-mentioned theory of Bandura [46], Helga Beimer can be considered as a positive role model for social change. In the episode she motivates her family, friends and her community to be a part of the event and to discuss and share opinions about climate change issues. As a platform for environmental discussions and exchanges, Helga Beimer planned and implemented a climate change quiz and a climate cycling rally.

The *Lindenstraße* episode aimed at making viewers think about climate change issues. It integrated various 'easy to understand' messages on more climate-friendly lifestyles in its storylines. For

[2] In 2007, *Lindenstraße* had on average around 5 Million viewers each episode [44].

instance, the character of Hans Beimer, the ex-husband of Helga Beimer, is very engaged in protecting the environment. In one storyline an energy consultant visits him at home to figure out where potentials for saving energy can arise. In the dialog with the consultant, Hans Beimer emphasizes using a steamer saves 60 percent of energy compared to using a regular cooking pot. Moreover, Hans Beimer mentioned that he buys regional products because frozen products consume four times more energy.

Another example for sending climate-friendly messages uses the character of Murat Dagdelen, a Turkish citizen who is less interested in environmental issues. He had been involved in criminal activities in the past, but during the last episodes he did quite a good job in re-building new life goals: He has successfully opened a doner snack bar and is happily married. However, in the episode he was not portrayed as an 'over-motivated' activist for saving the environment like Helga Beimer, but in one dialog he authentically remarked that saving the environment may have advantages for himself, his family and other people. On the environment day, he served burgers for the visitors of the event. In one storyline with Helga Beimer, he emphasized that he bought organic ingredients, and he argued that veggie burgers could be a good alternative to meat burgers. According to Banduras theory of social modeling, Murat Dagdelen can be considered as a transitional role model, first having a negative attitude according to climate protection in former episodes and then changing his mind to a more positive attitude.

7.5. Critical Discussion of the Approach

Using the governmental entertainment approach, win-win effects for both the TV and the governmental side can be created, because through collaborating (1) the governmental institution can reach a target group of millions of people with climate protection messages and (2) TV makers can produce accurate and authentic storylines with the help of certified expert knowledge. An essential condition for mainstreaming the approach may be to design a formalized collaboration setting (e.g. with a formal collaboration contract and a guidance of the partnership by effect research). Currently, implementing sporadically and informal collaborations is the foremost practice in Germany, in terms of 'let us do something together and let us see how it works out'. So, further development and improvement is needed in this field.

There are also disadvantages of the governmental entertainment approach. First and foremost, the fact that German governmental organizations and entertainment TV makers collaborate together for sending climate protection messages is reflected quite critically in Germany's public. Several reasons for this can be found. One major point may be that the Nazi regime misused entertainment TV for propaganda purpose [47] and out of this history legislation forbids governmental organizations to influence TV makers content in terms of 'we pay and you send the message'. Governmental organizations are only allowed to advise the work of the TV side, to sponsor it, and to guide the TV formats with participative events such as expert workshops or face-to-face meetings of soap celebrities and audience members, discussing climate change issues.

In consequence, when the collaboration partners come together, they undergo a thin red line between legality and illegality. Where do advising and sponsoring end? Where does influencing media content start? Germany's media legislation does not make a clear statement on that [48] and outsiders cannot comprehend, whether collaboration partners crossed boarders of legality while collaborating. The majority of the initiated collaborations suffer in public reputation, because the governmental part is suspected to influence the program content. Current discussions in collaboration practice deal with these areas of conflict and try to find solutions for future activities.

8. NGO/Association Entertainment Approach

8.1. Description/Definition

The NGO/Association entertainment approach refers to two important characteristics. The first alludes to the circle of actors who are sending these messages. These are NGO's (like *Greenpeace*, the *Red*

Cross etc.) and associations (like the *Central Marketing Association for German Agriculture*, the *German association for Healthy Sports and Sports Therapy* etc.). When comparing this group of agents to the governmental entertainment approach, it is not in any case clear, which societal and economic interests these organizations serve. For example, the German platform *Nutrition and Exercise* cooperated with the German commercial TV broadcaster *SuperRTL* in the children's program *Peb & Pebber* in 2007 to promote healthy and organic food. The platform includes non-profit as well as for-profit organizations in its portfolio. Interestingly, there is no information available (e.g. on the platform website) regarding the (social, commercial and economic) interests of the platform.

The second characteristic of this approach is that the sustainability messages are less accepted in society comparing them to the messages sent in the governmental approach. For example, the German commercial broadcaster, *RTL*, has aired six episodes of *Greenpeace TV* in 1997. In these episodes *Greenpeace* promoted specific messages that served their organizational concerns like "stop poison and gene technology in food". *Greenpeace* produced the six episodes on their own and offered sponsoring money to *RTL*. *RTL* was giving them the timeslot for airing.

Like in the governmental entertainment approach, impact research on messages developed in this approach is missing yet, so it is hard to say, which impact climate change messages implemented by a NGO/association and TV makers have on audience members so far. Impact research could investigate with a pre-/post design how the specific message "stop gene technology in food" influences audience members. The audience may be asked before watching the Greenpeace format, what their opinion about gene technology in food is and then after the show. In addition, research could investigate whether they remember the special message, and if they talked with family members or friends about the issues they saw in the episode.

8.2. Targeted Consumer Groups

The targeted consumer groups in the NGO/Association approach are quite similar to the governmental approach. The biggest difference is that this approach advocates more specific topics like promoting solar energy, no gene technology or German organic food. The consequences are that (1) the sustainable messages are more marginal and therefore (2) the consumer groups addressed are more specific comparing them to the governmental approach.

8.3. Media Approach

NGO's and associations also use an entertainment approach, which mixes up different media channels. However, their communication strategy is not as varied as in the governmental approach. It is more focused on special forums. For example, the TV program *Peb and Pebber* was only broadcasted by a TV channel targeted to children. The efforts to reach children with healthy and organic food messages were not combined in a bigger media strategy, i.e. sending the messages for example also via internet, radio or print. At least it was not incorporated in a special campaign. This goes with Volpers et al. [49] who suggest that many NGO's and associations particularly focus their media strategy on one or two channels, like collaborating with a television program or a youth magazine for promoting their pro-sustainability messages.

8.4. Case Example

A case example for this approach is the activities of the *CMA* (*Central Marketing Association for German Agriculture*). In 2003, *CMA* collaborated with the German music television channel *VIVA*. They launched the cooking show *The Last Judgment* together. The show aimed at making young people vocal regarding German agricultural products and to change their knowledge, attitude and consumption behavior towards these products. In a lively manner, the program depicted a variety of innovative recipes for regional and organic agricultural products. Additionally, celebrities from the music, sports and the entertainment professions were invited to cook together with the host of the show, the well-known *VIVA* moderator Tobi Schlegl.

Another collaboration of *CMA* in the field of entertainment television has been the co-production of the entertainment magazine, *Made in Germany – The Best from Countryman,* broadcasted by the commercial TV broadcaster *Terranova* from 2005 until 2007. The magazine focused on German agricultural products and tried to promote a positive image regarding these products. In the episode 09th May 2007, the president of the *German Meat Association*, who is also a member of the *CMA*, talked about German meat, emphasizing 'how healthy and organic' it is. He referred to German butchers who 'work with transparency' and who have 'nothing to hide' (*Terranova*, 09th May 2007, also see [50]).

8.5. Critical Discussion of the Approach

On one hand NGO's/associations have the advantage that they can work together with TV makers for promoting their topics to a mass audience. On the other hand, TV makers can receive (1) new ideas for the message content and (2) sponsoring money for implementing the program.

When talking about the disadvantages of the approach, two main points can be made. First, NGO's and associations do not find collaboration partners from the TV field as easy as governmental agencies, because their topics are more marginal and it is not easy to locate TV makers, who fully support more specific messages such as "no gene technology in food" or "solar energy is the best alternative". Second, it is to say that NGO's and associations regularly pay sponsoring money to financially support the TV makers. As described above in section 7.5., this accuses them even more in public that they actively influence the program content. For example in the case of Greenpeace mentioned above, German media scholars had criticized that *Greenpeace "had the possibility to report about their own concerns without any kind of control or authority"* [51]. Beyond doubt, the aims of *Greenpeace* can be considered as welfare-orientated. However, some media scholars argue that *Greenpeace* and other NGO's, as well as, associations do not serve the public welfare but their own interests, means, clients, members and their donors [52]. Thus, image or financial interests of these organizations dominate the welfare or sustainability interests. Obviously, future media legislation needs to find clear standards for resolving conflict areas arising in this collaboration approach [53].

9. Business Entertainment Approach

9.1. Description/Definition

Pratt and Sabapathy [54] note that business can play a key role in promoting sustainable consumption patterns and climate-friendly lifestyles. In recent years, companies discovered the potentials of the entertainment media to send pro-sustainability messages with commercial purposes. To summarize these developments in the business sector, this paper proposes the name business entertainment approach.

Michaelis [55] suggests that companies will need to go beyond their current business framework and engage in collaborations with civil society as well as with media in order to promote climate-friendly consumption patterns. Sending affective rather than cognitive, fact-based messages can be a powerful means of reaching consumers – especially from middle and lower classes with little knowledge and limited interest in sustainability matters.

However, when it comes to business communication of sustainability messages, especially via affective, entertaining media strategies, the risk of being accused of "Greenwashing", needs to be taken into account i.e. being accused of disseminating environmental disinformation in order to gain an environmentally responsible public image. Companies that make use of affective media strategies rather than communicating "hard facts" are especially at risk of these kinds of accusations. Therefore, these companies are well advised to be able to provide reliable facts and figures regarding sustainability aspects of their products and services when asked for it in order to back their sustainability message.

The furniture company IKEA can serve as an example here, which promotes a family image (abolishment of child labor in production etc.) on an affective base. The informative brochure *Social and Environmental Responsibility*, which contains facts and figures about their sustainability efforts [56], is not handed out in the stores, but in contrast to other IKEA brochures is only available for customers who express interest on how IKEA deals with sustainability issues. The reason is that the company is extremely cautious not to take any actions that could be misinterpreted as Greenwashing [57]. If the company promotes its sustainability efforts too pro-actively and markets its progress in concrete numbers and figures, it risks that environmentally concerned groups and other critics might point to shortcomings in sustainability still existing and accuse the company of Greenwashing. Corporations with strong environmental claims often remain prime targets of critics, who point to shortcomings of their measures, so that applause is seldom far from a Greenwashing charge [58]. The companies hence risk getting into what can be identified as a Reputation Trap [59].

9.2. Targeted Consumer Groups

The targeted consumer groups are current or potential customers of the respective company. Thereby affective media strategies are a valuable means for reaching customers from middle and lower socio-economic classes. As outlined above, these customers often have limited knowledge about sustainability so that fact-based, cognitive media messages (e.g. CO_2 emissions per km for a car) are less effective.

Furthermore, the sustainability interest of these consumers is often driven by the interest to save money and/or because sustainability is "trendy" and allows representation as well as discussions among family members, friends and community (participative function). The sustainability trends followed by these consumers are often those adopted by higher socio-economic groups that function as peer-groups for consumers of middle and lower socio-economic classes [60]. A main part of these latter consumers are what can be called LOHAS (lifestyle of health and sustainability)-affine consumers [61] who are interested in sustainable products, but whose knowledge about and commitment on sustainability remains clearly behind 'traditional' LOHAS or 'ecos'. LOHAS-affine consumers are driven by affective triggers, e.g. they favor products from local and national produce, products which they consider 'authentic' and related to their daily life, and they prefer new, sustainable products from existing, traditional companies which they have built trust in, rather than new sustainability start-ups [62].

These kinds of consumers, like LOHAS-affine consumers, are a fast growing group, which makes it interesting for companies as well as for researchers to investigate if and how these consumer groups can best be reached via affective media campaigns.

9.3. Media Approach

The business entertainment media approach very often uses a twofold strategy to target their consumers. The approach can be related to the Elaboration Likelihood Model (ELM) [63] presented above: Companies communicate a climate-friendly message with a high acceptance in society, for instance 'saving energy'. First an affective communication is used, e.g. by involving stars/idols or by using animals or small child patterns. This communication is initiated through a peripheral route, in which a motivation and a positive attitude towards the theme (e.g. saving energy) are created in an affective way. As a second step, a cognitive message based on facts and figures can be launched, e.g. amount of energy and hence CO_2 to be saved. This can motivate people to take the central route according to ELM and hence to sustain environmental behavior change (e.g. saving energy by purchasing energy efficient products). In order to further spur motivation, the climate-friendly message can be coupled with additional values relevant for the specific consumer group, e.g. saving money, being trendy or simply an entertaining, fun message.

The other main component of a business entertainment approach, besides the environmental message is the promotion of a specific product or service. Thereby the product or service is presented as a means for realizing the sustainability value – and the related additional value –, e.g. "if you use our energy system, you can not only save money, but you are climate-friendly too". This last layer provides specific and frequently limited messages, which seek to encourage consumers to buy commercial products or make use of commercial services.

It is important to note that in the business entertainment media approach, the affective component clearly is in the forefront. Facts and figures can be provided in addition within the media message or – as outlined above – can be made available separately 'on demand', for example within a special brochure (IKEA) or on a separate webpage (see RWE case example below). The affective component is predominant because the business entertainment media approach targets middle and lower social classes who are open to and interested in affective messages, but who often are not familiar with environmental topics such as climate change. Affective messages therefore can prove useful to spur motivation towards sustainability topics and environmentally friendly lifestyles.

9.4. Case Example

As a case example for the business entertainment approach serves the resource efficiency campaign of the German energy company RWE. As a major player in the European energy market, RWE is chosen as an example for the business approach because the business approach is discussed here with respect to E-E makes sense, especially for large, multinational companies, which target consumer mainstream (consumer masses).[3] As discussed previously, affective E-E strategies are especially well suited to address lower and middle class consumers.

In 2007, RWE launched an affective communication campaign, which used animals as 'efficiency experts'. The campaign aimed at identifying potentials for saving energy and money for 'the average' consumer and hence consumers from middle and lower socio-economic classes. To reach these consumer groups successfully, RWE launched a campaign with animals like dachshunds, hamsters and lizards, which give energy saving advices in print media, on television and radio. For example, in one television-advertising clip, RWE told the story of a speaking dachshund who walked through his house and found evidence that the house he lived in is not well insolated. For example, he discovered that the front door had a draft. He emphasized that energy saving hints from RWE are helpful to avoid energy loss and help to save costs and money. This story was told in a very emotional and funny/entertaining manner. At the end of the clip the information was given that RWE provides energy saving checks on their website. The evaluation of RWE's business entertainment approach paralleled effects of the animal stories on the consumers. Many consumers made the energy check and discovered more climate-friendly lifestyles for their every day life [64]. The company however only tracked the number of consumers, which run the check. Whereby data on consumers changing their behavior after the check was not tracked.

9.5. Critical Discussion of the Approach

Especially for companies, which serve mainstream consumers from middle and lower socio-economic classes, which are often less informed and interested in sustainability topics, it can prove useful to accompany their sustainable products by affective-emotional media strategies. In order to promote environmental behavior change, these strategies should go beyond mere marketing and should be based on E-E approaches, which are theoretically founded, e.g. which measure effectiveness and change of lifestyles rather than only amounts of products sold. So far this is hardly done. To realize such an endeavor, companies could engage in partnerships with researchers in this field.

A general pitfall for the business approach discussed can be put forward in form of a paradox: As some authors argue, it can prove contradictory to promote 'more consumption', as sustainability contains the concept of less consumption [65]. On an affective level excessive consumption may seem attractive, especially for lower class consumers as a hard or unobtainable dream, but the real excessive – and unsustainable – consumption is to a higher degree performed by those who can actually afford it. When higher socio-economic classes function as a peer group social climbing for lower classes, it is

[3] In contrast, small, genuine green companies usually address a smaller segment of well-educated consumers from higher economic classes and not the consumer mainstream.

exactly the opportunity to consume more – live in larger houses, have more cars etc. – that is attractive. But there is a smaller segment of higher classes, typically those with high cultural capital who may set a standard for moderation and simple living etc. Thus, companies have to reflect in detail, which kind of messages are useful to promote their products and services.

10. Critical Reflection

In order to map different collaboration approaches for media strategies within the field of E-E, this paper suggested a sender-oriented approach with a differentiation into governmental organizations, NGO's/associations and businesses collaborating with entertainment TV makers. Alternative mappings also would have been possible, such as mapping according to receivers (e.g. consumer types) of the messages or a differentiation regarding geographical partnerships, e.g. collaboration approaches in Western and/or Non-Western settings. As the senders are the ones, which need to apply media strategies within the E-E field in order to successfully communicate environmental topics, the authors mapped the different media strategies according to this group.

The mapping of different collaboration approaches depicted in this paper were chosen to highlight essential practice developments and to pinpoint (1) that theoretical foundation such as effect research is a necessity for future strategies in this field and (2) that legislative clarification and differentiation is needed to support the distribution of accurate and authentic sustainability information via entertainment TV and thus to foster 'social or sustainability change' [66].

When applying the three described approaches, implementers from governmental organizations, NGO's/associations and businesses are confronted with the fact that they are allowed to advise and sponsor, but they are not allowed to influence the program content. Indeed, most – if not all – collaboration implementers criticize that there are no standards and guidelines, which they can use as an orientation when initiating and implementing a collaboration and 'how far they can go' without 'contravene' media laws. For example, when an organization sponsors money to soap opera producers for a resource efficiency episode of a series and they strongly recommend sending a specific message in a storyline – for example that preheating an oven saves 20 percent of energy – can this be considered as an influential act? Can it be considered as an influential act when TV makers first have to 'get convinced'? And can it be defined as a non-influential act when TV makers agree on suggestions for messages coming from their sustainability partners right from the start? Visibly, future media legislation has to deal with these questions. Especially the differentiation between product-placement – often done by businesses – and (social/sustainability) topic placement needs to have more clarification. It could be recommended to establish specific legislative standards and guidelines for the three entertainment approaches of governmental organizations, NGO's/associations and business described above. Volpers et al. [67] supports this by emphasizing that entertainment collaborations in the field of sustainability topics will by trend increase, so Germany and other countries may do good to think about specific collaborations guidelines and ways of monitoring and sanctioning them. For example in the Netherlands, it is intensively discussed to introduce a media committee, (1) which supervises entertainment collaboration approaches from the initiation to the end and (2) which gives sanctions, when legislations are violated. Such a committee may contribute to transparency of entertainment collaboration approaches and a documentation of positive lighthouse projects could be possible, while – in contrast to that – negative examples also can be pointed out. In consequence, collaboration approaches would get more comprehensible to the public and not each and every collaboration would stand under the general accuse of 'violating legislation'. The mapping, according to the different types of senders (governmental, NGO or business), could be useful to sort the various legal issues, so that each sender has an overview of potential media strategies it can use, as well as, the according legal issues associated with it. The legal issues however are not dealt with in this paper as this would lead to far at the beginning, but if the mapping proves useful, there could be a further refinement of the mapping suggested.

11. Recommendations

During the past decade, several German governmental organizations, NGO's, associations and businesses are communicating climate change messages in a strategical, affective manner. These messages are mainly targeted to reach middle and lower socio-economic consumer groups. The paper suggests that these approaches can be related to E-E. Based on this, the paper provides three recommendations:

1. ***Different agents and media strategies***: It makes sense to differentiate the existing agents into three different organizational types of approaches that can recently be found in E-E field: (I) The Governmental Entertainment Approach, (II) the NGO/Association Entertainment Approach and (III) the Business Entertainment Approach. These three approaches differ in (A) the sender of the pro-sustainability messages, (B) in their targeted consumer groups, and (C) in their media approaches to reach lower socio-economic groups with climate-friendly messages.
2. ***Two paths for further research regarding Entertainment-Education ((1) Improve approaches through theory, (2) Generate new theory by analyzing approaches)***: In Germany, many approaches are existent, which can be related to the E-E strategy. So far most of these approaches are not yet theoretically founded. This paper suggests two different paths for further activities:
   a) *improve* existing approaches by using theoretical foundations from E-E theory. This comprises at its core (I) to do formative, process and summative effect research on the messages and (II) to use E-E theory from the field of social psychology, sociology and communication science for further improvement and
   b) *generate* new theories in the field of E-E by analyzing existing approaches used by government organizations, NGO's and businesses in the media to communicate climate change. Analyzing these paths may lead to new application areas and theories in the E-E field.
3. ***Learning from other countries***: Regarding the first path of the second recommendation (Improve approaches through theory), Germany can learn from the international field where a theoretical foundation of E-E projects is established (at least in the governmental arena). This could comprise formative and summative research for E-E in Germany.

Implementing theoretical foundations to the current approaches and analyzing the approaches for generating new theory may be important future steps in the field of E-E. This includes doing research on the effects that pro-sustainability messages have on the audience members' knowledge, attitudes and behaviors. These and other research results may contribute to deepen and extend the suggested threefold mapping and its characteristics for reaching middle and lower consumer classes with climate-friendly messages.

References

[1] European Environment Agency 2005 *Household consumption and the environment* EEA Report 11

Monagham P 2007 *What assures consumers on climate change?* June 2007

Pratt N 2009 forthcoming *Ich kaufe, also bin ich – gut? Nachhaltiger Konsum - eine Kontextbestimmung* (Wien: Transit-Neue Kritik)

[2] Lubjuhn S and Bouman M 2009 Entertainment-Education in den Niederlanden und den USA. Was kann Deutschalnd davon lernen *merz (medien + erziehung), Zeitschrift für Medienpädagogik*, **53/4** 44-49

Singhal A Pant S. et al. 2000 *Environmental Activism Through an Entertainment-Education Radio Soap Opera in India* ed M. Oepen and W. Hamacher Communicating the environment. Environmental Communication for Sustainable Development (Frankfurt: Peter Lang)

[3] Verlag 2005 OTC *Tools. Marken, Macher, Meinungen* (München: Gong Verlag)

[4] Porter Novelli Centers for Disease Control and Prevention et al. 2005 *TV Drama/Comedy Viewers and Health Information 2005* (Porter Novelli HealthStyles Survey)

[5] Stacey R D 1996 *Complexity and Creativity in Organizations (*San Francisco, CA: Berrett-Koehler Publishers)

Wheatley M J 2006 *Leadership and the New Science Discovering Order in a Chaotic World* (San Francisco, CA: Berrett-Koehler)

[6] Singhal A 2008 Where social change scholarship and practice went wrong? Might complexity Science provide a way out of the mess? **2** *Communication Development and Social Change*, retrieved from http://utminsers.utep.edu/asinghal/Articlesandchapters/singhal-throught-piece-CDSC-10-18-07-Where_Social_Change_Scho.pdf

[7] Papa M J, Singhal A and Papa W H 2006 *Organizing for Social Change. A Dialectic Journey of Theory and Praxis* (Thousand Oaks, CA London: Sage)

Waldrop M M 1992 *Complexity. The Emerging Science at the Edge of Order and Chaos* (New York, NY: Simon & Schuster)

Zimmerman B, Lindberg C et al. 1998 *Edgeware Lessons Learned from Complexity Science for Health Care Leaders* (Dallas, TX: VHA Inc)

[8] Bouman M P A 1999 *Collaboration for Prosocial Change. The turtle and the peacock. The Entertainment-Education Strategy on Television* (Wageningen, The Netherlands: Wageningen Agricultural University)

Liedtke C, Welfens J and Stengel O 2007 *Ressourcenschonung durch lebensstilorientierte Bildung* ed G Altner H Leitschuh and G Michelsen (München, C.H. Beck) Jahrbuch Ökologie 2008 pp 142-53

[9] Movius L, Cody M et al. 2007 Motivating television viewers to become organ donors *Cases in Public Health Communication & Marketing*, rerieved from http://www.casesjournal.org/volume1/peer-reviewed/cases_1_08.cfm.

Morgan S E, Movius L et al. 2009 in press The power of narratives The effect of entertainment television organ donation storylines on the attitudes, knowledge, and behaviors of donors and non-donors *Journal of Communication*

Whittier D K Kennedy M G et al. 2005 Embedding health messages into entertainment television: Effect on gay men's response to a syphilis outbreak **10** *Journal of Health Communication* 251-59

[10] Zillmann D 2003 *Theory of Affective Dynamics. Emotions and Moods* ed. J Bryant D Rosko-Ewoldsen and J Cantor (Mahweh, New Jersey, London: Lawrence Erlbaum Associates Publisher) pp 533-567

[11] Vorderer P (2003): Entertainment Theory. ed. J Bryant D Rosko-Ewoldsen and J Cantor (Mahweh, New Jersey, London: Lawrence Erlbaum Associates Publisher) pp 131-53

[12] Zillmann D 2003 *Theory of Affective Dynamics. Emotions and Moods* ed. J Bryant D Rosko-Ewoldsen and J Cantor (Mahweh, New Jersey, London: Lawrence Erlbaum Associates Publisher) pp 533-567

Zillmann D and Cantor J 1977 Affective responses to the emotions of a protagonist *Journal of Experimental Social Psychology* **13** 155-65

[13] Clobes H G, Hagedorn F 2008 *Nicht nur das Klima ändert sich. Fenseh-Programme auf dem Weg zur Nachhaltigkeit?* ed C Schwender F S Schulz and M Kreeb (Marburg: Metropolis Verlag) chaper 10 pp 221-34

Lichtl M 2007 *Markenwerbung und nachhaltiger Konsum - Faszination Nachhaltigkeit oder Bedrohung Nachhaltigkeit?* (Berlin: dissertation.de)

Reisch L A, Bietz S et al. 2006 How to communicate sustainable lifestyles to hard-to-reach consumers? A report on the large scale experiment „balance-f" ed M Charter and A Tucker Proceedings of the SCORE! Launched conference "sustainable consumption and production" Opportunities and Treats pp 39-52.

[14] Fietkau H-J and Kessel H 1981 *Umweltlernen. Veränderungsmöglichkeiten des Umweltbewußtseins Modelle Erfahrungen* (Königstein/Taunus: Hain)
[15] Fishbein M., Ajzen I. 1975 *Belief, Attitude, Intention and Behavior: An introduction to theory and Research* Reading: Addison-Wesley
[16] Schwartz S H, Howard J A 1981 *A normative decision Making model of altruism* ed J P Rushton R M Sorrentinno (Hillsdale: Erlbaum) pp 189-211.
[17] Matthies E 2005 Wie können PsychologInnen ihr Wissen besser an die PraktikerIn bringen? Vorschlag eines neuen, integrativen Einflussschemas umweltgerechten Alltagshandelns *Umweltpsychologie* **9/1** 62-81
[18] Klöckner C A 2005 Können wichtige Lebensereignisse die gewohnheitsmäßige Nutzung von Verkehrsmitteln verändern? *Umweltpsychologie* **9/1** 28-45
Klöckner C, A Matthies E and Hunecke M 2003 Problems of operationalizing habits and integrating habits in the normative decision-making model. *Journal of Applied Social Psychology* **33** 396-417
[19] Diekmann A and Preisendörfer D 2001 *Umweltsoziologie Eine Einführung* (Reinbek: Rowohlt).
[20] Sinus Sociovision 2007 Nachhaltigkeit geht! Rückschlüsse des Bio-Booms für den Non-Food Bereich, retrieved from: www.sociovision.de/uploads/tx_mpdownloadcenter/Nachhaltigkeit_geht.pdf
[21] Reisch L A and Bietz S 2008 Kommunikation des Nachhaltigen Konsums für die Massen? Ergebnisse der Konsumfrschung im Projekt balance [f] ed C Schwender F S Schulz M Kreeb (Marburg: metropolis) pp 315-36
[22] Reisch L A and Scherhorn G 2005 Kauf- und Konsumverhalten ed D Frey (Weinheim/Basel: Beltz Verlag) pp 180-87
[23] Hirsch-Hadorn G, Maier S and Wölfing Kast S (2002) *Transdiziplinäre Forschung in Aktion. Optionen und Restriktionen nachhaltiger Ernährung* (Zürich: vdf Verlag)
[24] Reisch L A and Bietz S 2008 Kommunikation des Nachhaltigen Konsums für die Massen? Ergebnisse der Konsumfrschung im Projekt balance [f] ed C Schwender F S Schulz M Kreeb (Marburg: metropolis) pp 315-36
[25] Ryerson W N 2008 *The Effectiveness of Entertainment Mass Media in Changing Behavior* (Shelburne, Vermont: Population Media Center)
Singhal A and Rogers E M 1999 *Entertainment-Education - A Communication Strategy for Social Change* (Mahwah, New Jersey, London: Lawrtence Erlbaum Associates)
[26] Singhal A and Rogers E M 1999 *Entertainment-Education - A Communication Strategy for Social Change* (Mahwah, New Jersey, London: Lawrtence Erlbaum Associates)
Wang H and Singhal A 2009 in press *Entertainment-Education through Digital Games* (New York: Routledge)
[27] Singhal A and Rogers E M 1999 *Entertainment-Education - A Communication Strategy for Social Change* (Mahwah, New Jersey, London: Lawrtence Erlbaum Associates)
[28] Singhal A and Rogers E M 1999 *Entertainment-Education - A Communication Strategy for Social Change* (Mahwah, New Jersey, London: Lawrtence Erlbaum Associates)
[29] Bandura A 1995 *Exercise of Personal and Collective Efficacy in Changing Societies* ed A Bandura Self-efficacy in changing societies (New York: Cambridge University Press) pp 1-45
Bandura A 2005 *Social Cognitive Theory for Personal and Social Change by Enabling Media* ed A Singhal, M J Cody, E M Rogers, M Sabido Entertainment-Education and Social Change History, Research and Practice (New Jersey: Lawrence Erlbaum Associates) pp 75-96
Schwarzer R 1992 Self-efficacy in the adoption and maintenance of health behaviors *Theoretical Approaches and a New Model* ed R Schwarzer Self-efficacy thought Control of Action (Washington DC: Hemisphere)
[30] Dutta M J 2005 Theory and practice in health communication campaigns - A critical

interrogation *Health Communication* **18/2** 103-22

Dutta M J 2006 Theoretical approaches to entertainment education campaigns - A subaltern critique *Health Communication* **20** 221-31

Singhal A and Rogers E M 2002 A theoretical agenda for Entertainment-Education *Communication Theory* **12/2** 117-35

Wang H and Singhal A 2009 in press *Entertainment-Education through Digital Games* (New York: Routledge)

[31] Wang H 2009 in press Narrative engagement and persuasion of Entertainment-Education for social change - A critical integration of theories and constructs *Journal of Communciation*

[32] Petty R E and Cacioppo J T 1986 The elaboration likelihood model of persuasion *Advances in Experimental Social Psychology* **19** 123-205

[33] Kaiser Family Foundation 2004 *Entertainment Education and Health in the United States* (Washington DC, Kaiser Family Foundation)

Kaiser Family Foundation 2004 *Hollywood & Health Health Content in popular TV* (Washington DC: Kaiser Family Foundation)

Kaiser Family Foundation 2008 *Television as a health educatior. A case study of Grey's Anatomy* (Washington DC: Kaiser Family Foundation)

Center Media & Health 2009 *Summative research on the project "Sound"* (Gouda: Centrum Media & Health).

[34] Lubjuhn S and Bouman M 2009 Entertainment-Education in den Niederlanden und den USA. Was kann Deutschalnd davon lernen *merz (medien + erziehung), Zeitschrift für Medienpädagogik*, **53/4** 44-49

[35] Kaiser Family Foundation 2008 *Television as a health educatior. A case study of Grey's Anatomy* (Washington DC: Kaiser Family Foundation)

[36] Bouman M P A 1999 *Collaboration for Prosocial Change. The turtle and the peacock. The Entertainment-Education Strategy on Television* (Wageningen, The Netherlands: Wageningen Agricultural University)

Haas A 2007 Medienmenüs *Der Zusammenhang zwischen Mediennutzung, SINUS-Milieus und Soziodemographie* (München: Verlag Reinhard Fischer)

[37] Schwender C Schulz F S and Kreeb M 2008 Medialisierung von Nachhaltigkeit. Das Forschungsprojekt balance(f): Emotionen und Ecotainment in den Massenmedien (Marburg: metropolis)

[38] Liedtke C, Welfens J and Stengel O 2007 *Ressourcenschonung durch lebensstilorientierte Bildung* ed G Altner H Leitschuh and G Michelsen (München, C.H. Beck) Jahrbuch Ökologie 2008 pp 142-53

[39] Schwender C Schulz F S and Kreeb M 2008 Medialisierung von Nachhaltigkeit. Das Forschungsprojekt balance(f): Emotionen und Ecotainment in den Massenmedien (Marburg: metropolis)

[40] Liedtke C, Welfens J and Stengel O 2007 *Ressourcenschonung durch lebensstilorientierte Bildung* ed G Altner H Leitschuh and G Michelsen (München, C.H. Beck) Jahrbuch Ökologie 2008 pp 142-53

[41] Liedtke C, Welfens J and Stengel O 2007 *Ressourcenschonung durch lebensstilorientierte Bildung* ed G Altner H Leitschuh and G Michelsen (München, C.H. Beck) Jahrbuch Ökologie 2008 pp 142-53

[42] Chatterjee J S, Bhanot A et al. 2009 in press The importance of interpersonal discussion and self-efficacy in knowledge, attitude, and practice models. *International Journal for Communication*

Singhal A 1988 *Prosocial Observational Learning from Television Soap Operas* (Los Angeles: Annenberg School for Communication, University of Southern California)

Singhal A and Dura L 2008 *Listening and Healing in the Peruvian Amazon - An Assessment of Minga Peru's intercultural Radio Educative Project to prevent and control Domestic*

Violence and HIV/AIDS (El Paso,TX: University of Texas)

Tuner J R 1993 Interpersonal and psychological predictors of parasocial interaction with different television performers *Communication Quarterly* **41** 443-53

Sood S and Rogers E M 2000 Dimensions of parasocial interaction by letter writers to a popular Entertainment Education soap opera in India. *Journal of Broadcasting and Electronic Media* **44** 386-414

Valente T W, Poppe P R et al. 1996 Mass-media-generated interpersonal communcation as a source of information about family planning *Journal for Health Communcation* **1/3** 247-66

Valente T W and Saba W P 2001 Campaigns exposure and interpersonal communciation as factors in contraceptive use in Bolivia *Journal for Health Communciation* **6/4** 303-22

[43] Porter Novelli Centers for Disease Control and Prevention et al. 2005 *TV Drama/Comedy Viewers and Health Information 2005* (Porter Novelli HealthStyles Survey)

[44] Bondy G 2007 1000. Sendung der Lindenstraße retrieved from http://www.br-online.de/wissen-bildung/kalenderblatt/2007/01/kb20070130.html

[45] Lindenstraße (2008): Umweltthemen in der Lindenstraße, eine interne Bestandsaufnahme (München: Lindenstraße)

[46] Bandura A 2005 *Social Cognitive Theory for Personal and Social Change by Enabling Media* ed A Singhal, M J Cody, E M Rogers, M Sabido Entertainment-Education and Social Change History, Research and Practice (New Jersey: Lawrence Erlbaum Associates) pp 75-96

[47] O'Brien M E 2004 *Nazi cinema as enchantment The politics of entertaiment in the Third Reich* (Woodbridge, Suffolk, UK: Camden House)

[48] Brahnal U 2009 Media legislation and Entertainment-Education (Dortmund: Institute for Journalism University of Dortmund) Interview January 20th 2009

[49] Volpers H, Bernhard U et al. 2008 *Public relations und werbliche Erscheinungsformen im Fernsehen - Eine Typlologie persuasiver Kommunikationsangebote des Fernsehens* (Berlin: Vistas)

[50] Volpers H, Bernhard U et al. 2008 *Public relations und werbliche Erscheinungsformen im Fernsehen - Eine Typlologie persuasiver Kommunikationsangebote des Fernsehens* (Berlin: Vistas)

[51] Loosen W and Meckel M 1999 Journalismus in eigener Sache - Veränderungen im Verhältnis von Journalismus und Public Relations am Beispiel von Greenpeace TV *Rundfunk und Fernsehen* **47** 379-392 p 376

[52] Volpers H, Bernhard U et al. 2008 *Public relations und werbliche Erscheinungsformen im Fernsehen - Eine Typlologie persuasiver Kommunikationsangebote des Fernsehens* (Berlin: Vistas)

Voss K 2007 *Öffentlichkeitsarbeit von Nichtregierungsorganisationen - Mittel, Ziele, interne Strukturen* (Wiesbaden: VS Verlag)

[53] Volpers H, Bernhard U et al. 2008 *Public relations und werbliche Erscheinungsformen im Fernsehen - Eine Typlologie persuasiver Kommunikationsangebote des Fernsehens* (Berlin: Vistas)

Volpers H, Holznagel B 2009 *Trennung von Werbung und Programm im Fernsehen. Zuschauerwahrnehmung und Regulierungsoptionen* (Berlin: Vistas)

[54] Pratt N 2009 forthcoming *Ich kaufe, also bin ich – gut? Nachhaltiger Konsum - eine Kontextbestimmung* (Wien: Transit-Neue Kritik)

Sabapathy J 2007 *Sustainable Consumption and Production – A Business Primer* (Cambridge: Programme for Industry University of Cambridge)

[55] Michaelis L 2003 The role of business in sustainable consumption *Journal of Cleaner Production* **11/8** 915-21

[56] IKEA 2004 *Social and Environmental Responsibility* Brochure (Leiden: IKEA Group)

[57] Nattrass B and Altomare M 2001 *Natural Step for Business - Wealth, Ecology and the*

Evolutionary Corporation vol 2 (British Columbia, Canada: New Society Publishers)

[58] Kramer A 2000 Greenwashing charges make businesses blue *Sustainable Industries Journal* Retrieved from http://www.sijournal.com

Vogel D J 2005 The low value of virtue *Harvard Business Review* **83/6** 26-36

Vogel D J 2005 *The Market for Virtue - The Potential and Limits of Corporate Social Responsibility* (Washington, D.C.: Brookings Institute)

[59] Pratt N 2007 A Systems Perspective on Sustainability - Contributing to Sustainability Science (Witten/Herecke: University Witten/Herdecke)

[60] Sociovision 2008 *Die Sinusmilieus* (Heidelberg: Sinus Institute)

[61] Häusler R and Kerns C 2008 *LOHAS – Mythos und Wirklichkeit* (Berlin: Stratum)

[62] Häusler R and Kerns C 2008 *LOHAS – Mythos und Wirklichkeit* (Berlin: Stratum)

[63] Petty R E and Cacioppo J T 1986 The elaboration likelihood model of persuasion *Advances in Experimental Social Psychology* **19** 123-205

[64] RWE 2007 *Mit Energieeffizienz Konsum nachhaltig gestalten* Global Compact Network Germany (Essen: RWE AG)

[65] WWF 2008 Weathercocks and Signposts - The environment Movement at a Crossroad, retrieved from http://www.wwf.org.uk/research_centre/research_centre_results.cfm?uNewsID=2224

Kinght A 2004 *Sustainable Consumption - The Retailing Paradox* paper for the sustainable development commission retrieved from http://www.sd-commission.org.uk/pages/sustainable-consumption-the-retailing-paradox.html

[66] Volpers H, Bernhard U et al. 2008 *Public relations und werbliche Erscheinungsformen im Fernsehen - Eine Typlologie persuasiver Kommunikationsangebote des Fernsehens* (Berlin: Vistas)

Volpers H, Holznagel B 2009 *Trennung von Werbung und Programm im Fernsehen. Zuschauerwahrnehmung und Regulierungsoptionen* (Berlin: Vistas)

Lubjuhn S and Bouman M 2009 Entertainment-Education in den Niederlanden und den USA. Was kann Deutschalnd davon lernen *merz (medien + erziehung), Zeitschrift für Medienpädagogik*, **53/4** 44-49

[67] Volpers H, Bernhard U et al. 2008 *Public relations und werbliche Erscheinungsformen im Fernsehen - Eine Typlologie persuasiver Kommunikationsangebote des Fernsehens* (Berlin: Vistas)

Public engagement in climate change – Disjunctions, tensions and blind spots in the UK

C Höppner

Environmental Change Institute ECI, Oxford University Centre for the Environment, Dyson Perrins Building, South Parks Road, Oxford OX1 3QY

corina.hoppner@ouce.ox.ac.uk, corina.hoeppner@wsl.ch

Abstract. *There is much talk about engaging the public in climate change mitigation and adaptation in the United Kingdom and elsewhere. Governments rush to demand greater engagement of the public in tackling climate change and delivering sustainable futures. The importance that public engagement has gained as part of the UK climate agenda begs the questions of what is actually behind this call and what are the implications. This paper analyses the rationale for public engagement as enshrined in major policy documents. This rationale is clearly instrumental in that citizens are expected to engage by adopting the 'right attitude', by performing prescribed behaviours, and by consenting to proposed measures. Using recent cases of climate change mitigation and adaptation practice the paper discusses the implications of such an approach to public engagement. The paper concludes that until the manifold disjunctions between climate related policy agendas and their rationales for engagement are explicitly addressed citizen engagement will be serving incumbent interests rather than contributing to socially sustainable and democratic decision-making*

1. Introduction

Climate change mitigation and adaptation intersects with many policy agendas and sectors [1]. Already the first UK Climate Change Programme in 2000 [2] emphasised the critical role of spatial planning and technology for climate protection as one critical aspect of sustainable development.

Calls to engage citizens in sustainable development have gained widespread prominence since the 1992 Rio Earth Summit, the 1992 Framework Convention on Climate Change [3] and the Aarhus Convention [4]. In fact, as Owens [5] points out, such appeals are now 'commonplace' in the UK's spatial development policies. In this spirit, efforts such as the Local Agenda 21 have endeavoured to involve citizens in addressing climate change in sustainable community development over the past two decades [6].

From a sustainability perspective, public engagement is usually justified by claiming that citizens in a democracy should be empowered in decision-making processes, that resulting decisions are of better quality, that citizens' consent and trust could be gained and that engagement contributes to favourably changing public attitudes and behaviours [7, 8, 9, 10, 11]. Closely connected to the first two reasons for public engagement is the vision of active citizens taking formative influence as vital ingredients of 'sustainable communities' in the UK [11, 12, 13].

These grounds for citizen engagement have also proliferated in UK science and technology (S&T) policies [14]. Over the past decade, citizen engagement in S&T has been advocated as a route to robust policymaking in times of uncertainty and contested certainties in science [15]. These efforts gradually nurtured a shift in official understandings of engagement from being a process of public education to being processes of public dialogue and mutual learning [16, 17]. Irwin [18] however argues that despite resounding rhetorical commitments, public engagement is still a matter of ongoing contention within government rather than a settled one. At present, practice indeed appears to lag behind the relatively new rhetoric of public dialogue. Wynne [17] and Irwin [18] even see signs for a re-emphasis on engagement exercises that solely serve to increase public trust and consent rather than facilitating substantial public dialogue as contribution to innovation.

Although the literature on public engagement is extensive in each of the aforementioned areas, public engagement at the intersection of climate change, sustainability, science, technology and spatial development remains under-researched. Kasemir and colleagues' [19] effort to evaluate the standing and the potential of public engagement for 'sustainable climate futures' in Europe is certainly an important exception. As part of that analysis Stoll-Kleemann et al. [20; p 248] maintain that 'there has never been a more propitious time to realise and extend the vision of public participation in sustainability science' and for a 'more imaginative and empowering approach to citizenship'.

Indeed, at first glance the prospects for increased public engagement in climate change issues appear to be good in the UK. The recent UK Climate Change Programme [21] dedicates an entire chapter as to why and how to engage the public and there is even a Citizen Agenda for tackling climate change [22]. However, the resounding call for public engagement raises at least two questions. Which understanding of engagement and 'the public' does actually underlie the official call and does this call promote an extended vision of public engagement in climate change mitigation and adaptation?

More specifically, this paper addresses the following questions:

What is the official approach to and rationale for public engagement in climate change mitigation and adaptation? What are the implications for mitigation and adaptation practice at the intersection with spatial and technology development as to vital fields for the implementation of climate policies?

2. Public engagement

Academics and practitioners likewise tend to use the terms 'public engagement' and 'public participation' synonymously. This common practice however makes it difficult to distinguish between public engagement as the practice of giving members of the public formative influence in the 'agenda-setting, decision-making, and policy-forming activities of organizations or institutions' [23; p 88] on one hand and as citizens' performance of particular behaviours to achieve prescribed policy goals on the other (e.g. reducing individual energy consumption to mitigate climate change).

Arguably, the study of public engagement in climate change mitigation and adaptation, thus far, struggles on this first definitional barrier. Depending on the specific perspective one takes, engagement in climate change issues is an end in itself (e.g. to involve the public in developing responses) or a means to an end (e.g. to widely implement prescribed actions). While the first understanding is intellectually grounded in literature on participatory and deliberative democracy and governance as well as empowerment and social capital theory, the latter has traditionally been the field of psychological behaviour change and communication research.

Scholars have started to harness concepts from the aforementioned fields to study public engagement in climate change [24, 25]. However, researchers have, as yet, not substantially engaged in analysing underlying official rationales for engaging the public in climate change and their wider implications. This paper employs Fiorino's [65] distinction between normative, substantive and instrumental rationales for citizen engagement and its adaptation by Stirling [27, 28] for such an analysis (see Table 1).

Table 1: Rationales for public engagement as adopted from Fiorino [26] and Stirling [27, 28]

	Rationale		
	Normative	Substantive	Instrumental
Reasons	- empowering citizens in agenda-setting and decision-making since it is their democratic right	- improving agendas and decisions through the inclusion of diverse views, kinds of knowledge, value and belief systems	- engagement is a means to endorse favoured decisions and favoured outcomes such as citizens' trust, consent and behaviour change
Role of engaged individuals	- active subjects with formative and reactive roles	- active subjects with formative and reactive roles	- passive objects with reactive roles
Room for input	- room for input, the extent and influence on agendas and decisions varies	- room for input, the extent and influence on agendas and decisions varies	- no room for input that challenges predefined favoured decisions and outcomes
Relationship to the exercise of power in and over engagement	- attentive to power and framing issues - quality criteria and measures to guarantee inclusive, fair and equal empowerment are central to contain the exercise of power	- typically blind to power and framing issues since the focus lies on the actual input and the quality of outcomes	- blind to the exercise of power or even supporting and justifying the exercise of power by participants if their interests match with favoured decisions and outcomes

Accordingly, a *normative rationale* aims at empowering citizens to partake in decision-making that affects them since it is their democratic right. Practice, research and policies committed to such normative arguments are particularly determined to develop and to ensure participatory exercises that are based on democratic principles such as inclusiveness, equity and equality. Under an *instrumental rationale*, engagement exercises and mechanisms are a means to endorse 'particular favoured decisions' and to achieve specific outcomes such as citizens' trust, consent and performance of prescribed actions. A *substantive rationale* sees the benefit of engagement in improved decisions through the in-depth learning about and the incorporation of diverse kinds of knowledge, belief systems and values. Engagement mechanisms are hence designed to substantially broaden the basis for social choice by shedding light onto possible long-term effects, the conditions under which they emerge, underlying interests, potential benefits and costs.

As real world engagement processes inevitably take place within specific constellations of agendas and interests, the underlying rationale for engagement is crucial as it determines to what extent processes are susceptible to the exercise of power and how outcomes are evaluated. Since equality, equity and inclusiveness are guiding principles of normative engagement, related exercises are attentive to and try to contain the exercise of power over and in the process. In comparison, engagement driven by a purely substantive rationale is typically blind to power issues [27]. Similarly, instrumentally motivated engagement does not critically engage with such questions and might even invite and legitimise instrumental agendas and powerful actors if they convincingly claim that their interests contribute to the favoured outcomes of engagement. Under such conditions, the use of engagement to endorse incumbent interests can easily be justified as public communication towards a favourable end rather than being criticisable as manipulation through the unrestricted exercise of power.

Unlike instrumentally motivated engagement, citizen involvement inspired by normative and substantive ideas leave participants with room for input. The nature and extent of this room can however considerably differ. Citizens might be invited to discuss only 'a single course or a very small subset of possible courses of action' [28; p 279]. In such 'closing-down' exercises citizens' input is usually sought to choose from a limited number of options, to perfect an option by including salient concerns and to legitimise a particular choice rather than to discuss alternatives and the 'conditioning assumptions' around choices. On the other hand, citizens' input can also serve to 'open-up' a wider debate on many alternative pathways by comprehensively exploring different perspectives, uncertainties, conditions and consequences [28]. From a developmental democracy perspective, an important asset of normative-substantive engagement for participating individuals as well as society as a whole is its potential to foster democratic values, to develop the capacities for and enlarge the practice of active citizenship and to increase people's sense of socio-political efficacy [29, 30, 31].

3. Public engagement in climate change as enshrined in the 2006 UK Climate Change Programme and the Citizen Agenda 2007

In the following the above outlined framework is used to analyse the rationale for public engagement as enshrined in the UK Climate Change Programme and the Citizen Agenda [21].

The Climate Change Programme, published by the Department for Environment, Food and Rural Affairs (Defra) in 2006, [21; p 4] '*...sets out the Government's commitments both at international and domestic levels to meet the challenge of climate change'*. It is thus a key document enshrining policies and priorities for action in the UK. In comparison to its precedent from 2000, it puts more emphasis on setting out the official approach to '*...strengthening the role that individuals can play'* [21; p 4]. The rationale for public engagement is probably best illustrated by the following excerpt: *'Engaging with the public on climate change is a vital part of the UK climate change programme both to encourage specific behaviours to reduce carbon dioxide emissions directly and to gain acceptance for more ambitious Government policy'* [21; p 121]. This clearly instrumental rationale dominates the mitigation section. Citizens are given a reactive role rather than a formative one.

The rationale is matched by advice on how to *'exemplify, enable, engage, and encourage the right attitude'* [21; p 118] and to help individuals *'...to understand their role and responsibility in tackling climate change'* [21; p 6]. To 'catalyse' the desired change in people's behaviour a four 'Es' model, originating from the UK Strategy for Sustainable Development [10], is presented. Under the label 'exemplify' the government recognises the importance of reciprocal action and commits itself to setting a good example. 'Enabling' behavioural change is the next concern: *'We need to help people make responsible choices by providing them with accessible alternatives and suitable infrastructure, and supporting them with the necessary skills and information'* [21; p 119]. Here, product labelling and voluntary agreements with industries and retailers are expected to facilitate alternative consumer behaviour. Additionally, a specific fund (Climate Change Challenge Fund) should support '*community groups in taking small simple actions on energy, waste and water which add up to make a big difference nationwide'* [21; p 119]. In comparison, the component 'Engage' appears rather weak and nebulous. What becomes clear, however, is that the document stresses the importance of proper communication materials and media channels for engaging people at the local level. Under 'Encourage – give the right signals' the document sets out how consumers' energy efficiency will be supported by the work of the Energy Saving Trust. As public engagement in the mitigation of climate change is clearly understood as a means to prescribed end, the communication of the end and how to achieve it are the focal points of therein suggested engagement mechanisms. Accordingly, the document sets out several 'climate change communication initiatives' and a new internet information platform as engagement tools [21; p 121].

While the mitigation section extensively deals with the role of individuals, words like individuals, citizens or consumers are hardly found in the adaptation section. In fact, the role of individuals other than local, regional and national decision-makers remains a blind spot. The rationale for engaging these key stakeholders is to help them planning their response to the impacts of climate change [21; p 139] and to build up their adaptive capacity. Engagement is a means to these ends. It is instrumental in that it aims at helping stakeholders to understand the need for adaptation to act accordingly. It is substantive in that stakeholders' input should lead to a better quality of specific adaptation decisions. Table 2 summarises rationales, roles and mechanisms of engagement as communicated in the programme.

The Citizen Agenda, published by Defra in 2007 [22] is dedicated to the question of *'how the ordinary citizen can change his or her lifestyle to minimise the impact of climate change and to mitigate its effects'* [22; p 5]. Similar to the Climate Change Programme, the Citizen Agenda adopts a focus on awareness raising and behaviour change. However, it more specifically deals with the *'barriers to uptake climate change mitigation strategies at the level of the individual'* and how barriers might be overcome [22; p 6]. The understanding of citizen engagement is again instrumental as a way to achieve voluntary emission reduction.

Table 2: Public engagement as enshrined in the UK Climate Change Programme 2006 and the Citizen Agenda 2007

	UK Climate Change Programme 2006		Citizen Agenda 2007
	Mitigation agenda	Adaptation agenda	
What is the rationale for engagement?	- instrumental	- instrumental, substantive	- instrumental
Who should engage?	- all individuals	- decision-makers as stakeholders	- all individuals
What is the role of those engaged?	- reactive	- reactive, formative	- reactive
What are the selecting preconditions and barriers to engagement (other than individual internal factors such as willingness, awareness and knowledge)?	- socio-economic characteristics, infrastructure provision	- role as decision-makers and/ or having salient knowledge, expertise, organisational power	- socio-economic characteristics, infrastructure provision
What mode of communication is required?	- one-way-communication	- one-way, two-way and dialogical communication	- one-way communication
What engagement mechanisms are endorsed?	- internet information service - communication initiatives to raise awareness and advise consumer choice - range of behaviour incentives	- stakeholder partnerships and national stakeholder forums - information and awareness raising initiatives	- information packages - economic and infrastructural incentives to reducing households' energy consumption, increasing their energy efficiency, to stimulate the use of low carbon alternatives and microgeneration

4. Implications of the official rationale for public engagement practice

The following examples shed light onto the implications for citizen engagement at the intersections of climate change mitigation and adaptation on one hand and spatial and technology development on the other. The selected cases show how the climate agenda's instrumental perspective on engagement can contain normative-substantive engagement at these intersections. Furthermore, examples illustrate how disjunctions between the sustainability and climate agenda easily result into tensions between citizen empowerment and an instrumental take on engagement.

4.1. Containment of actual normative-substantive citizen engagement

The case of the UK's most recent 'Nuclear Consultation' in 2007 is perhaps the most impressive case of public engagement at the intersection of climate change mitigation and S&T. Inspired by normative-substantive objectives for citizen engagement in S&T the consultation was designed to allow dialogue on the future of nuclear power and an alternative energy landscape in the UK. Around 1000 citizens attended deliberative events under the label 'Talking Energy: The Future of Nuclear Power'. The process design could be regarded as a remarkable innovation as it considerably deviated from the conventional consultation via written documents.

However, instead of opening-up a dialogue on alternative energy landscapes, officials framed and endorsed nuclear power as the only viable technological trajectory towards low carbon energy security [32, 33]. Consequently, the exercise turned out instrumental rather than substantive aiming at gaining public consent and increasing consumer demand for nuclear energy. In this way, the exercise fulfilled the climate agenda's rationale for engagement by advising consumers on a climate friendly energy option; it met the climate mitigation agenda itself with it's general openness to nuclear technology as a low carbon option and it worked for the official energy agenda with its explicit endorsement of nuclear energy [34]. In matching and supporting each other conveniently, the instrumental objectives for engagement on both agendas helped to legitimise the marginalisation of substantive dialogue. Through the instrumental lens of the mitigation agenda nuclear advocates' use of climate change to frame the conditions of appraisal and to construct a 'strong justification' [27] for nuclear power might be evaluated positively as an attempt to win citizens' consent to a low carbon energy option. Yet, judged against a normative-substantive yardstick the exercise might be deemed a farce for it uncritically allowed incumbent interests to use climate change to prevent debate on an alternative 'energy landscape' [28].

As analysed earlier, the official conceptualisation of public engagement in climate change adaptation differs from the one for mitigation in two ways. First, the instrumental rationale for engagement is complemented by substantive objectives. Second, the range of individuals to engage is much narrower in the adaptation agenda including only key stakeholders with specific knowledge, expertise, and roles as representatives of public interests. Whilst seeming plain sailing on paper, such a perspective on engagement can become quite problematic in adaptation practice. Two recent and well-researched cases of coastal management in Christchurch Bay, Dorset-Hampshire and the Orkney Islands serve as cases in point [35]. According to the official adaptation agenda, local managers and decision-makers are stakeholders that should ensure appropriate adaptation to rising sea levels. Whether and how stakeholders are expected to engage with local citizens is not outlined in the agenda. In these cases, the actual design of the adaptation projects was, however, inspired by a normative motive to empower local people in decision-making originating from a sustainable development perspective on engagement. The project designers' broad perspective on engagement stimulated the involvement of a wider circle of locals following the principle of democratic inclusiveness. Consequently, the odds that managers would encounter with people that not even shared their understanding of the necessity of anticipatory action increased. Not surprisingly, managers had a hard time with defending the need for anticipatory adaptation although it was in fact a non-negotiable strategic imperative. The project, of course, revealed considerable tensions between specific prescribed actions and local views. In fact, fundamental disjunctions between the engagement paradigms of the current climate and sustainability agenda had created conditions conducive to increasing mistrust, dissatisfaction and opposition.

4.2. Containment of future normative-substantive citizens engagement

The limited rationalisation of engagement in climate policies leaves normative-substantive citizen engagement at the intersection of climate change and technologies as a vacuum within and between these agendas. This is unfortunate since the potential scope for such engagement widens as official strategies to tackle climate change increasingly embrace technologies such as nuclear power, genetically modified food (GM), carbon capture and storage (CCS), and bio/geo-engineering [36, 37, 38, 39, 40, 41, 42].

The recently reinvigorated debate on GM food might serve as a telling case in point. In 2003 the future of GM technology was opened-up for debate by the 'GM nation?' citizen dialogue. The process revealed concerns with certain aspects of the technology (for a comprehensive discussion see Horlick-Jones et al. 2007 [43]). Recently, advocates in politics, business and science have started a new effort to boost GM food on its alleged contribution to food security under changing climatic conditions [44, 45]. Thus far, however, there are no signs of a new 'GM nation?' dialogue. In fact, a recent statement by the UK's science minister Ian Pearson suggests a rather instrumental take on citizen engagement: *'I don't think that the GM debate in 2000 was handled very well...If consumers see benefits from GM then I think a significant majority of them will want to choose GM. That's what we have to do. We have to show that there are benefits to the consumer of adopting GM technologies'* [46]. Addressing the public as consumers, who need to be convinced of the technology's advantages, establishes the future of GM food as a public communication issue rather than as a matter for dialogue between citizens, scientists, business and politicians. Such framing converges conveniently with instrumental understandings of engagement. In prominently presenting GM food as a superior means to food security, which the climate agenda highlights as an important challenge, the technology is portrayed as an end point rather than an option. In this way, the scope for normative-substantive engagement narrows, whereas the room for instrumental engagement to win citizens' consent and trust becomes exclusive. Of course, initiating a new 'GM nation?' dialogue is at best futile if the exercise would be uncritical to or even legitimising such preemptive framing that narrows the outcome of engagement to either public ignorance or acceptance of the technology's benefits. Considering that research has indicated people's personal trade-off between technologies and climate change action is based on inner conflicts and bounded to certain conditions [47] such containment is all the more serious.

4.3. Containment of empowerment

In an instrumentally limited engagement discourse citizens are objects of engagement rather than emancipated subjects *in* engagement [28]. Accordingly, the instrumental approach takes the attention off engagement as a process in which citizens are empowered subjects that develop and use their capacities to critically evaluate and shape decisions [31]. An instrumental perspective is typically insensitive to the reasons why locals challenge actual planning schemes (eg. housing and renewable energy schemes) and deems people's action as an unusual or unwanted outcome. To make the outcomes of engagement instrumentally controllable such empowerment is contained.

The recent case of widespread oppositions against eco-towns proposed for various sites in England by central government shows how containment can look in practice [48, 49, 50]. From an instrumental perspective engagement is a success if residents readily accept eco-towns. Engagement is a failure if residents, for whatever reason, oppose development proposals. The means of engagement thus primarily focus on facilitating acceptance through communication and on marginalising diverging or unusual outcomes. The consultation process on the proposed eco-town 'Weston Otmoor' in Oxfordshire, for example, was accompanied by a massive communication effort on the new town's contribution to climate change mitigation and sustainable community development. The PR company that the private developer had contracted to consult the public developed a 'comprehensive communication strategy' and mailed a one-page consultation questionnaire to local residents [51]. The consultation form asked residents to agree or disagree with a number of statements that described potential benefits of the new town (e.g. 'The East West Rail Link would benefit Bicester and the local area', 'I would be likely to use the new Park & Ride'). Interestingly, the question most prominently positioned in the form and the final report merged people's acceptance of the need for 'more housing in the area' with their intention to accept such in Weston Otmoor. It thus resembled a typical NIMBY ('not in my backyard') versus Non-NIMBY categorisation (see Wolsink 2007 for a definition of NIMBY attitudes [52]). The response option actually obscured that people might disagree with the project not because they did not see a need for more housing or because they did not want it in their 'backyards' but because they questioned the suitability of the specific planning proposal. In fact, a majority of citizens doubted that the project was reconcilable with a sustainable development of the area [51]. A subsequent expert assessment of the developer's proposal highlighted that it might indeed yield negative effects on local biodiversity, exacerbate congestion through encouraging commuting and prejudice the development of a neighboring municipality [53].

What the case illustrates though is that a normatively prescribed consultation process can actually devalue what is encouraged under a sustainability paradigm, namely the empowerment of citizens in the critical evaluation of community projects [13]. From a view on engagement as a means to democratic and substantive citizen empowerment, people's willingness to critically engage might have been harnessed as a starting point for constructive and transformative work between citizens and developers.

5. Discussion – What's the matter with public engagement in climate change?

This paper has addressed the above question by first pointing to the shallow use of the term 'public engagement' as a barrier to understanding current shortcomings in the discussion of public engagement in climate change mitigation and adaptation.

Public engagement is now a high priority on the UK climate agenda. The rationale behind and the role ascribed to citizens, however, are clearly instrumental. Accordingly, engagement is currently conceptualised as a means to alter citizens' consumption behaviour and to gain acceptance for specific courses of action. The implications are manifold and serious. To be sure, this work does not deny the need for efforts to raise awareness for climate change or for incentives to help people contribute to its mitigation. Rather, it aims at shedding light onto the limitations and implications of an instrumental rationale.

As has been shown, tensions in practice appear where instrumental understandings of engagement encounter normative-substantive perspectives on engagement. This is likely to happen where climate mitigation and adaptation intersect with spatial and technology development which have both been important testing fields and intellectual drivers of normative-substantive citizen engagement over the past decades. Practitioners with normative democratic and substantive commitments to the role of engagement will inevitably struggle with a narrow or lacking scope for such engagement under the current climate agenda. The presented schema helps to retrace the systemic roots of tensions identified in recent spatial development projects from eco-towns and coastal community adaptation to the planning of renewable schemes in the UK and elsewhere [35, 54, 55]. Without such bigger picture in mind the social contestation of these projects might easily be mistaken as a mere consequence of unfit participants or designers rather than stemming from systemic disjunctions and blind spots in official conceptualizations of public engagement.

Importantly, a lacking openness for the merits of engagement to empower citizens as subjects with formative influence on agendas and decisions can nourish insensitivity for forms and results of engagement different from those set out in the climate agenda. Using levels of behaviour change, and acceptance and opposition, inferences about the success of engagement practices and about responsible vs. irresponsible citizens are quickly drawn. Yet, such evaluations carry the risk to contain and discredit the empowerment of those who seek to find ways to integrate their visions of sustainability with responses to climate change. From a normative-substantive perspective such behaviour is not either 'obedient' or 'disobedient' but a practice of citizenship and opportunity to create dialogue that as Barry [56; p 40] puts it are critical in the struggle for 'sustainable, just, and democratic societies'. Policies that aim for such a society need thus to actively address the question as to the room for citizens that challenge proposed measures and schemes as an expression of empowerment.

Furthermore, a failure to address questions as to the scope of substantive citizen engagement leaves the potential of opening-up the debate on alternative technological responses unexplored. As many writers would argue, such a shortcoming poses one barrier to overcoming 'entrenched ways of thinking about possible policy pathways' that many writers dismiss as unable to deal with the 'unprecedented nature of predicted social and biophysical climate change impacts' [57; p 210]. As the cases discussed here exemplify, such entrenchment might even be nourished by instrumental insensitivity to formative roles of citizens and by allowing incumbent interests' typically instrumental acting to close-down the range of possible pathways [28].

However, it might be argued that normative-substantive engagement does not necessarily yield sustainable results as has been shown for three 'new democratic practices' (referendum, Local Agenda 21, open planning process) in Munich's transport policy creation [58]. Yet, as Hajer and Kesselring [58] note the potential of such spaces to contribute to a rethinking of sustainable mobility policy was compromised by poor design that did not allow deliberation to transcend sectoral divisions and to open-up debate on alternative concepts. Furthermore, referenda on such a complex issue only make sense if they follow in-depth deliberation on available alternatives. If options to vote on are not developed in a participatory way too, such referenda might force people into response options that do not correspond with their intentions. Indeed, standing alone, such quantitative exercises can, as shown for the eco-town consultation, conceal rather than reflect people's concerns. That combined discursive and quantitative participation (referenda, consultation) can push the case of sustainability has been highlighted by various authors in contexts such as renewable energy investment, environmental protection and traffic road restrictions [59, 60].

Given such evidence, this paper goes with Smith [60] arguing that the mechanisms for normative-substantive engagement in climate policy creation and planning processes need to be selected according to the question at hand and be combined to harness their strengths and balance out their specific weaknesses (combine opening-up and closing-down exercises). Moreover, the merit of discursive citizen participation in technology development should not be mistaken as additional voting exercise on whether one technology should be allowed to fix climate change. Rather they should be understood as tools to bring to light different values and meaning systems and to reflect on a number of choices as it was originally planned for the UK Nuclear Consultation.

To be sure, this paper is not to argue for opening all decisions to citizen participation or for replacing democratic institutions. Rather, it reminds that in a governance system in which decisions on technology and spatial development are increasingly shaped by stakeholder networks and advocacy coalitions [58], citizen involvement appears to be needed to inform and democratise such governance. However, inviting citizens to engage without giving them substantive influence or valuing their input ultimately diminishes rather than restores trust in inviting institutions and people's feelings of efficacy. Disenchanted with invited engagement people are likely to channel their engagement into opposition or cease to engage with the issue altogether [61]. In the face of dangerous climate change, citizen engagement needs to be instrumentally rationalised as a way to reduce emissions, increase energy efficiency and adaptation. However, it should not be a mere instrument to prescribe attitudes on specific pathways (planning schemes, technological fixes) and to restrict engagement to officially favoured forms (reactive engagement in the private sphere rather than also formative engagement in the public sphere).

6. Conclusions

'Creating national climate policy is challenging' [62; p 562]. Public engagement has as yet been oversimplified and underestimated as a crucial, multi-faceted and most troublesome challenge. Indeed, the general rush to call for climate action and sustainable development in one breath overlooks that when it comes to the scope for democratic and substantive citizen engagement and the practice of good citizenship the relationship between these imperatives is neither straightforward nor unproblematic. To reconcile instrumental agendas with normative and substantive rationales for engagement is all but a new challenge. However, the challenge becomes all the more serious with the new rhetoric of urgency in climate policies and with the re-domination of top-down planning approaches in the UK and elsewhere [54, 55, 63].

The prominence of instrumental understandings of engagement and citizens' role in the UK climate agenda is not too surprising given that current academic debates generally focus on engagement as a tool for 'cultivating 'green' attitudes and behaviours in individual citizens, leaving questions of democracy and collective action on the sidelines' [31; p 378]. Yet, if not critically reflected upon an instrumentalist approach can easily lead to premature inferences about a general incompatibility of democratic and substantive citizen engagement on one hand and 'effective' climate action on the other. Indeed, a purely instrumental understanding of the merits of citizen engagement will lead to the conclusion that in the case of climate change inclusive and substantive engagement is only possible where people are uncritically in favour of prescribed measures and do unconditionally trust those who implement them. Such a view however would negate much of what makes citizen engagement compelling from a sustainability and democratic perspective. Furthermore, a purely instrumental orientation distracts from the manifold challenges with normative-substantive engagement. For instance, how the 'translation quality' in the evaluation of processes and the feedback to participants can be improved [64], how representativeness can be combined with in-depth deliberation and how citizen and stakeholder involvement could be related to another to 'further the democratization of government in the context of the emergence of co-operative management regimes' [58; p 19].

Stoll-Kleemann et al. [20] have claimed that with the emerging ,vistas of governance' which aim at empowering citizens in 'sustainable climate futures' a more participatory form of democracy is beginning to appear. While this might be true for the involvement of key stakeholders and certain interest groups in multi-level governance, it is doubtful regarding ordinary citizens' normative-substantive participation. If it is actually an aim of governance to empower citizens as partners in democratic decision-making there must first be an overdue and open discussion on the manifold disjunctions, tensions and blind spots that have mounted up between agendas and their engagement rationales. Otherwise invited citizen engagement at the intersection of climate change, sustainability and technology might silently regress to an instrumental understanding since it is, in the UK, the most obvious and convenient conjunction between these agendas.

References

[1] Kok, M T J & De Conick, H C 2007 Widening the scope of policies to address climate change: directions of mainstreaming. Environmental Science & Policy 10 587-99.

[2] DETR 2000 Climate Change – The UK Programme. Department of the Environment, Transport and the Regions, London.

[3] UNFCCC 1992 United Nations Framework Convention on Climate Change United Nations.

[4] UNECE 1998 Convention On Access To Information, Public Participation In Decision-Making And Access To Justice In Environmental Matters. United Nations Economic Commission for Europe, Aarhus, Denmark.

[5] Owens, S 2000 'Engaging the public': information and deliberation in environmental policy. Environment and Planning A 32 1141-48.

[6] Bulkeley, H & Betsill, M 2005 Rethinking Sustainable Cities: Multilevel Governance and the 'Urban' Politics of Climate Change. Environmental Politics 14 42-63.

[7] Freeman, C, Littlewood, S & Whitey, D 1996 Local Government and Emerging Models of Participiction in the Local Agenda 21 process. Journal of Environmenta Planning and Management 39 65-78.

[8] Selman, P & Parker, J 1997 Citizenship, civicness and social capital in local agenda 21. Local Environment 2 171-84.

[9] Gough, C, Darier, É, De Marchi, B, Funtowicz, S, Grove-White, R, Guimarães Pereira, A, Shackley, S & Wynne, B 2003 Contexts of citizen participation. in Kasemir, B, Jäger, J, Jaeger, C C & Gardner, M T eds Public Participation in Sustainability Science. Cambridge University Press, Cambridge.

[10] Defra 2005 Securing the future – The UK Government Sustainable Development Strategy. Department for Environment, Food and Rural Affairs, London.

[11] DCLG 2007 What is a sustainable community? Department of Communities and Local Government. http://www.communities.gov.uk/communities/sustainablecommunities/whatis Accessed 3 February 2009.

[12] Office of the Deputy Prime Minister 2004 Community Involvement in Planning: The Governement's Objectives, London.

[13] McClymont, K & O'Hare, P 2008 "We're not NIMBYs!" Contrasting local protest groups with idealised conceptions of sustainable communities. Local Environment 13 321-35.

[14] Chilvers, J 2008 Deliberating Competence – Theoretical and Practitioner Perspectives on Effective Participatory Appraisal Practice. Science, Technology & Human Values 33 155-85.

[15] Irwin, A 2001 Constructing the scientific citizen: Science and democracy in the biosciences. Public Understanding of Science 10 1-18.

[16] Wynne, B 2005 Risk as globalizing 'democratic discourse? Framing subjects and citizens. in Leach, M, Scoones, I & Wynne, B eds Science and Citizens – Globalisation & The Challenge of Engagement. Zed Books, London/New York.

[17] Wynne, B 2006 Public Engagement as a Means of Restoring Public Trust in Science – Hitting the Notes, but Missing the Music? Community Genetics 9 211-20.

[18] Irwin, A 2006 The Politics of Talk: Coming to Terms with the 'New' Scientific Governance. Social Studies of Science 36 299-320.

[19] Kasemir, B, Jäger, J, Jaeger, C C & Gardner, M T eds 2003 Public Participation in Sustainability Science Cambridge University Press, Cambridge.

[20] Stoll-Kleemann, S, O'Riordan, T & Burns, T R 2003 Linking the citizen to governance for sustainable climate futures. in Kasemir, B, Jäger, J, Jaeger, C C & Gardner, M T eds Public Participation in Sustainability Science. A Handbook. Cambridge University Press, Cambridge.

[21] Defra 2006 Climate Change – The UK Programme 2006. Department for Environment, Food and Rural Affairs, London.

[22] Defra, 2007 Climate change: the "citizen's agenda". Department for Environment, Food and Rural Affairs, London.

[23] Rowe, G, Marsh, R & Frewer, L J 2004 Evaluation of a Deliberative Conference. Science, Technology, & Human Values 29 88-121

[24] Slocum, R 2004 Consumer citizens and the Cities for Climate Protection campaign. Environmental and Planning A 36 763-82.

[25] Hobson, K 2006 Environmental psychology and the geographies of ethical and sustainable consumption: aligning, triangulating, challenging? Area 38 292-300.

[26] Fiorino, D J 1990 Citizen Participation and Environmental Risk: A Survey of Institutional Mechanisms. Science, Technology, & Human Values 15 226-43.

[27] Stirling, A 2006 Analysis, participation and power: justification and closure in participatory multi-criteria analysis. Land Use Policy 23 95-107.

[28] Stirling, A 2008 "Opening Up" and "Closing Down": Power, Participation, and Pluralism in the Social Appraisal of Technology. Science, Technology, & Human Values 33 262-94.

[29] Holden, B 1988 Understanding Liberal Democracy The Alden Press, Oxford.

[30] Warren, M 1992 Democracy Theory and Self-Transformation. The American Political Science Review 86 8-23.
[31] Latta, A 2007 Locating democratic politics in ecological citizenship. Environmental Politics 16 377-93.
[32] Dorfman, P ed 2008 Framing the nuclear consultation. in Dorfman, P ed Nuclear Consultation – Public Trust in Government.
[33] Stirling, A 2008 Choosing Energy Futures: Framing, Lock-In, And Diversity. in Dorfman, P ed Nuclear Consultation – Public Trust in Government.
[34] DTI 2007 Meeting the Energy Challenge – A White Paper on Energy Department of Trade and Industry, London.
[35] Few, R, Brown, K & Tompkins, E L 2007 Public participation and climate change adaptation: avoiding the illusion of inclusion. Climate Policy 7 46-59
[36] Hoffert, M, Caldeira, K, Benford, G, Criswell, D R, Green, C, Herzog, H, Jain, A K, Kheshgi, H S, Lackner, K S, Lewis, J S, Lightfoot, H D, Manheimer, W, Manskins, J C, Mauel, M E, Perkins, L J, Schlesinger, M E, Volk, T & Wigley, T M L 2002 Advanced Technology Paths to Global Climate Stability: Energy for a Greenhouse Planet. Science 298 981-87.
[37] Wigley, T M L 2006 A Combined Mitigation/Geoengineering Approach to Climate Stabilization. Science 314 452-54.
[38] The Times, 2007 We need GM crops, says new chief scientist 13 December 2007. London.
[39] Pielke Jr., R 2007 Air capture of Carbon Dioxide: Economic and Political Considerations. Discussion paper presented at the Environmental & Governance Seminar, Environmental Change institute, Oxford University, Oxford.
[40] The Guardian, 2007 I am creating artificial life, declares US gene pioneer. 6 October 2007. London.
[41] The Guardian, 2008 Could US scientist's 'CO2 catcher' help to slow warming? 31 May 2008. London.
[42] Weisser, D, Howells, M & Rogner, H-H 2008 Nuclear power and post-2012 energy and climate change policies. Environmental Science & Policy 11 467-578.
[43] Horlick-Jones, T; Walls, J; Rowe, G; Pidgeon, N; Poortinga, W; Murdock, G & O'Riordan, T 2007 The GM Debate: Risk, politics and public engagement. Routledge, UK.
[44] The Guardian, 2007 GM food safer than normal food, government adviser says. 27 November 2007. London.
[45] The Independent, 2008 It's crazy to ignore benefits of GM food, says 'Country Life'. The 20 March 2007. London.
[46] The Guardian, 2008 Science minister attempts to reopen the debate on GM crops. 22 September 2008. London.
[47] Bickerstaff, K, Lorenzoni, I, Pidgeon, N F, Poortinga, W & Simmons, P 2008 Reframing nuclear power in the UK energy debate: nuclear power, climate change mitigation and radioactive waste. Public Understanding of Science 17 145-69.
[48] The Independent 2008 Eco-towns are just vandalism dressed up in a trendy green coat. 18 May 2008. London.
[49] The Sunday Times 2008 Whitehall to force through eco-towns. 18 May 2008. London.
[50] BBC 2008 Protest march over eco-town. BBC 1, http://news.bbc.co.uk/1/hi/programmes/politics_show/7215759.stm, Accessed 13 January 2009.
[51] MISTRAL 2008 Weston Otmoor – proposed eco-town http://www.westonotmoor.co.uk/inovem/gf2.ti/f/131234/2341221.1/pdf/-/Weston%20Otmoor%20community%20consultation%20report%20%20compressed.pdf. Accessed 27 January 2009.
[52] Wolsink, M 2007 Wind power implementation: The nature of public attitudes: Equity and fairness instead of public attitudes: Equity and fairness instead of 'backyard motives'. Renewable and Sustainable Energy Reviews 11, 1188-1207.

[53] DCLG 2008 Eco-Towns. Sustainability Appraisal and Habitats Regulations Assessment of the Eco-towns Programme. Weston Otmoor and Cherwell. Prepared by Scott Wilson for Communities and Local Government. http://www.communities.gov.uk/documents/planningandbuilding/pdf/westonotmoor.pdf, Accessed 10 June 2009.
[54] Cowell, R 2007 Wind Power and 'The Planning Problem': the Experience of Wales. European Environment 17 291-306.
[55] Wolsink, M in press Near-shore wind power–Protected seascapes, environmentalists' attitudes, and the technocratic planning perspective. Land Use Policy. Doi:10.1016/j.landusepol.2009.04.004
[56] Barry, J 2006 Resistance is fertile: from environmental sustainability to citizenship. in Dobson, A & Bell, D eds Environmental Citizenship. MIT Press, Cambridge, MA.
[57] Hobson, K 2008 Reasons to Be Cheerful: Thinking Sustainability in a (Climate) Changing World. Geography Compass 2 199-214.
[58] Hajer, M & Kesselring, S 1999 Democracy in the risk society? Learning from the new politics of mobility in Munich. Environmental Policy 8, 1-23.
[59] Fishkin, J S; Luskin, R & Jowell, R 2000 Deliberative Polling and Public Consultation. Parliamentary Affairs 53, 657-666.
[60] Smith, G. 2001 Taking Deliberation Seriously: Institutional Design and Green Politics. Environmental Politics 10, 72-93.
[61] Höppner, C; Frick, J & Buchecker, M 2008 What drives people to discuss local landscape development? Landscape Research 33, 605-622.
[62] Tompkins, E L & Adger, W N 2005 Defining response capacity to enhance climate change policy. Environmental Science & Policy 8 562-71.
[63] Breukers, S & Wolsink, M 2007 Wind power implementation in changing institutional landscapes: An international comparison. Energy Policy 35 2737-50.
[64] Horlick-Jones, T; Rowe, G & Walls, J 2005 Citizen engagement processes as information systems: the role of knowledge and the concept of translation quality. Public debate about transgenic crops. Public Understanding of Science 16, 259-278.

IOP Conf. Series: Earth and Environmental Science **8** (2009) 012011 doi:10.1088/1755-1315/8/1/012011

Climate change – a challenge for the arctic indigenous peoples – the inuit response*

Aqqaluk Lynge

President - ICC-Greenland, Vice Chair - Inuit Circumpolar Council (ICC), P.O. Box 204, 3900 Nuuk. Greenland

E-mail: aqqaluk@inuit.org

***Opinion article (not peer-reviewed)**

I would like to thank the University of Aarhus and the organizers of this very important conference for their kind invitation to speak here today. This conference is important for many reasons, not the least because it is helping us to prepare for one of the most important global meetings that Denmark has hosted. The Inuit Circumpolar Council, which represents Inuit from Russia to Greenland, is very keen on playing an active role in the events leading up to the post Kyoto meeting. And we are also very happy to be playing a role here as well.

We are gathered here for a conference about moving forward, about forging new partnerships among the public sector, the private sector, and civil society. Many of these partnerships will be centered on new technologies that we hope might help address the enormous problem of climate change.

But I am here today as a representative of the Inuit, the indigenous people of the Arctic, and our perspective on climate change is different. Our worldview is not focused on technology. Our worldview does, however, have a great deal in common with Gro Harlem Brundtland's groundbreaking work on sustainable development, which pointed us to a holistic view of development, an understanding of development as a process that involves much more than simply finding innovative, technological solutions to developmental problems like climate change.

When I think of "sustainable development," I think of traditions. I think of moving forward in ways that cherish the land, the sea and the Arctic flora and fauna. I think of how my ancestors lived for countless generations and how they gradually developed tools and technologies that were adapted to our changing circumstances. For me, adaptation is doing the same things my ancestors have done for centuries, because adaptability is at the core of what it means to be Inuit. My Inuit ancestors passed on the knowledge that adaptation means moving to follow where the wildlife is going, sharing a successful hunt with other families who might not have found the food they needed, adjusting our diet from year to year based on which animals or fish are plentiful at the time, and learning how to make warm clothing out of animal skins to keep our hunters safe and healthy out in the cold. I think of these old ways because they were developed precisely out of our need to continuously adapt. And these ways have been successful in helping my people to survive for hundreds of years.

Inuit are adapting to climate change. But rather than doing something radically new, we are mostly adapting in ways that are similar to the ways that Inuit have always adapted. The Inuit perspective on climate change is that we must find adaptation strategies that look both to the past and to the future.

When we look to the past, it is clear to us that the physical, mental, social, cultural and economic aspects of our well-being are all intimately connected to the Arctic ecosystem. Our health and well-being are utterly dependent on what we are able to harvest from the land and the sea around us.

Now, I am not trying to tell you that each and every Inuit person is out hunting or fishing every day. In Greenland, our full-time hunters share and sell what they have harvested, and many who participate in the wage economy go hunting several times a year. But whether they are out on the land themselves or not, the majority of Inuit place great importance on eating traditional foods. Traditional foods are part of our identity. They are more affordable than store-bought food. And they are healthier, in spite of the persistent organic pollutants, or POPs, that originate thousands of kilometres away and become embedded in our food chain.

Because our well-being is so dependent on our subsistence activities, anything that threatens our subsistence activities threatens our health. We have faced such threats before.

In the 1980s, the threat to Inuit health came from a political decision made in Europe to ban seal imports. The resulting collapse in the global seal market had an enormous impact on Inuit communities across the circumpolar Arctic. Our hunters were told not only that their work was worth very little, but also that it was morally wrong. Our artists, who made beautiful products from seal parts, found that their market too had collapsed. Inuit health was severely affected. Suicide rates increased, and a sense of worthlessness was more evident in many villages. Now the European Union is about to embark on another ban, this time on all seal products it seems. While they talk about an "Inuit exemption," the ban will still hurt us tremendously.

In the 1990s, we discovered that POPs (persistent organic pollutants) had invaded our pristine Arctic ecosystem from the south and attached themselves to the fatty tissue of the sea mammals that form the basis of the Inuit diet. These pollutants – such as DDT pesticides and PCBs (Polychrorinated biphenyls) – have found their way into our bodies and even into the breast milk of our nursing mothers at a concentration seven times that found in mothers in Aarhus or Toronto.

Now, in the past decade, Inuit health is being threatened by the effects of climate change. As most of us have heard by now, the Arctic is warming more quickly than any other part of the planet. Greenland's ice cap is melting. Our coasts are eroding, storms are becoming stronger and less predictable, the sea ice we travel on while hunting is becoming much more dangerous and difficult to read, and the species we rely upon for subsistence are changing their migration patterns.

The effects of climate change are a threat to Inuit health mainly because they make it more difficult and dangerous for us to harvest our food. However, climate change is also expected to bring new bacteria and other microorganisms to our region, with unknown effects. The melting permafrost will make it more difficult for us to freeze our food in the traditional ways, and the shifting ground is damaging our sanitation and water supply systems. All of these issues affect Inuit health and well-being, either physically because they threaten the safety of our food and water, or socially and mentally because they threaten the way of life upon which our identity is based.

Experts tell us that climate change may lead to increased economic development of the Arctic. But Inuit know that any economic advantages that may trickle down to us can never compensate for the hugely negative effects of climate change on our health and well-being. Other experts tell us that new technologies can help save us from the negative effects of climate change. Inuit are sceptical about this, but open to exploring the possibilities. We would welcome any new technology that would help us cope with the enormous challenges we face. We are an adaptable and pragmatic people. We have taken on new technologies very quickly when it was in our interest to do so.

But perhaps we are placing most of our hope for the future in effective partnerships that will allow us to make our own contributions to the solutions of world problems and at the same time to protect our way of life as a unique people. We welcome new partnerships with the public sector, the private sector, and other like-minded civil society groups. We recognize the value of working together with others who are knowledgeable in the fields of medicine, epidemiology, political science, and economic development.

How should these partnerships work? We need to consult each other at the early stages of our research. We need to share our knowledge and the results of our research with each other. And most importantly, we need to make a commitment to each other as partners, recognizing that we all have knowledge and skills that we can bring to our joint enterprises. The partnerships we forge should serve the needs of both the scientific community and the Inuit community as we all move forward into an uncertain future.

Recently, a new partnership was forged between Denmark and Greenland, which gives Greenland further autonomy as a nation. It is my hope that my country Greenland will use this new autonomy to very quickly adopt its own climate change policy – which not only reacts to the world's effects on us, but one that demonstrates responsible stewardship of the environment by Inuit as well.

Other partnerships exist and must be strengthened in our march towards the December summit in Copenhagen. The Inuit Circumpolar Council is an active partner with Denmark at the Arctic Council table, and hopes to work closely with the Danish government on matters of climate change especially as it assumes the role of Arctic Council chair this year. ICC's international involvement in many areas, such as the increasing debate over Arctic sovereignty, makes it an important organization for Denmark to work closely with.

To adapt to rapidly changing circumstances, while at the same time preserving important elements of our culture, we as Inuit need to keep one foot in the past and one foot in the present and balance on both feet as we walk into the future. We need to find a balance between old and new ways, between scientific and experience-based knowledge, between change and stability.

And I dare to suggest that the scientific community needs to find a balance too. I believe that scientists also need to have one foot in the present and one foot in the past, balancing their excitement over new knowledge with a respect for the ancient wisdom and experience of Inuit and other indigenous peoples around the world.

As we work together to develop new adaptation methods, I hope Inuit will be supported in our efforts to preserve our cultural heritage. We are not willing to throw out our traditions, and we do not believe that throwing them out would help us become more adaptable. In fact, it is our traditions that give us the strength and the tools we need to adapt. We have a rich, proud tradition of adaptation. And we want to use our own resources and our own methods of adapting whenever possible.

Denmark is considered a world leader in supporting indigenous peoples around the world. It has also been a strong supporter of the Inuit Circumpolar Council by working closely with us, and helping us in our work at the United Nations bodies, such as the UN Human Rights Council, the UN Permanent Forum on Indigenous Issues, and in drafting the UN Declaration on the Rights of Indigenous Peoples. ICC has also been an active player in the UNFCCC (United Nations Framework Convention on Climate Change) process. We hope that this partnership continues well into the future, but will also build upon the past.

The challenges being forced upon all of us by climate change are immense. We need solutions and we need them urgently. In our avid search for these solutions, I hope we'll remember that the best answers are not always found in the latest and greatest innovations. Sometimes the best answers are found by looking to the past to see what has worked well for hundreds and even thousands of years.

IOP Conf. Series: Earth and Environmental Science **8** (2009) 012012 doi:10.1088/1755-1315/8/1/012012

Comparing past and present climate – a tool to distinguish between natural and human-induced climate change

Marit-Solveig Seidenkrantz[1,4], Antoon Kuijpers[2] and Torben Schmith[3]

[1]Centre for Past Climate Studies, Department of Earth Sciences, Aarhus University, DK 8000 Aarhus C, Denmark.

[2]Geological Survey of Denmark and Greenland, DK 1350 Copenhagen K, Denmark

[3]Danish Meteorological Institute, 2100 Copenhagen Ø, Denmark

Email: mss@geo.au.dk

Abstract. Variations in climate over the last approx. 2000 years and a contrast in the winter climate regime of NW Europe and West Greenland may be ascribed to variations in the teleconnection pattern expressed by the North Atlantic Oscillation (NAO). This well-known climate phenomenon is, among others, reflected in an increased inflow of warm Atlantic Water into the Labrador Sea under conditions of a colder European winter regime (negative NAO indices). Cooling of the Labrador Sea appears to have been a feature typical for a milder European winter climate (positive NAO indices). This pattern is seen both on shorter, (sub)decadal timescales of the last approx. 60 years and on a longer, centennial timescale through the last millennia. During a significant part of the 20th century warming, episodes of milder winters in Europe may be related to Atlantic patterns of natural climate variability similarly as recorded in the past millennia. However, records for recent decades (since the mid 1980s) show a different pattern indicating a significant change in North Atlantic ocean- and atmospheric circulations. This seemingly changed behaviour of the climate system corresponds in timing to the episode of the most extensive global temperature increase, suggesting an increasing anthropogenic effect on the climate development of the past ca. 20 years.

1. Introduction

In the present debate on climate change the discussion is often limited to a comparison of climate records beginning in the mid 19th century, i.e. since the first instrumental observations. The instrumental records are, however, too short to truly display and explain longer scale (>100 years) climate variability patterns, and it is thus necessary to extend the climate record beyond the instrumental record. Furthermore, in order to distinguish natural and human-induced climate change and to improve predictions of future climate, a better understanding of processes, linkages, and time scales involved in climate change and ocean circulation is required.

Not taking into account more short-term variability during this period, the instrumental observations have documented a clear global increase in atmospheric temperature. However, there is not a simple correlation between the atmospheric greenhouse gas concentrations and the rate and pattern of climate change in the 20th century (figure 1). Furthermore, in the discussion it is sometimes forgotten that the start of the instrumental records coincided with the end of the so-called 'Little Ice Age', the coldest period in the European/North Atlantic region of the last approximately 6-8000 years; i.e. since the final melting of the large Northern Hemisphere ice sheets of the last glacial period. Furthermore, in particular the last about 2-3000 years have seen a pronounced climate variability, significantly affecting Northern Hemisphere atmospheric and ocean (figure 2) circulation; the most well-known phenomena being the 'Roman Warm Period' (RWP, ca. 500 BC - AD 400), the cold

[4] To whom any correspondence should be addressed.

'(European) Dark Ages' (DA; ca. AD 400-700), the 'Medieval Warm Period' (MWP, ca. AD 800-1200), the 'Little Ice Age' (LIA, ca. AD 1350-1850), and finally the 20th century 'Modern Warming' (MW) (figure 3a) [1]. The 20th century warming thus not simply represents the change from a stable 'normal' colder climate to an unprecedented warming regime, but may to a large extent be a consequence of generally unstable climate conditions of the last 2-3000 years. One of the outstanding challenges is thus to distinguish between natural climate variability and human impact on the climate system. We here present a comparison between geological (proxy) records of past climate change and the instrumental record from the North Atlantic region (figure 2, 3) in order to test whether such an approach is in fact possible. In the following discussion we will mainly focus on changes in ocean- and atmosphere circulation patterns of the North Atlantic region.

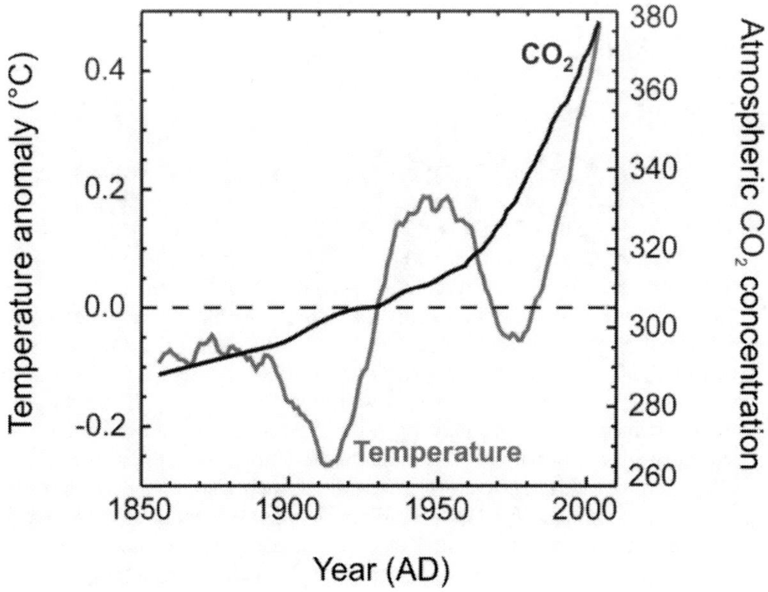

Figure 1. Change in North Atlantic sea-surface temperature (red curve) compared to atmospheric CO2 content (black) after 1856. Calculated after [2].

Several theories have been proposed to explain the possible cause of the above climate variability. Explanations include long-term solar variations that modulate the atmospheric circulation [e.g. 3], sulfate aerosols from volcanic eruptions and changes in large-scale ocean circulation [4, 5]. In the present discussion we will mainly focus on changes in ocean- and atmosphere circulation patterns of the North Atlantic region (figure 2).

2. The North Atlantic Oscillation – controlling past and present climate

When discussing climate variability it must also be taken into account that climate patterns may not be alike across larger geographical areas. *"In Greenland, all winters are severe, yet they are not alike. The Danes have noticed that when the winter in Denmark was severe, as we perceive it, the winter in Greenland in its manner was mild, and conversely"*. This statement (translated from Danish) was noted in the diary of Hans Egede Saabye, who was the Danish missionary in Greenland in the 1770s [6, 7]. This signified one of the first reports on the occurrence of such a contrasting climate pattern between Europe and West Greenland. Similar records of variations in temperature, precipitation and sea-level pressure are also found in other historical and instrumental records [8-10], documenting that coherent climate fluctuations reached from Siberia to North America and from the Arctic Ocean to the equatorial regions [11-12]. This now well-known pattern of climate variability has been recognized as being related to the so-called North Atlantic Oscillation (NAO) [13-16], which has been shown to influence both atmospheric conditions and ocean circulation patterns [e.g. 17]. The NAO is a subsystem of the Northern Annular Mode (NAM), which represents patterns in atmospheric pressure variations of the extra-tropical Northern Hemisphere. The NAO index is measured as the normalised sea-level pressure difference between Stykkisholmur, Iceland and Lisbon, Portugal [14]. The instrumentally-based NAO record may be reconstructed back to the mid 19th century, but few studies of actual temperature variations have hitherto been made.

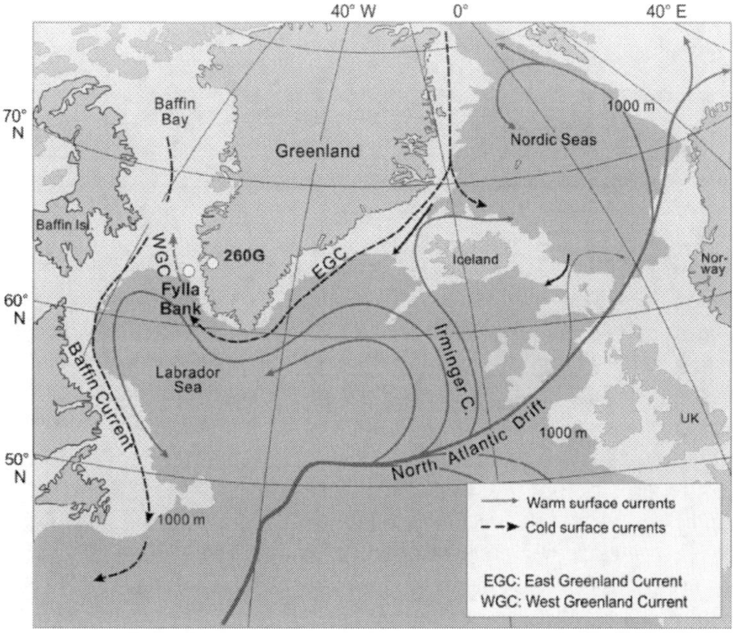

Figure 2. Map of the North Atlantic with location of gravity core 248260-2G from Ameralik Fjord, SW Greenland and the oceanographic monitoring station at Fylla Bank. Blue arrows indicate cold Polar Water transport pathways via the EGC and Baffin – Labrador Current system (WGC = West Greenland Current, EGC = East Greenland Current); red arrows show the main warm-water transport pathways.

Furthermore, most studies concentrate on the NAO signal of the instrumental records, although several recent studies have suggested that that the NAO signal may be traced further back in time using both atmospheric and marine records [e.g., 18-21]. This North Atlantic seesaw pattern appears, however, to be more complex on the longer (late Holocene) time scales [22, 19]. Thus, it has been suggested that the larger-scale climate variability of the last 2-3000 years (late Holocene) may have been linked to a general change in the overall NAO-mode implying that an oceanic-atmospheric seesaw mechanism is also operating on a centennial to millennial time scale [19-21, 23-24]. These authors, among others, found that via the Irminger Sea Water contribution to the West Greenland Current (i.e. figure 2) the Atlantic water inflow into the Labrador Sea increased during periods when climate in the eastern North Atlantic region was generally cold (i.e. 'Dark Ages' and 'Little Ice Age'), while a weaker Atlantic water entrainment into the Labrador Sea region occurred during periods of a warmer European climate ('Roman Warm Period' and 'Medieval Warm Period') (figure 3b). This suggests that centennial-scale ocean and climate fluctuations are related to an overall change in the NAO index. Within this context, it is noteworthy that there may be a link to large scale ocean-atmosphere circulation changes in the Southern Hemisphere (El Niño-Southern Oscillation), as it presumably was the case for the Medieval Warm Period [20]. It also implies that the 20th century warming, particularly in winter, may to a large extent be caused by a change to an increasingly positive NAO scenario with stronger westerly winds having prevailed during this period [e.g. 25].

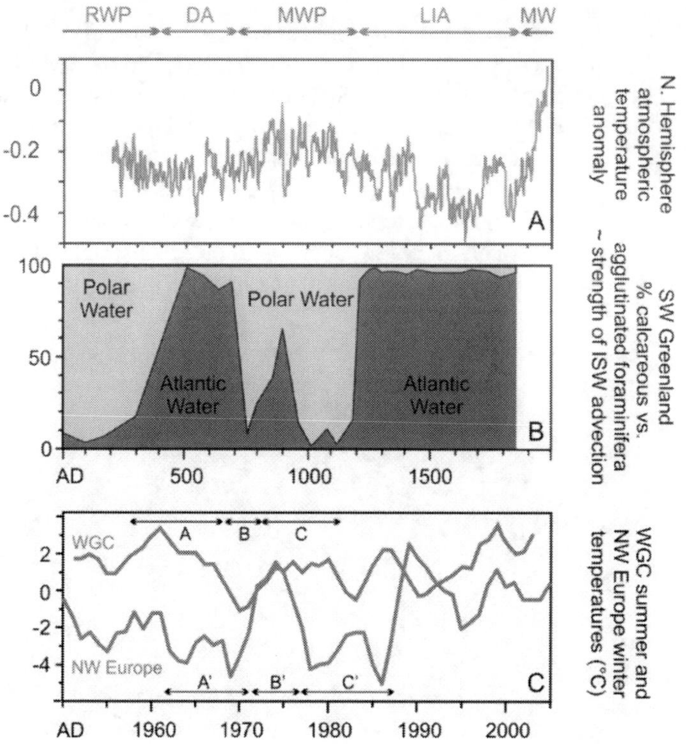

Figure 3. A) Northern Hemisphere atmospheric temperature anomaly from decadally-resolved multi-proxy based reconstruction (one standard error uncertainty 0.16). Data are shown as anomalies based on the 1961-1990 instrumental reference period [after 26]. RWP = Roman Warm Period, DA = (European) Dark Ages; MWP = Medieval Warm Period; LIA = Little Ice Age. **B)** Reconstruction of temperature of subsurface waters off West Greenland based on calcareous vs. agglutinated benthic foraminifera from core 260248-2G, Ameralik fjord in the Godthåbsfjord region, West Greenland [after 19, 20]. These data show warming of West Greenland subsurface waters during European cold periods (DA, LIA) and reduced water temperatures during warm periods (RWP, MWP) indicating that significant climate variability of the last 2000 years was linked to significant changes in ocean circulation. **C)** West Greenland Current (WGC) surface water (0-40 m water depth) temperature annually recorded (June) at Fylla Bank offshore SW Greenland (source: Danish Meteorological Institute and Northwest Atlantic Fisheries Organization, Canada) shown as 3-year running average (blue curve). Elevated temperatures in the 1950s and 1960s coincide with on average colder NW European (50-65°N, 20°W-20°E) mean winter (DJF) air temperatures [27] (green curve; 3-year running average). Multi-annual WGC anomalies are labeled A(+), B(-), and C(+), while (following) series of European winters are labeled A'(-), B'(+), and C'(-).

3. The 20th century warming

Enhanced subsurface advection of warm, saline Atlantic Water (Irminger Sea Water) by the West Greenland Current (WGC, figure 2) thus appears to be a characteristic feature of Subpolar gyre circulation during periods of colder European climate (e.g. Little Ice Age) in the past. In order to test this pattern in the modern instrumental records (starting 1950), we compared annual (June) temperature records of the WGC from offshore SW Greenland (Fylla Bank; figure 2, 3c) with NW European mean winter (December to February) air temperatures. The Fylla Bank sea-surface temperatures display elevated values until the end of the 1960s, followed by repeated, multi-annual episodes of surface water cooling, with renewed warming persisting after the mid 1990s. The prevalence of on average colder Northwest European winters during the 1950s and 1960s (figure 3b) coincided with a long-term (>decade) upper ocean warming off SW Greenland [cf. 27, 28]. In corroboration, positive salinity anomalies persisting in the subpolar gyre were recorded in the 1950s and 1960s [29], and are a typical feature associated with a negative NAO circulation mode. The records (figure 3c) furthermore show that prior to the mid 1980s both positive and negative multi-annual (<decade) temperature anomalies at Fylla Bank thus lead series of colder, respectively warmer winters in Northwest Europe by 4-5 years. The severity of Northwest European winters is mainly determined by the prevailing wind direction, i.e. variation in the strength of (westerly) zonal

circulation as indicated by the NAO index. A plausible explanation for the above ocean-climate link is therefore atmospheric circulation changes in response to sea-surface temperature anomalies in key areas in the north-western Atlantic, as demonstrated by many modelling studies.

4. Distinguishing natural and human-induced climate change

This pattern of opposite development between NW European air temperatures and W. Greenland sea-surface temperatures, however, does no longer apply after the mid 1980s, when global warming started to increase markedly, and at a 30-year average the last three decades (1973-2002) were likely the warmest period of the past 500 years [e.g. 30]. The persistence of generally warmer-than-normal NW European winters, although the Fylla Bank record has documented a series of uninterrupted positive temperature anomalies in the Labrador Sea since the mid 1990s (figure 2c), strongly suggests a fundamental change in the 'normal' ocean-atmosphere driven Atlantic climate pattern, presumably due to the effect of greenhouse-gas induced warming.

In summary, the palaeoceanographic records from the WGC region in combination with recent oceanographic evidence from that region suggest that we should have entered a period with more frequent colder Northwest European winters after substantial warming of the WGC region, which occurred after the mid 1990s. The severity of Northwest European winters is mainly determined by the prevailing wind direction, i.e. variation in the strength of (westerly) zonal circulation as indicated by the NAO index, which after the early 1990's indeed has shifted from positive towards more neutral or negative values. Despite of this change in the NAO index almost 15 years ago, we have not yet seen clear evidence of more frequent colder winter weather in Europe, a fact which we tentatively attribute to an increasing anthropogenic effect on the North Atlantic and European climate regime. This demonstrates that based on investigations of climate variability beyond the instrumental record it appears to be possible to distinguish between natural and human-induced climate change.

References

[1] Lamb H H 1995 *Climate history and the modern world.* (London: Routledge) p 433.
[2] Kaplan A, Cane M, Kushnir Y, Clement A, Blumenthal M, Rajagopalan B 1998 *J. Geophys. Res.* **103** 18,567-18,589.
[3] Shindell D T, Schmidt G A, Mann M E, Rind D and Waple A 2001 *Science* **294** 2149-52.
[4] Broecker W S 2000 *Proc. Natl. Acad. Sci. USA* **97** 1339-42.
[5] Crowley T J 2000 *Science* **289** 270-77.
[6] Saabye H E 1942 *Meddelelser om Grønland* **129(2)** p 45 and 103.
[7] van Loon H, Rogers J C 1978 *Monthly Weather Review* **106** 296-310.
[8] Cranz D 1765 *Histoire von Gronland enthaltend die Beschreibung des Landes and der Einwohner und. inbesonder die Gesichte der dortigen Mission der Evangelischen Bruder zu Neu-Herrnhut und Lichtenfels* (Leipzig: Heinrich Detlef Ebers) p 1132.
[9] Hann J 1890 *Meteorologische Zeitschrift* **7** 109-10.
[10] Defant A. 1924 *Geogr. Ann.* **6** 13-41.
[11] Walker G T 1924 *Memoirs of the Indian Meteorological Department* **24** 275-332.
[12] Walker G T and Bliss E M 1932. *Mem. Royal Meteorological Soc.* **4 (36)** 53-84.
[13] Rogers J C 1984 *Monthly Weather Rev.* **112** 1999-2015.
[14] Hurrell J W 1995 *Science* **269** 676-79.
[15] Portis D H, Walsh J E, Hamly, M E and Lamb, P J 2000. *Journ. Clim.* **13** 2069-78.
[16] Hannaz E and Cappelen J 2003 *Geophys. Res. Lett.* **30 (3)**, doi:10.10029/2002GL015797, 1132.
[17] Buch E 2002 *Danish Meteorological Institute, Scientific Report* **02-02** 1-39.
[18] Nesje A, Matthews J A, Dahl S O, Berrisford M S and Andersson C 2001. *The Holocene* **11** 267-80.
[19] Seidenkrantz, M-S, Aagaard-Sørensen, S, Sulsbrück, H, Kuijpers, A, Jensen, K G, Kunzendorf,

H 2007 The *Holocene* **17** 387-401.
[20] Kuijpers A, Malmgren, B A and Seidenkrantz, M-S 2009 *PAGES News* **17 (2)** 76-77.
[21] Trouet V, Esper J, Graham J N E, Baker A., Scourse J D, Frank D C 2009 *Science* **324** 78–80, doi: 10.1126/science.1166349
[22] Dawson A G, Elliott L, Mayewski P, Lockett P, Noone S, Hickey K, Holt T, Wadhams P and Foster I 2003 *The Holocene* **13** 381-92.
[23] Keigwin L D and Pickart, R S 1999 *Science* **286** 520-23.
[24] Seidenkrantz, M-S, Roncaglia, L, Fischel, A, Heilmann-Clausen, C, Kuijpers, A and Moros, M 2008 *Mar. Micropaleontol.* **68** 66-83, doi:10.1016/j.marmicro.2008.01.006
[25] Hurrell, J W, Kushnir, Y, Ottersen, G and Visbeck, M (Eds.) 2003. *Geophys. Monograph Ser.* **134**, 279pp.
[26] Mann, M E and Jones P D 2003 *Geophys. Res. Lett.* **30**, No. 15, 1820 doi: 10.1029/2003GL017814. Data citation: Mann, M.E. and P.D. Jones 2003, 2,000 Year Hemispheric Multi-proxy Temperature Reconstructions, IGBP PAGES/World Data Center for Paleoclimatology Data Contribution Series #2003-051. NOAA/NGDC Paleoclimatology Program, Boulder CO, USA.
[27] Jones, P D and Moberg, A. 2003. *J. Clim.* **16**, 206-23.
[28] Levitus, S, Antonov, J I and Boyer, T P 1994 *Science* **266** 96-99.
[29] Hátún, H, Sandø, A B, Drange, H, Hansen, B, Valdimarsson, H 2005 *Science* **309** 1841-44, doi: 10.1126/science.1114777.
[30] Luterbacher J, Dietrich D, Xoplaki E, Grosjean M and Wanner H 2004 *Science* **303** 1499-1503, doi: 10.1126/science.1093877.

Acknowledgements
We acknowledge the German Research Council for shiptime (RV *A. v. Humboldt*, 2002) funding as well as the The Danish Council for Independent Research | Natural Sciences (FNU 21-04-0336) and 'Kommisionen for Videnskabelige Undersøgelser af Grønland' (2008-2009) for supporting this study.

IOP Conf. Series: Earth and Environmental Science **8** (2009) 012013 doi:10.1088/1755-1315/8/1/012013

Systems in peril: climate change, agriculture and biodiversity in Australia

Chris Cocklin[1,3] and Jacqui Dibden[2]

[1] Division of Research & Innovation, James Cook University, Townsville, Australia

[2] School of Geography & Environmental Science, Monash University, Australia

Abstract. This paper reflects on the interplay amongst three closely linked systems – climate, agriculture and biodiversity – in the Australian context. The advance of a European style of agriculture has imperilled Australian biodiversity. The loss and degradation of biodiversity has, in turn, had negative consequences for agriculture. Climate change is imposing new pressures on both agriculture and biodiversity. From a policy and management perspective, though, it is possible to envisage mitigation and adaptation responses that would alleviate pressures on all three systems (climate, agriculture, biodiversity). In this way, the paper seeks to make explicit the important connections between science and policy. The paper outlines the distinctive features of both biodiversity and agriculture in the Australian context. The discussion then addresses the impacts of agriculture on biodiversity, followed by an overview of how climate change is impacting on both of these systems. The final section of the paper offers some commentary on current policy and management strategies that are targeted at mitigating the loss of biodiversity and which may also have benefits in terms of climate change.

1. Introduction

Climate change poses a global threat to biodiversity and to human settlements. In the case of the island continent of Australia, the advance of a European style of agriculture over the past two centuries has already imperilled Australia's unique biodiversity, including many endemic species. The loss and degradation of biodiversity has, in turn, had negative consequences for agriculture and hence for Australia's economy, which is highly dependent on the export of commodities. While the Australian government has made commitments to protect and enhance biodiversity and to ecologically sustainable development, the reality is that environmental decline has continued and increasing numbers of indigenous species are threatened or have become extinct. Climate change is bringing new pressures to bear on both agriculture and biodiversity, with drought, floods and other extreme weather events imposing additional pressures on already stressed production and natural systems. Indeed, climate change threatens to administer the *coup de grace* to many unique and fragile ecosystems already damaged by human activity.

In this paper, we outline the characteristics of the environment and biodiversity of Australia; summarise the impacts of settlement of the continent on the natural landscape, particularly through land clearing and introduction of European agricultural practices; examine the various threats to

[3] Corresponding author. Email: Chris.Cocklin@jcu.edu.au

biodiversity and agriculture, and the exacerbating effects of climate change; and finally consider policy and management responses to environmental and climate change. While both land and maritime environments are under threat, our focus in this paper is on terrestrial biodiversity and natural and agricultural ecosystems. We conclude that, while these are all imperilled by both past and current practices and climate change, it is possible to envisage mitigation and adaptation responses that would alleviate pressures on all three systems (climate, agriculture, biodiversity).

2. Biodiversity and settlement in Australia

2.1 International significance of Australia's biodiversity

The distinctive features of Australia's biodiversity have been attributed to "the country's size, isolation, naturally fragmented landscapes and long-term climate variability" (Beeton et al. 2006). Australia is included in a group of 17 countries identified as being 'megadiverse' (Mittermeier et al. 1997), because of the large number and high proportion of endemic species to be found within these countries. Although the group of megadiverse countries occupy less than 10% of the earth's surface, they are thought to support more than 70% of all known life forms. Of the megadiverse countries, only Australia and the United States are in the high income category, leading Williams et al. (2001) to suggest that a high level of biodiversity protection is a reasonable expectation of these countries. In practice, this has proved difficult to achieve.

Australia is home to an estimated 600-700,000 species, many of which are endemic. These include distinctively Australian species such as kangaroos and koalas, but - even more remarkably - "it is likely that about 90% of its native arthropods and plants are endemic to Australia, which is a level of endemism as high as that recorded for any part of the world, including small, remote islands" (Raven and Yeates 2007: 179). It has been estimated that 83% of mammals, 45% of birds, 89% of reptiles, 93% of amphibia and 90% of fishes are endemic to Australia (Cork et al., 2006), but many of the species native to Australia have not yet been described. This is particularly the case with organisms that are difficult to detect or of little direct interest to people (Chapman 2006: 2).

2.2 The impact of human settlement

When the Indigenous Australians arrived in the continent approximately 50,000 years ago, they used fire in many areas to transform the landscape from forests to open savannas in order to increase the availability of food (Dovers 1992), but their hunting and gathering activities were relatively gentle on the land. By contrast, European settlement from 1788 brought about a profound transformation, and one that was not for the better. As Raven and Yeates argue:

> *The antiquity and isolation of Australia, its ancient and often infertile soils, and its overall aridity, have given its environment a unique fragility that has not responded well to methods of cultivation and grazing developed in other better-watered and more fertile parts of the world. ... [The] settlement of virtually all of the arable land during the 19th century without regard to the peculiar properties of the Australian environment led rapidly to many of the environmental problems faced today.*
> (Raven and Yeates 2007: 182)

In the course of 200 years (from 1788 to 1988), 17% of total land area was converted from forest, woodland or shrub cover to grassland or pasture. Mackey et al. (2008: 13) estimate that approximately 50% of woodland and forest ecosystems have been cleared, and in productive bioregions as much as 95% of all native vegetation has been modified or destroyed (see also ABS 2006; NLWRA 2001a). Clearing was undertaken to open up land for cropping, forestry, grazing and mining, as well as urban settlements. Today, about 60% of the land area is occupied by agricultural and pastoral enterprises (SOE, 1996). Grazing has resulted in environmental deterioration, exacerbated by recurrent droughts, and the spread of introduced pests. Artificial pastures have replaced native grasses and more than 1700 species of weeds have become established (Raven and Yeates 2007: 183).

Many of the changes introduced by European settlers have had adverse effects on biodiversity, which quickly became apparent. Mackey et al. (1998: 13) report that as early as the 1920s transformation of the landscape had led to many species becoming extinct or threatened. The abandonment of indigenous fire management practices altered fire regimes in ways which damaged fire-sensitive species and ecosystems (Bowman 2000). Furthermore, altered fire regimes and introduced pasture species have interacted with adverse ecological effects, particularly in northern regions (Soule et al., 2004: 268). Cork et al. (2006) report that there are now less than 2% of the original native grasslands in the temperate zone, which has severely reduced habitats for native fauna (see also Beeton et al. 2006: 37, 39). In many cases, the harmful practices of settlers in the early days were the result of "government policies that encouraged (and even forced) farmers to clear natural vegetation from their properties under terms of conditional purchase" (Tonts 2005: 197; Bolton 1981) and later were promoted through "financial support from government agencies and new land clearing technologies developed by CSIRO [the Commonwealth Scientific and Industrial Research Organisation]" (Beeton et al. 2006: 72). In interviews, farmers not infrequently raise the issue of government responsibility for instigating environmentally damaging practices in the past.

European farming practices have proved environmentally damaging in a number of ways. Destruction of native vegetation was initially encouraged by a desire to create a landscape that was more like 'home', with British crops and animals (Dovers 1992). Several of these introduced plants and animals subsequently became notorious pests. Settlers lacked the skills and knowledge of the environment to realise that the introduced European agricultural practices were often unsuited to the Australian environment (Gray and Lawrence 2001). The adoption of high-input 'productivist' farming techniques, from the 1950s, increased production by privileging economic returns over land stewardship (Cocklin, 2005). The emphasis on ever-increasing production has become even more pronounced since the 1970s/early 1980s due to a shift in government ideology. The adoption of a free-market policy stance by Australian governments has led to the expectation that farmers will become more competitive through stripping away the price supports, import restrictions and subsidies that had previously buffered them against the fluctuations of global commodity prices (Dibden and Cocklin, 2005, 2009; Tonts 2005). Falling prices for agricultural produce, together with rising prices for inputs (such as fertilisers, livestock feed, fuel), have pressured farmers to increase both the intensity and scale of production through farm expansion – in some areas through further land clearing – with predictable consequences for the environment:

> *... market liberalisation, the cost-prize squeeze, and the increasing influences of multinational agri-food companies have further underscored long-standing patterns of environmental degradation. ... Significantly, it is not only the environment that suffers – literally millions of dollars of production are lost each year, impacting adversely on individual farmers and their communities* (Cocklin 2005: 174).

3. Biodiversity in decline

Ecosystems continue to be destroyed in many regions. The 2006 national *State of the Environment Report* reported that 17 million hectares have been cleared since 1973, with 1.5 million hectares of that deforestation occurring between 2001 and 2004 (Beeton et al. 2006). Because many of these changes in land use are quite recent, Australia stands out as being one of only a few developed countries "that remains a leading contributor to the current human-induced global mass extinction event" (Mackey et al., 2008: 11). Data collected in the Australian Bureau of Statistics *Measures of Australia's Progress* (ABS 2009a) showed that, between 2000 and 2008, "the number of terrestrial bird and mammal species assessed as extinct, endangered or vulnerable rose by 14% from 154 to 175 (of which 69 were birds and 106 were mammals)", with 18% of these presumed extinct. Johnson (2006) found that nearly half of all mammal extinctions globally have occurred in Australia in the last 200 hundred years. Native species have also declined in range, with some experiencing serious but often unacknowledged regional declines – what Mackey et al. (2008: 11) describe as "secret extinctions".

Despite efforts to reverse this trend, declines in biodiversity are continuing because many of the threats that have caused past extinctions or declines are still present. These include the impact of invasive organisms – estimated to be increasing at the rate of 20 new pests or diseases each year; changed fire regimes; widespread drought conditions, now into the thirteenth year in south-eastern Australia; declining native vegetation, water quality and quantity (BDWG 2005; Cork et al. 2006; Mackey et al. 1998; O'Brien, 1990). Some regions are also affected by ongoing land clearing, pollution of waterways, soil salinity and modified waterways (Cork et al. 2006). The National Biodiversity Strategy Review Task Group (NBSRTG, 2009) identified several long-standing threats to Australia's biodiversity:

- invasive species
- loss, fragmentation and degradation of habitat
- unsustainable use of natural resources
- changes to the aquatic environment and water flows
- inappropriate fire regimes.

The Task Force also added a new but increasingly important threat: "climate change (resulting in conditions such as prolonged drought)".

4. Agriculture and the environment

Agriculture has been seen by conservationists and others "as contributing significantly to erosion of biodiversity (and possibly ecological function), through the clearance of natural vegetation … and the intensification of farming practices on land already highly modified" (House et al., 2008: 153). However, environmental degradation does not in general result from deliberate action but through the "tyranny of small decisions" - "decisions that were never consciously made" but may nonetheless have severe impacts (Odum, 1982: 728; see also Kahn, 1966). This description is well exemplified by agriculture, where "individual farmers rarely set out to diminish the quality of the environment, and generally the actions of each have a relatively minor impact in themselves, but cumulatively their 'decisions' impact in significant ways upon the environment" (Cocklin, 2005: 173; see also Cocklin, 1993; Spaling and Smit, 1995). Moreover, the decisions of farmers have not been made in isolation. They have been influenced by government policies, which have until recently been unsympathetic to environmental concerns, and more recently have been largely determined by what might be called the 'tyranny of the market'. As Tonts (2005: 197) points out: "In order to remain profitable, farmers are often forced to use large quantities of fertilisers and pesticides, and in some cases to overstock their paddocks or clear more native vegetation, despite an acute awareness that land degradation is a serious problem."

Numerous reports attest to the degraded state of the Australian environment, particularly its land and water resources (e.g., Australian State of the Environment Committee, 2001; Beeton et al. 2006; Conacher and Conacher, 2000; Yencken and Wilkinson, 2000). Land clearing remains a particularly significant threat to biodiversity, which "destroys and degrades the habitat on which native species rely", "allows weeds and invasive animals to spread, contributes to greenhouse gas emissions and can lead to soil degradation, such as erosion and salinity, which in turn can affect water quality" (ABS 2009a).

Australia's old, shallow soils are readily damaged by agricultural activities (ABS 2009b). In consequence, land degradation has afflicted Australian agriculture in many areas for more than a century, and has been recognised as a matter for state government attention since the 1930s (Vanclay and Lawrence 1995: xiii). Yet, despite increasing evidence of the severity of problems such as wind and water erosion, acidification and salinisation, it took a long time for the full gravity of the environmental damage to be recognised and for an awareness to dawn that some – perhaps most – of the land available for agriculture cannot safely be used for this purpose. As late as 1916, for example, land susceptible to salinisation at Nyah (in the mid-Murray region of northern Victoria) was distributed to returning soldiers as part of a soldier settlement program, with disastrous consequences both for the aspiring farmers and the environment (Barr and Cary 1992: 261).

Soil salinity (which is the result of both irrigation and tree clearing) is a major contributor to land degradation (ABS 2002; Beeton et al. 2006). The pervasiveness of salinisation resulting from irrigation has caused some researchers to question whether this land management strategy can be sustained (Haw et al. 2000: 156). Dryland salinity is a result of clearing native species whose adaptive root systems help to ensure salts in groundwater are not transported to the surface (Brennan 2004). It has gained great significance in Australia due its extent, economic cost and irreversibility (LWRRDC 1999). An assessment of dryland salinity indicated that approximately 5.7 million hectares were at risk or affected by dryland salinity and that within 50 years this might increase to more than 17 million hectares (NLWRA, 2001b). The non-farm effects of dryland salinity represent a significant cost for local and state governments: these effects include declining water quality, infrastructure damage to roads and buildings, habitat decline, increased flood potential, social costs and increased public expenditure on drainage schemes (LWRRDC 1999: 7-8). It is equally damaging for the natural environment, threatening biodiversity through the degradation of habitats and its effects on water quality (ABS 2009b). In 2001, it was estimated that up to 20,000 km of streams could be significantly salinity-affected by 2050 and that about 630,000 hectares of native vegetation and associated ecosystems were at risk; this could increase to 2 million hectares over the next 50 years (NLWRA 2001b).

Land clearing and intensive agriculture have resulted in both water and wind erosion in many areas (Beeton et al. 2006). For example, broadacre cropping regions, such as the Wimmera and Mallee in western Victoria, have experienced severe dust storms which stripped and carried away large quantities of topsoil over considerable distances. The remedy for this problem has been sought in 'no-till' or 'conservation' farming, a benign sounding practice which depends on application of herbicides to avoid the necessity of ploughing. A study of adoption of sustainable farming practices found that use of these potentially hazardous and expensive chemicals was "seen by farmers and others to place farmers on a technological as well as economic treadmill … which delivers certain long-term costs but doubtful long-term benefits" (Gray el al. 2000: 35).[4] The impacts on biodiversity have not, to our knowledge, been studied, perhaps reflecting a general tendency in Australia to under-value biodiversity on *farmed* (contrasted with *wild*) lands.

5. Impacts of climate change

5.1 Climate variability and change

Australia's climate has long been subject to extreme variations, both between climatic zones and over time. Indeed, variability and extremes are such established characteristics of the climate that a popular Australian poem ('My Country') celebrates Australia as a land of "droughts and flooding rains". Despite "severe climatic oscillations … over the last 500,000 years", Mackey et al. (2008: 12)

[4] This reluctance to adopt "zero tillage" appears to have been overcome, with no till practices used on "nearly two-thirds of land prepared for crops and pastures" in 2007-08 (ABS 2009b).

conclude that "all native species extant when Europeans first occupied the continent 220 years ago had persisted through all these long term trends and oscillations in temperature, CO2 concentration and wetness." Thus, Australian flora and fauna are to a large extent adapted to a highly variable climate (Soulé et al. 2004). What is different now is the scale of the problem and the fact that climate change is affecting a landscape which is already severely damaged and "in which the habitats have been fragmented, invaded by alien species, and otherwise disturbed, all of these factors increasing the impact of global warming greatly" (Raven and Yeates, 2007: 183; Mackey et al. 2008).

There is now scientific consensus that profound, human-induced climate changes can be anticipated in coming decades, and indeed have already have begun. Climate change projections for Australia (CSIRO and BOM 2007) indicate that significant increases in temperature can be anticipated: annual warming is predicted to increase on average by 1.0°C by 2030 relative to 1990 levels, with an increase in the frequency of very hot days (over 35.0°C). Rainfall projections are even more concerning, with declining rainfall predicted for all regions except the far north, and major changes in seasonal rainfall patterns in the east, south-east and south-west, i.e., in the major agricultural areas. Reduced rainfall will result in even larger decreases in stream flows (Garnaut 2008: 109). Drought may be expected to occur more frequently across Australia, particularly in south-western Australia, with up to 20 per cent more agricultural drought months[5] over most of Australia. Decreases in rainfall due to climate change are likely to be exacerbated by the influence of El Niño weather events, which reduce rainfall dramatically across southern Australia: the last El Niño in 2006 reduced rainfall to the lowest level every recorded in parts of southern Victoria and northern Tasmania. As a result of climate change, extreme events – heatwaves, bushfires, tropical cyclones and floods – will also become more common in some areas (CSIRO and BOM 2007; Garnaut 2008). Two new categories of fire weather have been defined – "very extreme" and "catastrophic" (Garnaut 2008: 120); an instance of catastrophic fire weather in south-eastern Australia recently resulted in the deaths of 173 people in one day.

5.2 Environmental impacts of climate change

A number of studies have identified the serious threat which climate change poses to Australia's fragile environment and ecosystems. This threat is amplified by the extent of damage already brought about over two centuries of human-induced environmental changes. Thus, a *Climate Change Review,* commissioned by the Australian government, came to the view that:

> *... added stressors from climate change would exacerbate existing environmental problems, such as widespread loss of native vegetation, over-harvesting of water and reduction of water quality, isolation of habitats and ecosystems, and the influence of introduced plant and animal pests* (Garnaut 2008: 141; see also Mackey et al. 2008).

Similarly, Cork et al. (2006) argued that: "Climate change will further compound and intensify pressures on biodiversity, especially by affecting rainfall patterns, and hence fire frequency, affecting regeneration of vegetation, and changing where plants and animals can live." Raven and Yeates (2007: 183) went further, concluding that: 'Perhaps the most important problem of all affecting the survival of Australia's unique plants and animals is global warming".

Morton et al. (2009: 3) identified various direct and indirect effects of climate change including:

- Range shifts of species, range fragmentation or shrinkage, extinctions and changes in the structure of ecosystems
- Potentially greater vulnerability to biological invasions

[5] These are months when agricultural production is constrained by extremely low soil moisture as a result of seasonal temperatures and rainfall (CSIRO and BOM 2007).

- Additional alteration to fire regimes
- Changes in linkages between ecological and socio-economic systems.

Already there are some convincing examples that climate-change induced shifts have occurred in the Australian Alps since the 1960s, such as the encroachment of snow gums (*Eucalyptus pauciflora*) on sub-alpine grasslands and changes in the ranges of vertebrates (Cork et al. 2006; Hughes 2003). For some species occupying specialized ranges, changes to their habitat due to climate change are likely to be fatal. The future of the flora and fauna of the Australian Alps is particularly hard to resolve, given the lack of higher terrain anywhere in Australia for species to migrate or be relocated to (NBSRTG 2009: 65). Similar impacts on higher altitude species have been detected in the Australian tropical rainforests (Cork et al. 2006; Krockenberger et al., 2003; Williams and Hilbert 2006, Williams et al. 2008).

Also facing a very uncertain future are numerous wetlands, due to declining rainfall and, on the coast, the prospect of saltwater intrusion and inundation (Hughes 2003). The RAMSAR-listed Coorong wetlands and adjacent lakes at the mouth of the Murray River have already been damaged, possibly beyond repair, by salinisation and acidification resulting from a combination of drought and over-extraction of water preventing environmental flows in the river (Office for Water Security 2009). Without global mitigation of climate change, Garnaut (2008: 125) found that by mid-century another major natural icon, the Great Barrier Reef, would be effectively destroyed, together with other reef systems, with "serious ramifications for marine biodiversity", as well as associated activities, such as tourism (see also Johnson and Marshall, 2007).

Agricultural areas in much of the country will also be damaged by climate change. According to Gunasekera et al. (2007: 657), "Australia is projected to be one of the most adversely affected regions from future changes in climate in terms of reductions in agricultural production and exports." The Garnaut report found that, without effective global mitigation of climate change, agriculture would be adversely affected by changes in temperature and water availability with the result that:

> *By mid-century, there would be major declines in agricultural production across much of the country. Irrigated agriculture in the Murray-Darling Basin would be likely to lose half of its annual output. This would lead to changes in our capacity to export food and a growing reliance on food imports.* (Garnaut 2008: 125)

Indeed, drought in much of the Murray-Darling Basin (MDB) has already reduced availability of water for irrigation-dependent agriculture and resulted in severe stress for farmers. Given that this area has been the food bowl of Australia, producing more than 40% of the total gross value of agricultural production, this is a serious matter. Without mitigation, it is predicted that impacts will be profound by the close of the century with Garnaut forecasting that:

> *The increased frequency of drought, combined with decreased median rainfall and a nearly complete absence of runoff in the Murray-Darling Basin, is likely to have ended irrigated agriculture for this region. Depopulation will be under way.* (Garnaut 2008: 129)

Other impacts on agricultural productivity – crop yields and livestock carrying capacity – are likely to include increased average temperature, changed rainfall patterns and increased climate variability; reduced soil moisture and increased evaporation; more severe weather events, including bushfires and flooding; an increased risk of soil erosion as a result of reduced ground cover and increased frequency of extreme wind and rain events; and increased incidence and occurrence of pests and diseases (AGO 2007; DCC 2008; Garnaut 2008). Weeds are already a major problem for Australian farmers, but this problem is likely to escalate as a result of changes in temperature combined with financial pressures on farmers which limit their ability to effectively manage them (DCC 2008). Some positive effects could occur, although generally only in the short run: in some regions, there could be a decrease in pests and diseases, while wheat production could increase in some parts of Western Australia because decreasing rainfall is likely to reduce soil salinity (Garnaut 2008: 132). Effects of climate change may include shifts in geographical location of production (Morton et al. 2009), with some farming groups and politicians expressing enthusiasm for the prospect of agriculture moving from the increasingly desiccated farming areas of southern Australia to the better watered north, ignoring the climatic extremes, and the pests and diseases, to which the northern region is prone. Some farming areas have become marginal, and many farmers have reduced production, due to drought. Perhaps because of this experience of drought, a substantial proportion of farmers appear to have accepted that the climate has already changed: a survey of all agricultural businesses in 2006-07 (when many regions experienced a severe drought) found that "65.6% reported that they considered the climate affecting their holding had changed" (ABS 2008: 4).

6. Environment and biodiversity protection policies

The imminent prospect of dangerous climate change is not only raising concerns about the probable environmental impacts but also about the adequacy and effectiveness of current and proposed strategies for biodiversity conservation and natural resource management. Before considering strategies specifically addressing impacts of climate change, we outline the development and characteristics of existing policies and programmes for biodiversity protection and sustainable land management.

The genesis of environmental conservation in Australia was associated with the growing appreciation, mainly amongst urban residents, of the natural environment, and particularly of 'wilderness' areas: This led to the establishment of nature reserves, primarily for recreation (Mosley 1978). During the 1970s and 1980s, an increasingly well-supported nature conservation movement engaged in a series of often contentious campaigns to protect wilderness areas from development. Subsequently, a number of states enacted wilderness protection legislation (Mackey et al. 1998). In Australia, as in other 'settler societies', a far higher valuation has been placed on natural areas than on the farmed countryside (Figgis, 2003). By contrast, in Europe, people have tended to "value the managed landscape that has evolved over the centuries" (Dobbs and Pretty, 2004: 225). The notion of protecting 'multifunctional' farmed landscapes, appreciated for their beauty and heritage attributes, has played an important role in Europe but has little resonance in Australian environmental and rural policies (Bjørkhaug and Richards 2008; Dibden and Cocklin 2009). Nonetheless, it has become increasingly apparent that the environmental values and problems of agricultural areas can no longer be neglected, given that farmers manage around two-thirds of Australia's land area, including landscapes with important conservation values (Davidson et al. 2006: 6); for example, "50 per cent of Victoria's threatened vegetation types are found almost entirely on private land" (Beeton et al. 2006: 43)

From the late 1980s, a number of farmer-led environmental groups sprang up in the state of Victoria, partly with state government encouragement, to tackle issues connected with salinity (Salt Action groups) and tree clearing (Farm Trees and Landcare groups). A joint initiative of the Australian Conservation Foundation and the National Farmers Federation persuaded the Australian government to launch a national Landcare programme in 1989/90. Another potentially significant initiative related to the Murray-Darling Basin (MDB), which covers over one million km^2 of south-eastern Australia including 75% of the country's irrigated land (ABS 2007; Garnaut 2008: 129): the Murray-Darling Basin Agreement, signed in 1992, promised state collaboration in promoting the integrated management of the MDB (although the agreement has signally failed to deliver urgently needed environmental improvements).

Some important international initiatives in the late 1980s also propelled the Australian government in the direction of greater concern both for the natural environment and for broader issues of sustainability. Of particular importance was the report of the United Nation's World Commission on Environment and Development (1987), *Our Common Future*, popularly known as the Brundtland Report. According to Tonts:

> *This report had a particularly important legitimising role within government. With increasing global attention being given to environmental issues, governments could no longer continue to dismiss the problems facing natural environments. Furthermore, by emphasising the connections between ecosystems, economies and societies, the Brundtland Report forced governments to begin to think about the issues in a more integrated and systematic way.*
> (Tonts 2005: 198)

In 1992, the Australian government responded to the Brundtland Report by launching a *National Strategy for Ecologically Sustainable Development* (ESD), with ESD defined as "using, conserving and enhancing the community's resources so that ecological processes, on which life depends, are maintained, and the total quality of life, now and in the future, can be maintained" (Commonwealth of Australia, 1992: 6). This was accompanied by the signing of an agreement which committed all levels of government to the principles of sustainability. In the same remarkably busy year, Australia made a further international commitment, this time to biodiversity conservation, as a signatory of the 1992 International Convention on Biodiversity; it also committed to five-yearly independent state of the environment reporting (Cork et al. 2006; NBSRTG 2009), the most recent being undertaken in 2006 (Beeton et al. 2006).

Commitments to improved natural resource management were backed up by two major Australian government funding programmes. The National Heritage Trust was established in 1996 with the goal of stimulating "activities in the national interest to achieve the conservation, sustainable use and repair of Australia's natural environment" (http://www.nht.gov.au): A$3 billion was allocated for projects that addressed major natural resource management issues. In 2000, a further A$700 million was made available through the National Action Plan on Salinity and Water Quality (NAP) to address the problems of regions in the Murray-Darling Basin. Responsibility for planning and administering environmental programmes was assigned to 56 regional 'community'-based catchment management groups. This model, based on 'civic regionalism' (Lane 2006), has been criticized on the grounds that it devolves responsibility without power (Dibden and Cocklin 2005), and its effectiveness has been questioned. For example, Beeton et al. (2006: 98) have pointed to: "the ephemeral character of groups receiving financial assistance, the short-term benefits of their work, the lack of scientific information combined with local knowledge as it is replaced with citizens' views and experiences, and the lack of a wider strategic direction in their work." They confront the core issue in natural resource management in Australia, attempting to mediate "the tension between conservation and production in agricultural landscapes" (House et al. 2008: 153). Many of the regional boards (e.g. in Victoria) are dominated by farmers, which creates the risk that producer interests will be privileged over the natural environment. They now find themselves facing a new challenge, incorporating responses to climate change into their planning and management responses (Campbell 2008; DCC 2008).

7. Australian policies in relation to the agriculture and the environment

The capacity of the Australian government to respond to environmental issues has been constrained by the important of exports of primary products to the country's economy. This dependence on exports underlies both an interest in encouraging the production of high volumes of commodities as efficiently and cheaply as possible and a commitment to trade liberalisation, which has been promoted actively through the World Trade Organisation (Dibden and Cocklin 2009, Dibden et al. 2009; Pritchard 2005). In particular, Australia has strongly opposed subsidization of agriculture, since this is a battle it cannot win against the giant economies of its competitors in Europe and North America. In support of this position, subsidies to farmers have been reduced to the second lowest amongst OECD countries (OECD 2006: 44), and European support for 'multifunctional' agriculture, including payment of agri-environmental incentives, has been strongly criticised in the WTO. Australian governments have also deregulated agricultural marketing arrangements which had to some extent provided a buffer against low and often wildly fluctuating global commodity prices. As Dibden and Cocklin (2009: 164) argue, this "unwillingness to provide more than token support for farming and for agriculture-dependent areas has created a disjuncture between entrenched political positions … and the reality of a growing environmental and social crisis in the countryside".

In attempting to improve environmental management *without subsidising* landholders, Australian governments at both federal and state levels have relied on a limited range of instruments, primarily information provision, voluntary action and regulation. The federal government's preferred model for promoting better natural resource management practices has been "the participatory, self-help approach epitomized by the National Landcare Programme (Dibden and Cocklin 2005: 138). This programme has sought to develop a national network of voluntary, community-based groups to address local environmental problems with limited financial support and advice from government agencies (Higgins and Lockie 2002; Lockie and Higgins 2007). While Landcare groups have been established in all states (Hodges and Goesch 2006; Nelson et al. 2004), they have never been able to attract more than a minority of landholders and the outcomes have been patchy. Not surprisingly, farmers tend to prefer to undertake environmental work which will improve production, such as planting trees on farm for shade and shelter for livestock or eradicating pest plants and animals. As a result, there is doubt that real ecological benefits have been achieved, particularly when a wider landscape perspective is taken.

In moving to a system of regional governance from 1996, more emphasis was placed on planning, setting targets and monitoring, at least partly as a response to criticisms that the Landcare Programme had failed to evaluate environmental outcomes to ensure that funds were well spent. Although improvements in environmental performance were in the main "to be accomplished through persuasion, education, information and incentives, some coercive power is provided by state-level legislation regulating water use and protecting native vegetation" (Dibden and Cocklin 2005: 140). In practice, much of the effort to conserve biodiversity has focused on preventing further clearing of native vegetation through regulation. Most states introduced legislation to regulate clearing of native vegetation, but "the legacies of past vegetation clearing—including changed hydrology, habitat loss and fragmentation, and impacts on seed supplies and regeneration of native vegetation—remain strong drivers of biodiversity decline" (Cork et al. 2006). Moreover, enforcement of restrictions on clearing has often been ineffective. As Brennan (2004: 127) points out, over the past decade the rate of vegetation clearing has increased, in spite of regulation and policy prescriptions. Only Brazil, Indonesia, the Democratic Republic of the Congo, and Bolivia have cleared more forest than Australia. Efforts to reverse this decline have often been counter-productive: for example, the *State of the Environment 2006* report pointed to "a recent increase in some states in advance of stronger clearing legislation" (Beeton et al. 2006: 70). In states where uncompensated restrictions on clearing both extant vegetation and 'regrowth' have occasioned much bitterness, farmers' groups have pointed to the unfairness of the Australian government claiming credit, and a reduction in emissions targets, in its response to the Kyoto Protocol, on the basis of decreased clearing and hence at the expense of farmers (e.g., Skuthorp 2008). Studies of the effects of clearing restrictions on farmers have revealed that significant costs were borne by landholders while the benefits are experienced by the wider society (Davidson et al. 2006), a finding which prompted the Productivity Commission (2004) to recommend the development of native vegetation policies that provide landholders with positive incentives to retain and manage native vegetation.

This shift in position by the Productivity Commission forms part of a wider recognition that previous approaches have not worked. It has become apparent that voluntary programmes, such as Landcare, do not have the capacity or breadth of vision to overcome environmental problems at a wider 'landscape' scale. Moreover, it has also been recognised that farmers – trapped between the requirements to produce more and yet somehow protect the environment at their own expense – are not able to undertake the work required, particularly at a time when many are struggling as a result of low prices for agricultural produce and drought conditions. This has resulted in a new willingness to consider strategies, supported by government funding, which recognize the public benefits of farmers' environmental 'stewardship' activities (Dibden and Cockin 2009).

Federal government policies on environmental issues have in recent years tried to resolve the conflict between promoting agricultural production and the need to tackle environmental (and social) issues through a shift from reliance on voluntary activities to promotion and funding of market strategies (Dibden and Cocklin 2009). These include providing federal funding to encourage adoption of:

- environmental management systems (EMS), which aim to gain market advantages for landholders practising 'environmental-friendly' farming;
- market-based initiatives, such as 'auction' schemes in which landholders tender to provide ecosystem services. These auction systems have been directed to specific purposes, such as biodiversity conservation or salinity mitigation.

These programmes have been linked to Australia's commitment to ecologically sustainable development (ESD). Thus the National Framework for EMS asserts that:

> *... the principles for ESD developed for Australia apply to EMS. In return, implementation of this Framework and voluntary adoption of EMS will help Australia to deliver on these ESD principles.* (NRMMC 2002: 5)

Similarly, the Victorian state government has identified EMS as a tool for incorporating ESD principles, such as maintenance of biodiversity, into "performance standards" for agriculture (Anderson et al. 2001: 1). Victoria has been particularly active in establishing auction schemes, which primarily serve environmental purposes – such as biodiversity conservation or salinity mitigation (achieved through reafforestation), while one project aims to preserve an endangered species with great symbolic importance, the Red-tailed Black Cockatoo, which was the mascot for the 2006 Commonwealth Games in Melbourne.

8. Climate change policies

The natural environment and agriculture in Australia are both likely to be threatened by climate change and its associated impacts. The agricultural sector in Australia has been judged by the IPCC as "particularly vulnerable to climate changes" (Stokes et al. 2008: 1). The vulnerability of the agricultural sector to climate change has already been demonstrated in the past by the impact of drought – not only on farmers and rural areas but on the Australian economy: for example, as a result of a severe drought in 2002-03, the gross value of agricultural production for that year fell by 19% (to around A$32 billion) which reduced GDP by approximately 1% (ABS 2004; DCC 2008). Agriculture faces an additional challenge if the Australian government introduces a proposed emissions trading scheme, since the sector accounts for a substantial proportion of Australia's greenhouse gas emissions – indeed, it is the second largest source (after energy) with 16.3% of the total in 2007 (DCC 2009: 4). However, as we argue below, agriculture's losses and difficulties could represent a gain for biodiversity.

The many threats to agriculture outlined above (section 5) will require farmers to make a variety of adjustments. A number of studies have discussed two types of responses to climate change (and to any potential emissions trading scheme for agriculture) – adaptation and mitigation (Gunasekera et al. 2007a; Jiang et al., 2009). Adaptation will be required whether an emissions trading scheme is applied to agriculture or not, since there will be no escaping the need to adjust to a changing climate. Mitigation refers to efforts to reduce greenhouse gas emissions, which is most likely to occur in conjunction with an emissions trading scheme. In practice, the two approaches are bound to overlap, with mitigation strategies underpinning adaptation (especially where a carbon price is applied), and many adaptation initiatives simultaneously contributing to mitigation efforts.

8.1 Adaptation strategies

Campbell (2008: 16) considers that it is helpful to conceptualise adaptation – at least for farm sectors and at regional or catchment scales – as "reducing vulnerability and/or increasing resilience." Change will be inevitable in many areas, as rising temperatures and decreases in rainfall make current farming practices untenable. Regions which will be particularly hard hit are those dependent on irrigation, such as the Murray-Darling Basin, and marginal farming areas, where farmers may already be operating beyond the desirable climatic limits for their productive activity. Garnaut (2008: 532) predicts optimistically that "land managers will respond to these dual challenges by pursuing new opportunities in carbon removal (or sequestration), energy production from biomass and low-emissions livestock production." Other forms of adjustment may require farmers to adapt current farming practices, adopt new technologies, switch to other forms of commodity production, combine on- and off-farm income earning activities or seek alternative sources of income outside of agriculture (FFRPG 2009; Gunasekera et al. 2007a).

The extent to which a majority of farmers will be willing to make 'adjustments' is at present uncertain. Stokes et al. (2008: 2) caution that adoption depends both on "confidence that the climate really is changing" and on "the motivation to change to avoid risks or use opportunities". A recent paper by Gray et al. (forthcoming) questions "the capacity of those currently managing Australia's farming and grazing lands to make the appropriate 'adjustments' that will, at one and the same time, deliver increased output (a key goal in terms of export income, and the key to continued economic viability) while protecting the environment – and all in the context of predicted climate changes." They point to a case study by Richards of the grazing industry in Central Queensland during a period of drought and falling prices. Graziers responded to this crisis situation by using more inputs, maintaining stocking rates, and clearing native vegetation, "despite the view of many observers (including some graziers) that this approach was unsustainable, and would lead to considerable environmental degradation" (Gray et al. forthcoming: 14-15; see also Bjørkhaug and Richards 2008). The lack of alternative options may have been equally important for graziers in this remote area.

Studies of farmers in the more closely settled state of Victoria indicate that in some areas adaptations are already occurring. For example, in our research in the Wimmera region of western Victoria, we found that, as the climate became drier, cropping was shifting southwards into former sheep country, which had previously been too subject to water-logging to be suitable for crops. A large, corporate farmer was already following another suggested strategy – spreading risk by growing crops in "multiple holdings in different climatic regions" (Gunasekera et al. 2007a: 500). Some grain growers are hoping to improve their future prospects by experimenting with genetically modified canola, although varieties which are drought tolerant – the most desirable trait for Australia – are not yet available (Glover et al. 2008). In more marginal country, an auction scheme (Catchment Tender), which paid farmers to plant trees for salinity mitigation, was quickly over-subscribed – a result which appears to have been partly a response to drought conditions (Dibden and Cocklin 2009).

8.2 Mitigation strategies

Agriculture has not at present been included in the Australian government's proposed emissions trading scheme (the 'Climate Pollution Reduction Scheme' or CPRS), largely because of the difficulty in measuring or estimating net emissions cost-effectively (Garnaut 2008: 558). However, the inclusion of agriculture is considered highly desirable, since agriculture currently accounts for about 16% of national greenhouse gas emissions (Jiang et al. 2009). Livestock production is particularly damaging, with methane from sheep and cattle comprising 12% of total greenhouse gas emissions – "the third highest single emissions source" (AGO 2005: 6). As a result, agriculture – particularly the livestock sector – would be "the hardest hit industry" if included in the CPRS (Jiang et al., 2009: xiv). Even without being included, farming would incur increased costs for inputs, such as fuel, electricity, farm chemicals and fertilisers. This would cause considerable hardship. As Jiang et al. (2009: xiv) point out: "Because most Australian agricultural products are exported and Australia is a small player in the world market, it is extremely difficult for Australian farmers to pass on the additional costs to their customers and end users."

The exclusion of agriculture form the CPRS is hence a mixed blessing – it means that farmers do not as yet incur charges for emissions from farming operations, yet at the same time does not relieve them of all costs resulting from the scheme and prevents them from taking advantage of potential opportunities. These opportunities include earning credits by reducing emissions from farming activities or for carbon sequestration in trees or – less certainly – in soil (Garnaut 2008; Gunasekera et al. 2007b). Some farmers are already taking action to guard against the risks and make the most of the opportunities which a future emissions trading scheme will bring. For example, the 'Resource Management Group' in the Northern Gulf region of far north Queensland is interested in opportunities to participate in environmental incentive schemes "related to carbon storage and sequestration"; potentially beneficial actions might include reducing carbon emissions through improved fire management in savanna landscapes (Greiner et al. 2009: 57; see also TS-CRC 2009). A similar scheme is already under way in northern Australia, using traditional Indigenous fire management skills (Garnaut Climate Change Review 2008).

Some major problems stand in the way of enabling farmers to participate in emissions trading, relating to difficulties in measuring or estimating net emissions from agriculture, particularly in view of the large number of agricultural establishments (around 130,000) in Australia, with a wide diversity of emission profiles (Gunasekera et al. 2007a: 505). Garnaut (2008:559) has proposed "a large role for collective action among farmers, or private broking functions, to reduce the costs of individual farmers' interaction with an emissions trading scheme." This possibility is already being explored by some potential participants. For example, in Victoria, a review suggested a number of possible options for regional catchment management bodies in relation to emissions trading, including taking on responsibility for measurement and quality assurance or, more ambitiously, "becoming a direct player in the market by 'bulking up' revegetation projects and managing and marketing a carbon pool" (Campbell 2008: 29; see also Campbell 2007). Another proposal has been put forward by Australia's largest dairy co-operative, which recently obtained federal government funding to assist farmers to implement best management practices to reduce and mitigate emissions while maintaining productivity. The outcomes will be measured through the co-operative's existing intranet information management system for farmers, which currently integrates milk production and financial information but will be modified to incorporate tools to measure nutrient loss, nitrogen utilisation, energy use and the effects of greenhouse gas abatement strategies. The accounting tools will be set up in such a way that farmers can feed in information relating to emissions, including bills for energy, fertilisers etc. from third party vendors, and transmit these data to the co-operative. This scheme not only gives the co-op a means of mitigating the substantial emissions from dairy farming but may also, in future, enable it to act as a broker gaining credits for farmer members for their actions in reducing emissions.

8.3 Alternative futures

The future for both farmers and ecosystems will be very different but not necessarily worse than the past. The need for farmers to adjust simultaneously to the biophysical outcomes of climate change, the impacts of an emissions trading scheme and/or the threat of penalties for greenhouse intensive activities could produce positive changes for the environment, for example by requiring shifts in land use and farming practices. Climate mitigation strategies have been proposed which serve a dual purpose, both rewarding or compensating farmers and helping the environment, including biodiversity. Examples include:

- the proposed use of drought assistance to promote improved environmental management (exit of marginal farmers, less intensive land use, areas retired from production) rather than simply propping up farmers who are struggling to survive prolonged drought (Botterill and Fisher 2003; FFRPG 2009);
- carbon credits for tree planting on farmland ('carbon farming') – which could also enhance biodiversity;
- carbon offsets for vegetation management on farms – this "could provide several other private and social benefits, including shelter for livestock and crops, reducing the incidence of salinity and waterlogging and conserving biodiversity" (Gunasekera et al. 2007a: 507);
- especially in marginal areas, new crops (such as oil mallees or other eucalypts) could be grown with multiple benefits – generating income for landholders, sequestering carbon, producing a range of products (essential oils, charcoal, biofuels), and providing ecosystem services (AGO 2005).

Many studies of future outcomes from climate change envisage a dual or multi-track pattern of land use, with land allocated to *either* productivist agriculture *or* multifunctional land uses *or* complete withdrawal from farming uses. A Victorian government group considering the future of farming in the state (FFRPG 2009: 22) came to the view that: "With the majority of the state subject to some form of landscape stress, it is increasingly difficult to keep land in production and, in some cases, alternative futures or interim support may need to be considered if productive uses are no longer viable." The future thus looks grim for farmers in areas no longer considered suitable for intensive farming or indeed for any farming. There may be limited prospects for them to sell ecosystem services, but government schemes have thus far been limited in scope. Agri-environmental schemes similar to those found in Europe have been proposed as a means of combining support for farmers with conservation objectives (e.g., Attwood et al. 2009), but this would depend on a willingness by governments to make substantial financial commitments, and run counter to Australia's present free-trade policy position (Dibden and Cocklin 2009; Dibden et al. 2009). Probably the best prospect of adequately compensating farmers for contributions to the natural environment lies in creation of an active carbon market, with prices far higher than those currently contemplated (Campbell 2008). If the price for carbon was sufficient, this would make land under trees, shrubs, even grasslands, potentially more valuable – and allow carbon storage to be combined with conserving land for biodiversity (Lindenmayer 2007; Mackey et al. 2008).

9. Conclusion

From an environmental perspective, the Australian continent is an ancient land. For most of the last 50,000 years, the native flora and fauna have been able to evolve gradually in the presence of the Indigenous Australians. This is not to suggest that the first humans did not transform the landscape – their use of fire in particular had quite significant environmental consequences. However, it was Australian settler society and especially the introduction of a European style of agriculture over the past 200 years that has really imperilled Australian biodiversity. Agriculture has impacted heavily on Australian biodiversity, driven historically by a 'frontier' ethic and more recently by free-market policies that emphasise production over stewardship. The losses have been massive, made all the more significant by the fact that so many of the Australian species are endemic.

At the same time, the loss and degradation of biodiversity has, in turn, had negative consequences for agriculture; there is a negative feedback loop, in that the destruction of biodiversity is affecting agricultural productivity. Dryland salinity is one of the most graphic examples. Vast tracts of the Australian landscape that have been cleared for agriculture have become useless as a result of soil salinity levels increasing to levels that are toxic to all but the most hardy of plants. Not only has the original fauna and flora been lost from these areas, but they are no longer suitable for production either.

Climate change is imposing new pressures on both agriculture and biodiversity and these impacts will intensify significantly. A majority of farmers now acknowledge that their farm enterprises have been impacted by climate change, a belief underpinned no doubt by the drought that has affected much of the southern agricultural landscapes for more than a decade. At the same time, there are discernible impacts on biodiversity particularly amongst 'indicator species' – e.g., vulnerable higher altitude species in both the temperate and tropical zones. Every indication is that Australian biodiversity will suffer even more, including iconic natural assets such as the Great Barrier Reef.

It is possible to envisage mitigation and adaptation responses that would alleviate pressures on all three systems (climate, agriculture, biodiversity). For example, innovation in public policy tools gives some hope for biodiversity management and climate change mitigation, but the reality is that a much more decisive and aggressive policy platform is required. Despite the rhetoric, it is as yet unclear as to whether Australian governments – federal and state – really have the political will to do what is required to protect the nation's unique flora and fauna from the combined threats of both agriculture and climate change.

10. References

Anderson, S., Lowe, K., Preece, K. and Crouch, A. (2001) *Incorporating Biodiversity into Environmental Management Systems for Victorian Agriculture: A discussion paper on developing a methodology for linking performance standards and management systems.* Melbourne: Parks, Flora and Fauna Division, Department of Natural Resources and Environment, State of Victoria.

Attwood, S.J., Park, E., Maron, M., Collard, S.J., Robinson, D., Reardon-Smith, K.M. and Cockfield, G. (2009) Declining birds in Australian agricultural landscapes may benefit from aspects of the European agri-environment model, *Biological Conservation*, 142(10): 1981-91.

Australian Bureau of Statistics [ABS] (2002) *Salinity on Australian Farms 2002.* Canberra: ABS.

--- (2004) Impact of the drought on Australian production in 2002-03, *Year Book Australia, 2004* feature article. Available on: www.abs.gov.au/ausstats/abs@.nsf/mf/1301.0

--- (2007) *Year Book Australia, 2007*, cat. no. 1301.0, Canberra: ABS.

--- (2008) *Farm Management and Climate 2006-07*. Available on: www.abs.gov.au

--- (2009a) *Measures of Australia's Progress.* Canberra: Commonwealth of Australia.

--- (2009b) *Land Management and Farming in Australia 2007-08.* Canberra: Commonwealth of Australia.

Australian Greenhouse Office [AGO] (2005) *Climate Change in Rural and Regional Australia.* Canberra: AGO, Department of the Environment and Heritage.

--- (2007) *Australia's Agriculture – Impacts of Climate Change.* Canberra: Australian Government Department of the Environment and Heritage. Available on: www.greenhouse.gov.au/impacts/agriculture.html

Australian State of the Environment Committee (2001) *Australia State of the Environment 2001.* Independent Report to the Commonwealth Minister for the Environment and Heritage. Canberra: CSIRO Publishing.

Barr, N. and Cary, J. (1992) *Greening a Brown Land: The Search for Sustainable Land Use,* South Melbourne: Macmillan Education.

Beeton R.J.S., Buckley, K.I., Jones, G.J., Morgan, D., Reichelt, R.E, and Trewin, D. (2006) *Australia State of the Environment 2006.* Independent report to the Australian Government Minister for the Environment and Heritage from the 2006 Australian State of the Environment Committee.

Biodiversity Decline Working Group [BDWG] (2005) *A National Approach to Biodiversity Decline. Report to the Natural Resource Management Ministerial Council.* Accessed 16.6.09 on: http://www.environment.gov.au/biodiversity/publications/biodiversity-decline.html

Bjørkhaug, H. and Richards, C.A. (2008) Multifunctional agriculture in policy and practice? A comparative analysis of Norway and Australia, *Journal of Rural Studies,* 24(1): 98-111.

Bolton, G (1981) *Spoils and Spoilers: Australians Make Their Environment, 1788-1980.* Sydney: George Allen and Unwin.

Botterill L. and Fisher, M. (eds., 2003) *Beyond Drought in Australia: People, Policy and Perspectives,* Collingwood, Victoria: CSIRO Publishing.

Bowman, D. (2000). *Australian Rainforests: Islands of Green in the Land of Fire,* Cambridge: Cambridge University Press.

Brennan, A. (2004) Biodiversity and agricultural landscapes', *Pacific Conservation Biology,* 10: 124-142.

Campbell, A. (2007) *Options for Catchment Management Authorities in the Carbon Market.* Report for Victorian Catchment Management Authorities. Queanbeyan: Triple Helix Consulting.

Campbell, A. (2008) *Managing Australian Landscapes in a Changing Climate: A Climate Change Primer for Regional Natural Resource Management Bodies*, Canberra: Department of Climate Change.

Chapman, A.D. (2006) *Numbers of Living Species in Australia and the World*, Report for the Department of the Environment and Heritage, Canberra, September 2005. Canberra: Department of the Environment and Heritage.

Cocklin, C (1993) What does cumulative environmental effects assessment have to do with sustainable development? *Canadian Journal of Regional Science* 16(3): 453-479.

Cocklin, C. (2005) Natural capital and the sustainability of rural communities. In C. Cocklin and J. Dibden (eds.), *Sustainability and Change in Rural Australia*, Sydney: UNSW Press, 171-191.

Commonwealth of Australia (1992) *National Strategy for Ecologically Sustainable Development*, Canberra: AGPS.

Commonwealth Scientific and Industrial Research Organisation [CSIRO] and Bureau of Meteorology [BOM] (2007) *Climate Change in Australia Technical Report*, Canberra: CSIRO.

Conacher, A. and Conacher, J. (2000) *Environmental Planning and Management in Australia*, Oxford University Press, Melbourne.

Cork, S., Sattler, P. and Alexandra, J. (2006) *SOE 2006 Biodiversity Theme Commentary*, prepared for the 2006 Australian State of the Environment Committee. Accessed 16.6.09 on: http://www.environment.gov.au/soe/2006/publications/commentaries/biodiversity/executive-summary.html

Davidson, A., Beare, S., Gooday, P., Kokic, P., Lawson, K. and Elliston, L. (2006) *Native Vegetation: Public Conservation on Private Land – Cost of Foregone Rangelands Development in Southern and Western Queensland*, ABARE Research Report 06.13, Canberra: ABARE.

Department of Climate Change [DCC] (2008) *Australia's Agriculture - Impacts of Climate Change*, Australian Government. Accessed 1.7.09 on: http://www.climatechange.gov.au/impacts/agriculture.html

--- (2009) *National Greenhouse Gas Inventory 2007*, Canberra: DCC.

Dibden, J. and Cocklin, C. (2005) Sustainability and agri-environmental governance. In: V. Higgins and G. Lawrence (eds.), *Agricultural Governance: Globalization and the New Politics of Regulation*, London: Routledge, pp. 136-152.

Dibden, J. and Cocklin, C. (2009). 'Multifunctionality': Trade protectionism or a new way forward? *Environment and Planning A*, 41(1): 163- 182.

Dibden, J., Potter, C. and Cocklin, C. (2009) Contesting the Neoliberal Project for Agriculture: Productivist and multifunctional trajectories in the European Union and Australia, *Journal of Rural Studies*, 25(3): 299-308.

Dobbs, T. and Pretty, J., 2004. Agri-environmental stewardship schemes and multifunctionality. *Review of Agricultural Economics*, 26: 221-237.

Dovers, S. (1992) The history of natural resource use in rural Australia: Practicalities and ideologies, In: G. Lawrence, F. Vanclay and B. Furze (eds.) *Agriculture, Environment and Society: Contemporary Issues in Australia*, Melbourne: Macmillan, 1-18.

Figgis, P. (2003) The changing face of nature conservation: reflections on the Australian experience. In: W. Adams and M. Mulligan (eds.), *Decolonizing Nature: Strategies for Conservation in a Post-colonial Era*, London: Earthscan, 197-219.

Future Farming Rural Planning Group [FFRPG] (2009) *Future Farms: Providing for Victoria's future rural land use,* East Melbourne: Dept of Planning and Community Development. Accessed 4.6.09 on: http://www.dse.vic.gov.au/CA256F310024B628/0/EA669A1013245490CA2575AE000F2697/$File/Future.Farms.Planning.Group.Discussion.Paper.pdf

Garnaut, R. (2008) *Climate Change Review: Final Report.* Accessed 1.7.09 on: http://www.garnautreview.org.au

Garnaut Climate Change Review (2008) Abating greenhouse gas emissions through strategic management of savanna fires - Northern Territory, *Climate Change Review: Case Study.* Accessed 1.7.09 on: http://www.garnautreview.org.au/CA25734E0016A131/pages/all-reports--resources

Glover, J., Johnson, H., Lizzio, J., Wesley, V., Hattersley, P. and Knight, C. (2008) *Australia's crops and pastures in a changing climate - can biotechnology help?* Canberra: Australian Government Bureau of Rural Sciences.

Gunasekera, D., Ford, M. and Tulloh, C. (2007a) Climate change: Issues and challenges for Australian agriculture, *Australian commodities,* 14(3): 493-515.

Gunasekera, D., Kim, Y., Tulloh, C. and Ford, M. (2007b) Climate change: Impacts on Australian agriculture, *Australian commodities,* 14(4): 657-673.

Gray, I. and Lawrence, G. (2001) *A Future for Regional Australia: Escaping Global Misfortune,* Cambridge: Cambridge University Press.

Gray, I., Lawrence, G. and Sinclair, P. (forthcoming) The sociology of climate change for regional Australia: Considering farmer capacity for change. In: J. Martin (ed.) *Climate Change Responses across Regional Australia: Social Learning and Adaptation.*

Gray, I., Phillips, E. and Dunn, A. (2000) Aspects of rural culture and the use of conservation farming. In: A. D. Shulman and R. J. Price (eds.) *Case studies in increasing the adoption of sustainable resource management practices,* Canberra: LWRRDC, 34-47.

Greiner, R., Gordon, and Cocklin, C. (2009). Ecosystem services from tropical savannas: economic opportunities through payments for environmental services. *The Rangeland Journal* 31, 51–59.

Haw, M., Cocklin, C. and Mercer, D. (2000) A pinch of salt: Landowner perception and adjustment to the salinity hazard in Victoria, Australia, *Journal of Rural Studies,* 16: 155-169.

Higgins, V. and Lockie, S. (2002) Re-discovering the social: neoliberalism and hybrid practices of governing in rural natural resource management. *Journal of Rural Studies,* 18: 419-428.

Hodges, A. and Goesch, T. (2006) Natural resource management: results for 2004-05 from an ABARE survey of Australian farmers. *Australian Commodities* 13 (3): 569-573.

Hughes, L. (2003) Climate change and Australia: Trends, projections and impacts, *Austral Ecology,* 28: 423-443.

Jiang, T., K. Hanslow, K. and Pearce, D. (2009). *On farm impacts of an Australian ETS: Economic analysis.* RIRDC Publication No. 09/064, Canberra: RIRDC.

Johnson, C. (2006) *Australia's Mammal Extinctions*, Cambridge: Cambridge University Press.

Johnson, J.E. and Marshall, P.A. (2007) (eds.) *Climate Change and the Great Barrier Reef: A Vulnerability Assessment*, Canberra: Great Barrier Reef Marine Park Authority and the Australian Greenhouse Office.

Krockenberger, A Kitching R and Turton, S (2003) Environmental Crisis: Climate Change and Terrestrial Biodiversity in Queensland. Cooperative Research Centre for Tropical Rainforest Ecology and Management, Cairns.

Land and Water Resources Research and Development Corporation [LWRRDC] (1999) *Issues in natural resource management*, LWRRDC, Canberra. Available: <http://www.lwrrdc.gov.au>.

Lane, M. (2006) Critical issues in regional natural resource management. Paper prepared for the 2006 Australian State of the Environment Committee, Department of the Environment and Heritage, Canberra: <http://www.environment.gov.au/soe/2006/publications/integrative/nrm-issues/index.html>.

Lindenmayer, David (2007) *On borrowed time: Australia's environmental crisis and what we must do about it*. Camberwell, Vic.: Penguin Books in association with CSIRO Publishing.

Lockie, S. and Higgins, V. (2007) Roll-out neoliberalism and hybrid practices of regulation in Australian agri-environmental governance, *Journal of Rural Studies* 23(1): 1-11.

Mackey, B.G., Lesslie, R.G., Lindenmayer, D.B., Nix, H.A. and Incoll, R.D. (1998) *The Role of Wilderness in Nature Conservation*, Canberra: School of Resource Management and Environmental Science, ANU.

Mackey, B.G., Watson, J.E.M., Hope, G. and Gilmore, S. (2008) Climate change, biodiversity conservation, and the role of protected areas: An Australian perspective, *Biodiversity*, 9(3 & 4): 11-18.

Morton, S.R., Hoegh-Guldberg, O., Lindenmayer, D.B., et al. (2009) The big ecological questions inhibiting effective environmental management in Australia, *Austral Ecology*, 34: 1-9.

Mosley, G. (1978) A history of the wilderness reserve idea in Australia, in Mosley, G. (ed.) *Australian Wilderness: Conservation Progress and Plans*, Australian Conservation Foundation.

National Biodiversity Strategy Review Task Group [NBSRTG] (2009) Australia's Biodiversity Conservation Strategy 2010-2020 - Consultation draft. Prepared by the National Biodiversity Strategy Review Task Group. Convened under the Natural Resource Management Ministerial Council. Accessed 16.6.09 on:
http://www.environment.gov.au/biodiversity/strategy/draft-strategy.html

National Land and Water Resources Audit [NLWRA] (2001a) *Australian Native Vegetation Assessment 2001*, Canberra: Commonwealth of Australia.

--- (2001b) *Australian Dryland Salinity Assessment 2000*, Canberra: Commonwealth of Australia.

Natural Resource Management Ministerial Council [NRMMC] (2002) *Australia's National Framework for Environmental Management Systems in Agriculture*, Canberra: Australian Government.

Nelson, R., Alexander, F., Elliston, L. and Blias, A. (2004) *Natural Resource Management on Australian Farms*, ABARE eReport 04.7. Canberra: Australian Government.

O'Brien, P. (1990) Managing Australian wildlife, *Search*, 21: 24-27.

Office for Water Security (2009) *Water for Good: A plan to ensure our water future to 2050.* Adelaide: Government of South Australia. Available on: http://www.waterforgood.sa.gov.au/wp-content/uploads/2009/06/complete-water-for-good-plan.pdf

Odum, W. (1982) Environmental degradation and the tyranny of small decisions, *BioScience* 2(9): 728-729.

Organisation for Economic Co-operation and Development [OECD] (2006) *Agricultural Policies in OECD Countries at a Glance 2006*, Paris: Organisation for Economic Co-operation and Development.

Pritchard B. (2005) "Implementing and maintaining neoliberal agriculture in Australia: Strategies for securing neoliberalism" *International Journal of the Sociology of Agriculture,* 13: 1-14

Productivity Commission (2004) *Impacts of Native Vegetation and Biodiversity Regulations: Productivity Commission Inquiry Report*. Accessed 10.11.05 on: http//www.pc.gov.au/inquiry/nativevegetation/finalreport/nativevegetation.pdf

Raven, P. H., and Yeates, D. K. (2007). Australian biodiversity: threats for the present, opportunities for the future. *Australian Journal of Entomology*, 46(3), 177-187.

Skuthorp, L. (2008) Spencer takes carbon credit case to WA farmers. *The Land*, 6 March.

Soulé , M.E., Mackey, B.G. et al. (2004) The role of connectivity in Australian conservation, *Pacific Conservation Biology,* 10: 266-79.

Spaling, H and Smit, B (1995) A conceptual model of cumulative environmental effects of agricultural land drainage. *Agriculture, Ecosystems and Environment* 53(2): 99-108.

Tropical Savannas CRC [TS-CRC] (2009) The West Arnhem Land Fire Abatement Project (WALFA). Accessed 15.1.09 on: http://savanna.cdu.edu.au/information/arnhem_fire_project.html

Tonts, M. (2005) Government policy and rural sustainability. In C. Cocklin and J. Dibden (eds.), *Sustainability and Change in Rural Australia*. Sydney: UNSW Press, 194-211.

Stokes, C.J., Howden, S.M., Gifford, R.G., et al. (2008). *An overview of climate change adaptation in Australian primary industries – impacts, options and priorities*. Canberra: CSIRO.

Vanclay, F. and Lawrence, G. (1995) *The Environmental Imperative: Ecosocial Concerns for Australian Agriculture*. Rockhampton: Central Queensland University Press,.

Williams, S.E. and Hilbert, D. (2006) Climate change threats to the biodiversity of tropical rainforests in Australia. In Laurance, W.F. and Peres, C. (eds.) *Emerging Threats to Tropical Forests*. Chicago: Chicago University Press, 33-52.

Williams S.E., Isaac, J.L. and Shoo, L.P. (2008) The impact of climate change on the biodiversity and ecosystem functions of the Wet Tropics. In: N. Stork and S. Turton (eds.), *Living in a Dynamic Tropical Forest Landscape*, Oxford: Wiley-Blackwell Publishing,.

World Commission on Environment and Development (1987) *Our Common Future* (the Brundtland Report), New York: Oxford University Press.

Yencken, D. and Wilkinson, D. (2000) *Resetting the Compass: Australia's Journey towards Sustainability*, Melbourne: CSIRO Publishing.

Climate change sensitivity of the African ivory nut palm, *Hyphaene petersiana* Klotzsch ex Mart. (Arecaceae) - a key-stone species in SE Africa

A Blach-Overgaard[1], J-C Svenning and H Balslev

Ecoinformatics and Biodiversity group, Department of Biological Sciences, Aarhus University, Ny Munkegade 114, DK-8000 Aarhus C, Denmark

Email: anne.overgaard@biology.au.dk

Abstract. Africa is the most vulnerable continent to future climate change. Profound changes are projected for southwestern Africa with increased drying, notably with delayed onset of the rainy season in September-November, and temperature increases in all seasons. The projected climate changes combined with land-use changes are thought to constitute the main threats to biodiversity in the 21st century. To be able to predict the potential impact on biodiversity, it is crucial to achieve a better insight into the controls of contemporary species ranges. Using species distribution modeling, we assessed the climate sensitivity of the key-stone palm species *Hyphaene petersiana* (African ivory nut palm) in southern Africa. We tested the relative roles of climate vs. non-climatic range-controls and found that climate had a clear effect on the range of *H. petersiana* and that especially water-related variables (annual precipitation and precipitation driest quarter) were of high importance. Nevertheless, latitude was the overall most dominant variable, reflecting spatial constraints on the continental-scale distribution. Of the remaining non-climatic factors, soil type and human influence were as important as the climatic factors. A future decrease in annual precipitation below 400 mm and hydrological changes towards drier conditions could cause a dramatic decline in *H. petersiana* populations, while the influence of temperature changes is less clear. The ongoing, unsustainable utilization pressures on this palm species by humans and livestock are likely to exacerbate the negative effect of future climate changes on its populations, especially, given the expected human population increase in Africa.

1. Introduction

The global environment is rapidly changing, and the fourth report of the Intergovernmental Panel on Climate Change (IPCC) has established that observed climate changes are likely induced by anthropogenic activities, notably increases in anthropogenic greenhouse gas concentrations [1]. Over the past 50 years, the global average temperature has increased more over land areas than over the oceans. Concurrently, changes in precipitation patterns has occurred between 1900 and the present with increases in precipitation in some parts of the World, but drying in the Mediterranean region, parts of southern Asia, and in the Sahel and southern Africa [1]. Drought events have affected ecosystems and people since the end of the 1960s in Africa [2], and the African continent has experienced gradual warming in the past century with larger warming in June-August (JJA) and

[1] To whom any correspondence should be addressed.

September-November (SON) [3]. It has been stated that Africa is one of the most vulnerable continents to projected future climate changes due to lack of adaptive ability and interactions of additional confounding challenges [2]. Poverty, infrastructural, and technological challenges combined with disasters, political conflicts, and degradation of ecosystem functioning constitute key problems in Africa, exacerbating future climate-linked stresses and straining on the continent's adaptability, and response to future climate change [2]. It has been estimated from 21 multi-model projections of IPCC's A1B emission scenario that regions across Africa will likely experience median temperature increases between 3-4°C in all seasons (*e.g.*, December-February; March-May) at the end of the current century – an increase which is higher than the projected global mean [4]. However, the projected temperature changes depend on the different emission scenarios (of which only the median of the A1B emission scenario is provided here), and regional differences, notably for the latter part of the century where the differences in projections between the scenarios become more marked [1]. For example, up to 9°C temperature change has been projected for North Africa and 7°C change for southern Africa by the A1F1 emission scenario in certain seasons at the end of the current century [2]. In general, the drier subtropical regions will likely experience more warming than the moister tropics [4]. The most robust findings of current projections of the 21st century precipitation patterns are drying in the Mediterranean region and southern Africa, while eastern Africa is likely to experience increases in rainfall. The predicted southern African drying is more pronounced in the southwest in JJA (winter), which constitutes the dry season in most of the region, except for the winter-rainfall regions in the southwest. However, the contribution of winter drying to the overall annual mean drying is less than that of the reduction projected for SON (spring). Less precipitation in spring can be regarded as a delay of the rainy season and contribute to the maximum temperature response due to suppressed evapotranspiration [4]. Several previous studies have already addressed the potential consequences of the future climate changes on biodiversity in Africa (summarized in [5]): predicting that ~5000 African plant species will experience losses of climatically suitable area by 2085 [6]; while up to 25-40% of African mammals will potentially be critically endangered by 2080 due to loss of climatically suitable area [7]. Climate change is not the only foreseeable, future challenge for the biodiversity: land-use change will probably continue its increasing pressure on biodiversity in the 21st century [8].

In order to assess the potential impact of future climate and land-use changes on the biodiversity, it is critical to determine what mechanisms control the contemporary ranges of species [9]. Although a key issue, the scientific understanding of species ranges and their determinants is still limited [10, 11]. Here, we assessed the controls of the continental-scale distribution of the African ivory nut palm (*Hyphaene petersiana*) in order to estimate the climate sensitivity of this key-stone species. Palms constitute an important component of tropical and subtropical ecosystems around the World, with only a few species occurring in warm-temperate environments [12, 13]. Palms are sensitive to climatic gradients, and they are therefore good objects for studies that attempt to achieve insights into the 21st century prospects for tropical and subtropical organisms [14]. Recently, it has been found that climate, notably water availability, is an important control of the distribution and diversity patterns of Neotropical palms [14-16]. Hence, palms and the associated ecosystem services are likely to be sensitive to climate changes. Palms have been an important resource for human societies for >10,000 years [17], and cultivation of palms has taken place as early as 4000 years BC [18]. Today, cultivated as well as numerous wild palm species are still an integral element of the livelihoods of many local communities [19, 20] where they are utilized for numerous products and applications, *e.g.*, the leaves and petioles for building materials and basketry, and the fruits, sap, oil, and palm hearts for consumption [13, 21], with palms constituting an important nutritional source in many communities [22]. Besides their key role in the daily lives of local communities, cultivated palms provide widely used commodities and are important for national economies, notably the African oil palm, *Elaeis guineensis* and the coconut palm, *Cocos nucifera* [13]. The diversity of palms across the World is high: most palm species are found in Asia, followed by the New World; however, the palm diversity in Africa is comparatively low with only 65 recognized species [23].

Tomlinson [24] divided the African palms into two broad ecological groups: 1) species that tolerate dry and exposed habitats, and which are never found in dense forests; and 2) forest palms, which are intolerant of drought and light exposure. However, there are palms associated with intermediate habitats [12]. The genus *Hyphaene,* from Tomlinson's first group, is found in a variety of savanna types from the arid Sahel/Sudano savannas (*H. thebaica*) to the coastal savannas *(H. coriacea).* Restricted to the eastern and southern parts of Africa, *H. petersiana* (figure 1) is found primarily inland away from the coast in areas with high water tables, or in streamside riparian vegetation (figure 2) [12]. In Angola, Botswana, and Zimbabwe, *H. petersiana* occurs in the grasslands with high water tables, as well as, on the delta islands of the Okavango Delta, Botswana [25]. In southern Africa, *H. petersiana* is utilized in many ways by rural communities, and also by livestock and wildlife (table 1). Depletion of *H. petersiana* populations have already been observed [25-28] mainly due to over-utilization and bush fires [12, 27], destructions by elephants, and potentially due to lower rainfall and higher temperatures, as observed by local communities dependent on *H. petersiana* for daily needs [27].

Here, we used species distribution modeling (SDM) to examine the climatic and non-climatic limitations on the distribution of *H. petersiana.* We specifically wanted to 1) examine if climatic factors constitute important constraints on the distribution of *H. petersiana*, and 2) assess the relative role of climatic and non-climatic factors (soil type, human influence and spatial constraints) as range constraints.

Table 1. List of *Hyphaene petersiana* uses in southern Africa.

Basket–weaving: the leaves are used for creating baskets, the baskets being constructed using palm leaf pinnae, often from unopened young leaves [25, 27-29].
Timber/construction material: in many areas the thick petioles collected from tall palm individuals are used for building palisades and in other areas the felled stems are used for fencing posts or water troughs [28-30].
Human consumption: palm hearts (edible terminal shoots) and edible fruits are harvested for consumption. The fruits are further utilized for making palm liquor through fermentation, and the sap is used for palm wine [28-30].
Fodder for livestock: the palm is often browsed by domesticated livestock (cattle, goats and donkeys) [25, 28, 30].
Utilization by wildlife: especially elephants and baboons consume fruits of *H. petersiana,* and also act as seed dispersal agents [29, 31]; however, it has been noted that also hippopotami browse this palm species [25].

2. Methods

2.1 *Palm data and predictor variables*

The localities (n = 40) for *H. petersiana* came from herbarium collections, provided through the Royal Botanic Gardens, Kew, the Nationaal Herbarium Nederland (NHN) and Botanic Garden and Botanical Museum Berlin-Dahlem (the latter accessed through the Global Biodiversity Information Facility (GBIF) data portal (www.gbif.org)); literature surveys, and through Google Earth. *Hyphaene petersiana* is confined to eastern and southern Africa in hot, dry sub-tropical climate, normally found in areas with high water tables, or in riparian vegetation (figure 1, 2). It possesses very distinct grayish-green foliage, and it is a tall erect palm with fan-shaped leaves [12], thus identifiable on the satellite images provided through Google Earth.

We used the following four climatic variables obtained from the Worldclim dataset [32] to represent potential range controls for *H. petersiana*: 1) annual mean temperature (AMT), 2) annual precipitation (PREC), 3) precipitation seasonality (PSEA), computed as the coefficient of variation (SD/mean) of monthly precipitation averages, and 4) precipitation of driest quarter (PDRY). We also used the following three non-climatic variables to represent habitat characteristics, and one spatial variable to account for major spatial constraints on the distribution of *H. petersiana*: 1) soil type (SOIL), obtained from the Harmonized World Soil Database (HWSD), which for Africa is derived from the regional SOTER (soil and terrain) studies and the Digital Soil Map of The World (DSMW) at 30'' [33]; 2) human influence index (HMNINFL), which is an index of values representing low (0) to high (72) human influence, estimated for each grid cell based on several proxy data layers for human influence, such as population density, land transformation, accessibility from roads, rivers and coastlines, and electric power infrastructure [34]; 3) closeness to rivers (RIVERS), which is a categorical data layer with two classes: > 2 km (0) or ≤ 2 km (1) from major rivers or water bodies, computed in ArcGIS 9.2 (ESRI, Redlands, CA, USA) from the SRTM River-Surface Water Bodies data layer [35, 36]; and 4) latitude (LAT) was included to represent a spatial constraint on the distribution of *H. petersiana,* as it occurs only in southern parts of Africa. LAT was computed in ArcGIS 9.2 from a Worldclim data layer using the sampling tool in Spatial Analyst. The resulting table was converted to a point shapefile, and subsequently converted to raster using the Conversion tool using Y (latitude) as priority field. All raster data layers were reprojected to the Lambert Azimuthal equal area projection and resampled to 1-km grid size using the nearest neighbor or bilinear resampling techniques for categorical and continuous variables, respectively, using ArcGIS 9.2.

Figure 1. *Hyphaene petersiana* in dry savanna habitat in northern Namibia. Note the distinct grayish-green, fan-shaped leaves, which make this species easy to identify even on satellite images (photo by Marco Schmidt).

Figure 2. Typical savanna habitat for *Hyphaene petersiana* in close proximity to water courses, Okavango Delta, Botswana (photo by Teo Gómez).

2.2 *Predictive modeling and model evaluation*
A state-of-the-art SDM algorithm, Maxent [37], was implemented to model the distribution of *H. petersiana*. Maxent was chosen among a range of SDM techniques given its rank among the best performing predictive algorithms in recent comparative methodological studies [38-41]. To address the study questions, we set up four nested models of increasing complexity. The CL model consisted of the four climatic variables and LAT, in the CLS model, SOIL was added, in the CLSH model, HMNINFL was added, while finally in the CLSHR model, consisting of all eight predictors, RIVERS was added. The default settings for Maxent were used to allow feature classes (linear, quadratic, product, threshold and hinge functions of the variables) to be selected based on the number of point localities as this has been shown to provide good predictive performance over a range of datasets [42].

To assess the relative role of the variables in the study, we examined the regularized training gain attained by the variables when all variables of the particular model were included in the Maxent runs. In addition, the climatic variables were subsequently used in isolation in a model using all point localities to estimate their overall contribution to the prediction for *H. petersiana*, thus avoiding any interactions among the remaining predictor variables. Model performance was assessed by dividing the presence localities into random training (80%) and test (20%) datasets, and subsequently using the threshold independent assessment metric, the Area Under the receiver operating Curve (AUC) for model evaluation [43]. The models were calibrated on the training dataset and tested using the remaining 20% test data points.

3. Results

The Maxent modeling of the four nested models all provided good model predictions with AUC values ≥ 0.947 (figure 3). They all managed to predict the species' eastern and southern range (figure

3a) and its absence from the extreme south and northwest of Africa (figure 3b-d). Based on the AUC values, the model with superior predictive ability was the CLSH model (figure 3d), followed by the CLSHR model (figure 3e) and the CL model (figure 3b). PREC and PDRY were the two most influential climatic variables in the CL model while the effects of AMT and PSEA were much smaller (table 2). The more complex models provided similar results providing evidence of the strong precipitation-related climatic effects on the distribution of *H. petersiana* regardless of whether additional habitat variables were included or not (table 2).

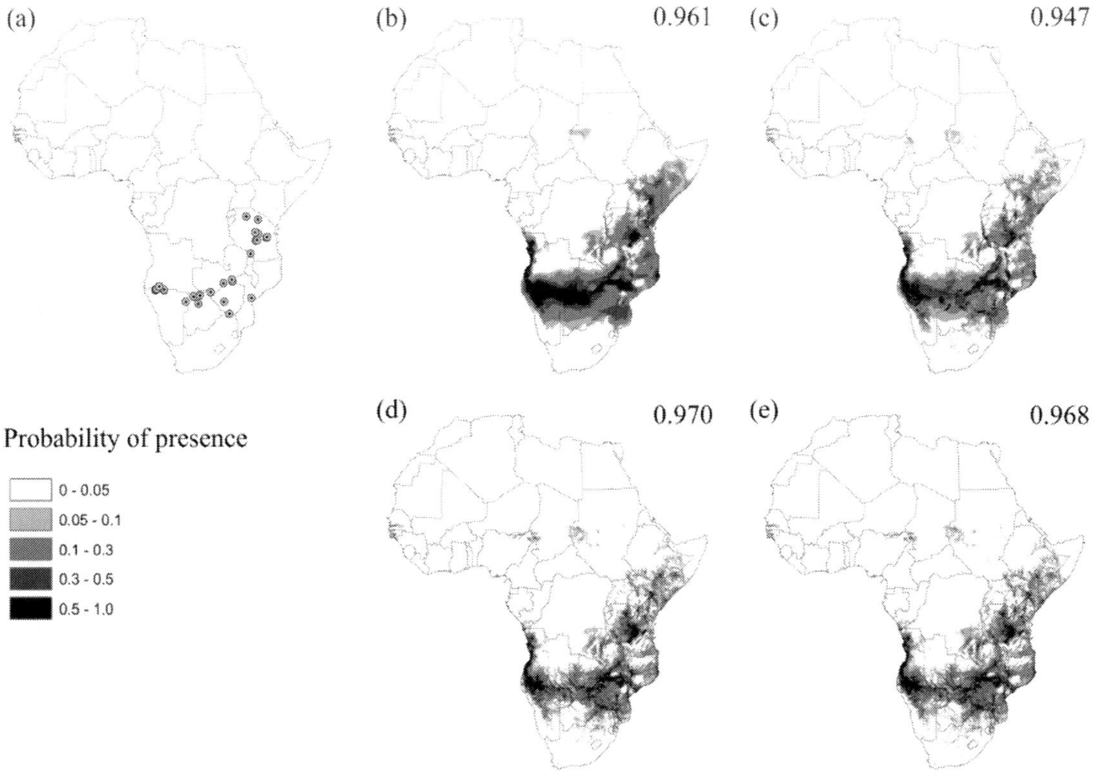

Figure 3. (a) The presence localities (n = 40), and (b-e) predicted distributions for *H. petersiana* based on Maxent models fitted to all presence localities using the (b) CL model, (c) CLS model, (d) CLSH model, and (e) CLSHR model. The AUC values are given in the upper right-hand corner of each predicted distribution map. The lowest legend interval is set to approximately correspond to the minimum training presence threshold (the lowest predicted probability corresponding with a presence record used in the model) across models.

Considering the response curves to the four climatic variables (figure 4), *H. petersiana* has a strong response to annual precipitation occurring primarily in areas with approximately 400 mm annual precipitation, and the likelihood of occurrence declining rapidly below this threshold, and more gradually above (figure 4a). In addition, it primarily occurs where there is pronounced rainfall seasonality (figure 4c), and 1 or more months with very little precipitation (figure 4b). With respect to temperature, the likelihood of *H. petersiana* occurrence peaks sharply between 22-25°C (figure 4d).

Regarding the non-climatic range-controls, LAT was the single most contributing predictor variable for the occurrence of *H. petersiana*, accounting for 43.3-59.4% of the model gain (table 2).

This reflects that the continental range of *H. petersiana* is limited by geographic constraints, with the species missing from climatically suitable areas outside southern and eastern Africa. The CLS, CLSH, and CLSHR models all had SOIL being the second-most important predictor (table 2). *Hyphaene petersiana* had a high probability of presence on six of the 31 soil types, namely, Fluvisols, Leptosols, Luvisols, Planosols, Solonetz, and soils in areas of inland waters [33]. HMNINFL was of similar importance as PREC and PDRY when included (CLSH and CLSHR models) (table 2). Overall the model response to HMNINFL was positive *i.e.*, the stronger human influence, the higher probability of finding *H. petersiana*. The influence of RIVERS was small (table 2), however, there was an elevated probability of presence ≤ 2km to rivers or other water bodies.

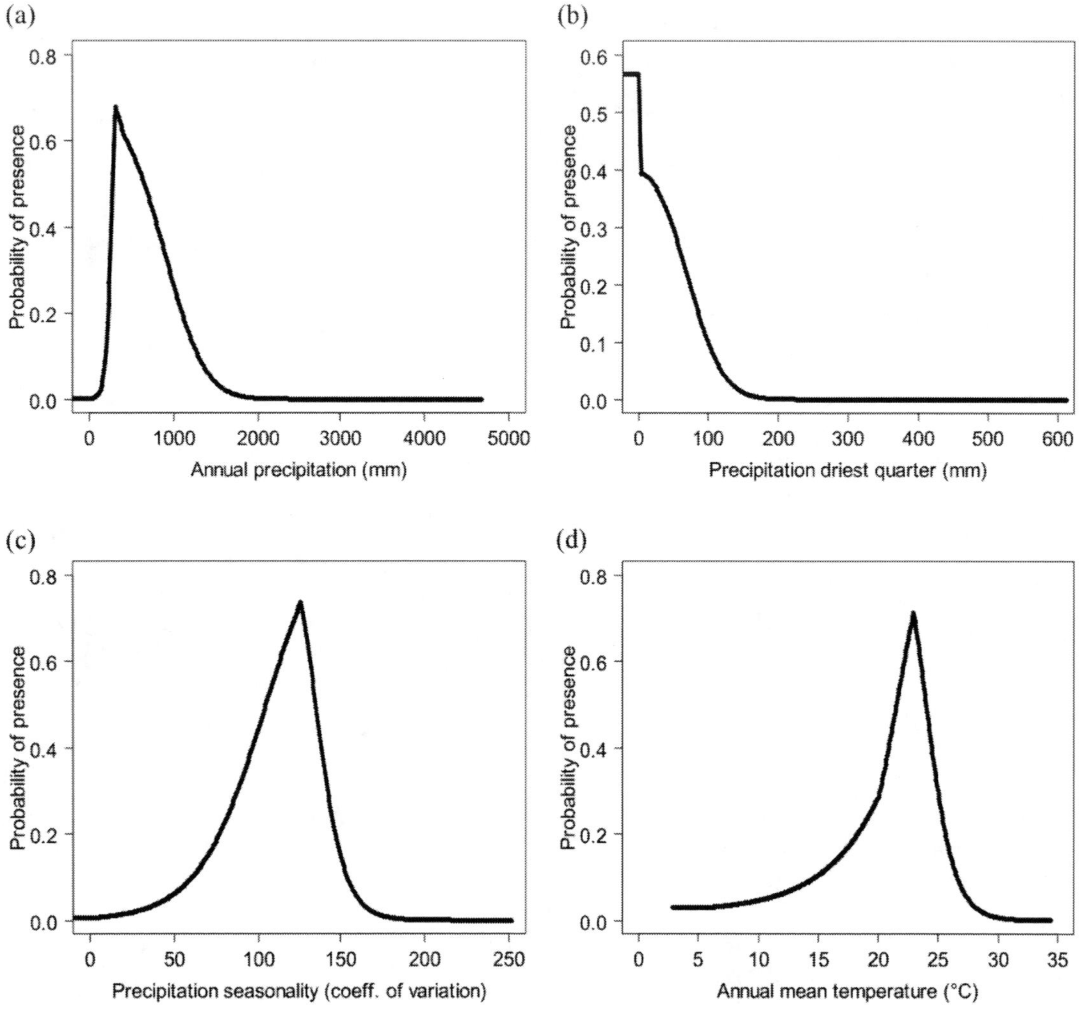

Figure 4. *Hyphaene petersiana* response curves to the climatic predictors. The response curves were estimated by Maxent modeling based on all presence localities using the particular climatic variable in isolation. Precipitation seasonality represents the coefficient of variation (SD/mean) of the monthly precipitation averages.

4. Discussion

Our analyses showed that the continental-scale distribution of *H. petersiana* is controlled partly by climate and partly by non-climatic factors, in particular geographical constraints, but also habitat characteristics such as soil type and human influence. Considering the climatic variables, it was clear that *H. petersiana* is very dependent on water-related factors (figure 4, table 2). The response to precipitation below 400 mm annually was marked, with the probability of presence decreasing drastically below this threshold, as well as, more gradually above 400 mm (figure 4a). In terms of temperature, it was similarly clear that above 23°C, the probability of *H. petersiana* presence decreased rapidly (figure 4d). The study also provided evidence that the distribution of *H. petersiana* is strongly influenced by non-climatic factors as well (table 2).

Table 2. Relative contribution of each predictor to each of the four Maxent models, as measured by its contribution to the final regularized training gain for a given model.

Variable	Model Acronym	CL[a] (%)	CLS[b] (%)	CLSH[c] (%)	CLSHR[d] (%)
Annual mean temperature	AMT	4.7	3.4	2.2	2.0
Annual precipitation	PREC	18.4	12.2	11.2	11.4
Precipitation seasonality	PSEA	1.8	0.3	0.2	0.1
Precipitation driest quarter	PDRY	15.7	12.7	12.9	13.0
Latitude	LAT	59.4	51.6	45.0	43.3
Soil type	SOIL		19.7	16.6	15.9
Human influence index	HMNINFL			12.0	11.2
Closeness to rivers	RIVERS				3.2

[a] AMT+PREC+PSEA+PDRY+LAT
[b] AMT+PREC+PSEA+PDRY+LAT+SOIL
[c] AMT+PREC+PSEA+PDRY+LAT+SOIL+HMNINFL
[d] AMT+PREC+PSEA+PDRY+LAT+SOIL+HMNINFL+RIVERS

It has recently been found that the geographic variation in palm species richness in the Americas is mainly controlled by the availability of water with palm species richness peaking in ever-wet regions [14, 15]. Our results show that the distribution of *H. petersiana* is also strongly dependent on the climatic water regime with a sharply defined optimum in relation to annual precipitation and rainfall seasonality achieving its peak occurrence at low-to-intermediate overall precipitation, pronounced rainfall seasonality, and ≥ 1 very dry months (figure 4a-c). The projected drying and temperature increase by the A1B emission scenario for the current century, especially in southwestern regions of Africa [4], could therefore be a potential threat to the survival of *H. petersiana* populations in the future. Given the narrow climatic niche requirements as indicated by the response curves to both PREC and AMT (figure 4a,d), it seems likely that even a slight change in annual means could potentially influence the maintenance of viable *H. petersiana* populations in southern Africa. Increasing scarcity of *H. petersiana* currently observed by rural communities in Botswana has been partly attributed to increased temperatures and reduced rainfall [27]. It has been projected that in southern Africa, the annual precipitation is likely to decrease by 5-15% and in some extreme western region up to 20% according to the A1B scenario by 2080-2099 compared to 1980-1999 [4]. However, the most pronounced decrease in precipitation is projected in the JJA quarter (up to 40% [2]) which extends into the early spring (SON) [4]. In terms of temperature changes, the A1B scenario projects temperature increases across all seasons, but notably in JJA and SON, the temperature increase will be at or above the annual mean (3.7°C and 3.4°C, respectively) for the same time period as above [4] in line with observations over the 20th century [3]. In the light of these projected changes there are potential risks that the precipitation will fall below 400 mm and the temperature will increase above 23°C, creating a stressful environment for *H. petersiana* within its current range. However, as noted by

Svenning and Condit [44], caution should be made when inferring detrimental effects of the projected temperature increases since some species may be able to cope with the projected changes due to inherent abilities to tolerate warmer climates than observed today. Apart from the climatic water regime, the soil relationship also highlights the importance of hydrology, particularly an ample soil water supply for the occurrence of *H. petersiana*. Certain palm species like *H. petersiana* are capable of persisting in areas of low annual precipitation and high day-time temperatures given a sufficient supply of ground water, *i.e.*, high ground-water levels or high soil-water retention [12]. Yet, climate change may well change the hydrology, notably increasing drought would diminish the occurrence of wet soil conditions, and hence, potentially influence the persistence of *H. petersiana*.

Non-climatic factors were important determinants of the distribution of *H. petersiana*, notably latitude, soil type and human influence, while the effect of the closeness to rivers or other water bodies was minimal. The pronounced predictive strength of LAT most likely reflects that climatically suitable regions for *H. petersiana* are found outside its current range, notably in the savanna swath stretching from west to east across northern Africa. Hence, the strong effect of LAT probably reflects that the range of *H. petersiana* is not in equilibrium with the environment at the continental scale (*i.e.*, the species is currently only partially filling its potential range [45]). A number of factors may cause such non-equilibrium ranges: time-limited expansion from place of origin or refugia [46-48], physical barriers [49], intrinsic range limitations [50], or interspecific competition [51]. Such non-environmental constraints on the range of *H. petersiana* may constitute yet another challenge for the 21st century fate of this palm species, notably by limiting its ability to track the changing climate. Additionally, land-use changes and the contingent fragmentation of suitable habitats [8] may further strongly impede the dispersal of *H. petersiana* [52]. The response to the SOIL variable showed a pronounced association of *H. petersiana* with the soil types, classified by FAO as: Fluvisols, Leptosols, Luvisols, Planosols, Solonetz, and soils in areas of inland waters. These findings corroborate the known observation of the species' association with the riparian zone and its association with sodic/alkaline alluvial soils with high water tables [12], being particularly characteristic of the young alluvial soils (Fluvisols), other river soils (Leptosols, Planosols), and soils associated with grasslands in semi-arid, sub-tropical regions (Solonetz) (www.fao.org, www.isric.org). The response to the HMNINFL factor was positive. This response could indicate several things, notably, reflect the sampling patterns of the palm collections, or the association of *H. petersiana* with waterways, which are also preferred sites for human settlements. The human influence index, as mentioned in the method section, is computed from a range of human impact proxies. Particularly, (1) the accessibility proxy (15-km buffer surrounding *e.g.,* rivers or roads); (2) gridded population of the world; (3) land transformations (built-up areas, population settlements, land use) could all explain the positive effect of HMNINFL in the CLSH and CLSHR models: It is generally known that plant collections tend to be made in accessible areas, *e.g.*, often along rivers and roads [53], thus, the positive HMNINFL relationship may reflect this sampling bias. *Hyphaene petersiana* is in some instances found near waterways [12, 29]; however, the RIVERS variable provided less explanatory power than HMNINFL. Alternatively, it has been noted that palms are dispersed by humans in close proximity to human settlements: palm seeds are often discarded after consumption of the palm fruits and traditionally villagers tend to manage fruit trees, as observed in Botswana for *H. petersiana* [29], *i.e.*, *H. petersiana* might indeed be more abundant in the more influenced areas. Given the association of *H. petersiana* to rivers and riparian vegetation, the lack of influence of the RIVERS variable was unanticipated; yet, one explanation could be that the SOIL variable provides a more detailed description of the habitat in which *H. petersiana* is found and thereby limit the effect of the RIVERS variable. In addition, the RIVERS variable is computed from only larger water bodies in Africa [35, 36] and, consequently, may not capture the true relationship between the occurrence of *H. petersiana* and closeness to rivers, even though, category 1 was dominant.

Despite the fact that *H. petersiana* is likely facing future climatic stresses, other factors are likely to further increase the strain on this key-stone species as well. As can be seen in table 1, *H. petersiana* is a much used wild palm species, and the anthropogenic utilization of the palm has already been noted to have a negative effect on its population densities in Botswana and Namibia [25, 27, 28]. An ethnobotanical study has shown that the increased utilization of *H. petersiana* leaves for basket weaving in Botswana has lead to the disappearance of the palm in certain areas around human settlements. Hence, some communities have over time noted that they need to travel further and further to harvest palm leaves. Human population increase in the Okavango Delta in Botswana since the late 1960s has changed the utilization pattern of useful plants in the region from subsistence to consumerism [25]. Basket weaving using *H. petersiana* leaves became industrialized by 1984 with 50% of all women from the rural communities producing baskets for local and overseas markets [25]. Studies have shown that the heavy over-utilization and notably, the destructive harvesting methods have lead to the decline of *H. petersiana* in this area [25, 27]. The harvesting methods often damage the apical meristems preventing the suckers from reaching reproductive maturity, and thereby preventing regeneration from seed production [25]. A recent study has observed even further depletion of the *H. petersiana* resource since the 1980s in the same basket weaving areas in Botswana despite the fact that advice had been given to rural villagers not to utilize palm leaves harvested in unsustainable ways [27]. Unfortunately, the destructive harvesting of palms is common place across Africa. Additional problematic harvesting methods have been observed for other palm species in Africa which as a consequence are likewise threatened with local extinction [22, 54]. In Namibia, similar palm population declines of *H. petersiana* have been observed, especially the skewed population structure with less mature individuals than expected [28], as found in Botswana [27]. The decline in Botswana has been attributed to the over-utilization of unopened leaves for basket weaving; nevertheless, the harvest of *H. petersiana* for basket production seems to be within sustainable limits in Namibia [30]. However, unsustainable browsing by livestock and felling of adult individuals for building materials have been recorded in both countries, notably livestock browsing in Namibia [25, 28, 30]. Sullivan et al. [28] also noted that the decline of *H. petersiana* was more marked in areas of higher population densities of villagers and livestock corroborating findings of an additional study where *H. petersiana* has been found to decrease in numbers as the environment becomes more and more degraded due to an increasing human population [55]. This indicates that the pressures on *H. petersiana* will be continuing during the 21st century given the high current [56, 57], and projected human population growth in Africa [57]. *Hyphaene petersiana* is associated with semi-arid regions with sufficient water availability from high water tables [29], and it has been projected that semi-arid parts of southern Africa are at moderate to high risk of desertification [2]. This may imply that there might be even higher future utilization pressures on the products of this palm species if other useful plants fail to cope with 21st century environmental change in the region.

5. Conclusions

The current distribution of *H. petersiana* is controlled by a combination of climate and non-climatic factors. The palm has a very narrow response to several climatic factors, notably annual means of temperature and precipitation as well as precipitation seasonality. The strong effect of soil types associated with wet conditions implies that hydrology limits its distribution as well. The projected climatic changes for southern Africa, with strong precipitation decreases in the dry season (JJA) and spring months (SON) may therefore have a strong negative impact on the distribution and abundance of this key-stone palm species. Moreover, the future climate-driven stresses may be exacerbated by the rapidly growing human populations in Africa. The unsustainable utilization of *H. petersiana* is likely to continue and increase with increasing human population unless marked changes in the management of *H. petersiana* occur.

Acknowledgements

This work was financed by the Danish Natural Science Research Council through grant #272-07-0242 to JCS, #272-06-0476 to HB, and the Faculty of Science at Aarhus University to ABO. We are grateful to professor emeritus John Dransfield for thorough evaluation of the palm data and helpful comments on the predicted palm distributions, and to Peder Klith Bøcher for computing the RIVERS variable. Moreover, we would like to thank the data contributors for making their palm data available for the project: The Royal Botanic Gardens, Kew (Dr. William Baker); The Nationaal Herbarium Nederland (NHN) (Dr. Jan Wieringa & Dr. Johan van Valkenburg); and the herbarium collections accessed through the GBIF data portal (Botanic Garden and Botanical Museum Berlin-Dahlem). Finally, we thank Marco Schmidt for permission to use figure 1.

References

[1] IPCC 2007 *Summary for Policymakers* in *Climate Change 2007: The Physical Science Basis. Contribution of Working Group I to the Fourth Assessment Report of the Intergovernmental Panel on Climate Change* ed Solomon S *et al* (Cambridge: Cambridge University Press) pp 1-18
[2] Boko M, Niang I, Nyong A, Vogel C, Githeko A, Medany M, Osman-Elasha B, Tabo R and Yanda P 2007 *Africa. Climate change 2007: Impacts, Adaptation and Vulnerability. Contribution of Working Group II to the Forth assessment Report of the Intergovernmental Panel on Climate Change* ed Parry M L *et al* (Cambridge: Cambridge University Press) pp 433-67
[3] Hulme M, Doherty R, Ngara T, New M and Lister D 2001 *Clim. Res.* **17** 145-68
[4] Christensen J H *et al* 2007 *Regional Climate Projections* in *Climate change 2007: The physical Science Basis. Contribution of Working Group I to the Fourth Assessment Report on the Intergovernmental Panel on Climate Change* ed Solomon S *et al* (Cambridge: Cambridge University Press) pp 846-71
[5] Fischlin A, Midgley G F, Price J T, Leemans R, Gopal B, Turley C, Rounsevell M D A, Dube O P, Tarazona J and Velichko A A 2007 *Ecosystems, their properties, goods, and services. Climate Change 2007: Impacts, Adaptation and Vulnerability. Contribution of Working Group II to the Fourth Assessment Report of the Intergovernmental Panel on Climate Change* ed Parry M L *et al* (Cambridge: Cambridge University Press) pp 211-72
[6] McClean C J, Lovett J C, Kuper W, Hannah L, Sommer J H, Barthlott W, Termansen M, Smith G E, Tokamine S and Taplin J R D 2005 *An. Mo. Bot. Gard.* **92** 139-52
[7] Thuiller W, Broennimann O, Hughes G, Alkemade J R M, Midgley G F and Corsi F 2006 *Global Change Biol.* **12** 424-40
[8] Sala O E *et al* 2000 *Science* **287** 1770-4.
[9] Midgley G F and Thuiller W 2005 *New Phytol.* **167** 638-41
[10] Gaston K 2009 *Proc. R. Soc. B.* **276** 1395-406
[11] Gaston K 2009 *Proc. R. Soc. B.* **276** 1391-3
[12] Tuley P 1995 *The palms of Africa* (Cornwall: The Trendrine Press) pp 1-189
[13] Bates D M 1988 Utilization pools: A framework for comparing and evaluating the economic importance of palms in *The Palm - Tree of Life: Biology, Utilization and Conservation* ed Balick M J (New York: The New York Botanical Garden) pp 56-64
[14] Bjorholm S, Svenning J-C, Skov F and Balslev H 2005 *Global Ecol. Biogeogr.* **14** 423-9
[15] Bjorholm S, Svenning J-C, Baker W J, Skov F and Balslev H 2006 *Bot. J. Linn. Soc.* **151** 113-25
[16] Svenning J-C, Harlev D, Sorensen M and Balslev H 2009 *Biodivers. Conserv.* **18** 219-28
[17] Roosevelt A C *et al* 1996 *Science* **272** 373-84
[18] DeMason D A and Sekhar K N C 1988 The breeding system in the Date palm (*Phoenic dactylifera* L.) and its recognition by early cultivators in *The Palm - Tree of Life: Biology, Utilization and Conservation* ed Balick M J (New York: The New York Botanical Garden) pp 20-35

[19] Balick M J 1988 The use of palms by the Apinayé and Guajajara Indians in northeastern Brazil in *The Palm - Tree of Life: Biology, Utilization and Conservation* ed Balick M J (New York: The New York Botanical Garden) pp 65-90
[20] Boom B M 1988 The Chácobo Indians and their palms in *The Palm - Tree of Life: Biology, Utilization and Conservation* ed Balick M J (New York: The New York Botanical Garden) pp 91-7
[21] Kahn F 1988 Ecology of economically important palms in Peruvian Amazonia in *The Palm - Tree of Life: Biology, Utilization and Conservation* ed Balick M J (New York: The New York Botanical Garden) pp 42-9
[22] Mollet M, Herzog F, Behi Y E N and Farah Z 2000 *Environ. Dev. Sustain.* **2** 45-59
[23] Dransfield J, Uhl N W, Asmussen C B, Baker W J, Harley M M and Lewis C E 2008 *Genera Palmarum. The Evolution and Classification of Palms* (Kew: Kew Publishing, Royal Botanic Gardens, Kew) pp 1-732
[24] Tomlinson P B 1962 *Principes* **6** 96-103
[25] Cunningham A B and Milton S J 1987 *Econ. Bot.* **41** 386-402
[26] Strohbach B J 2000 *Vegetation degradation in Namibia.* (Windhoek, Namibia: Namibia Wissenschaftliche Gesellschaft / Namibia Scientific Society)
[27] Kgathi D L, Mmopelwa G and Mosepele K 2005 *Nat. Resour. Forum* **29** 70-81
[28] Sullivan S, Konstant T L and Cunningham A B 1995 *Econ. Bot.* **49** 357-70
[29] Fujioka Y 2005 *African Study Monographs, Suppl.* **30** 89-105
[30] Konstant T, Sullivan S and Cunningham A 1995 *Econ. Bot.* **49** 345-56
[31] Sneed M W 1983 *Principes* **27** 143-51
[32] Hijmans R J, Cameron S E, Parra J L, Jones P G and Jarvis A 2005 *Int. J. Climatol.* **25** 1965-78
[33] Fisher G, Nachtergaele F, Prieler S, van Velthuizen H T, Verelst L and Wiberg D 2008 *Global Agro-ecological Zones Assessment for Agriculture (GAEZ 2008).* (Laxenburg: IIASA and Rome: FAO)
[34] Sanderson E W, Jaiteh M, Levy M A, Redford K H, Wannebo A V and Woolmer G 2002 *Bioscience* **52** 891
[35] Jenness J, Dooley J, Aguilar-Manjarrez J and Riva C 2007 *African Water Resource Database. GIS-based tools for inland aquatic resource management. 1. Concepts and applications case studies CIFA Technical Paper. No. 33, Part 1.* (Rome: FAO)
[36] Jenness J, Dooley J, Aguilar-Manjarrez J and Riva C 2007 *African Water Resource database. GIS-based tools for inland aquatic resource management. 2. Technical manual and workbook. CIFA Technical Paper. NO 33, Part 2.* (Rome: FAO)
[37] Phillips S J, Anderson R P and Schapire R E 2006 *Ecol. Model.* **190** 231-59
[38] Elith J *et al* 2006 *Ecography* **29** 129-51
[39] Graham C H, Elith J, Hijmans R J, Guisan A, Townsend Peterson A and Loiselle B A 2008 *J. Appl. Ecol.* **45** 239-47
[40] Guisan A, Graham C H, Elith J and Huettmann F 2007 *Divers. Distrib.* **13** 332-40
[41] Wisz M S, Hijmans R J, Li J, Peterson A T, Graham C H and Guisan A 2008 *Divers. Distrib.* **14** 763-73
[42] Phillips S J and Dudik M 2008 *Ecography* **31** 161-75
[43] Allouche O, Tsoar A and Kadmon R 2006 *J. Appl. Ecol.* **43** 1223-32
[44] Svenning J C and Condit R 2008 *Science* **322** 206-7
[45] Svenning J-C and Skov F 2004 *Ecol. Lett.* **7** 565-73
[46] Svenning J-C and Skov F 2007 *Global Ecol. Biogeogr.* **16** 234-45
[47] Svenning J-C and Skov F 2007 *Ecol. Lett.* **10** 453-60
[48] Svenning J-C, Normand S and Kageyama M 2008 *J. Ecol.* **96** 1117-27
[49] Brown J H, Stevens G C and Kaufman D M 1996 *Annu. Rev. Ecol. Syst.* **27** 597-623
[50] Holt R D, Keitt T H, Lewis M A, Maurer B A and Taper M L 2005 *Oikos* **108** 18-27
[51] Case T J, Holt R D, McPeek M A and Keitt T H 2005 *Oikos* **108** 28-46

[52] Neilson R P, Pitelka L F, Solomon A M, Nathan R, Midgley G F, Fragoso J M V, Lischke H and Thompson K 2005 *Bioscience* **55** 749-59
[53] Graham C H, Ferrier S, Huettman F, Moritz C and Peterson A T 2004 *Trends Ecol. Evol.* **19** 497-503
[54] Sambou B, Goudiaby A, Ervik F, Diallo D and Camara M C 2002 *Biodivers. Conserv.* **11** 1149-61
[55] Strohbach B J 2000 *Dinteria* **26** 45-62
[56] Tiffen M 2003 *World Dev.* **31** 1343-66
[57] Bengtsson M, Shen Y J and Oki T 2006 *Popul. Environ.* **28** 113-31

IOP Conf. Series: Earth and Environmental Science **8** (2009) 012015 doi:10.1088/1755-1315/8/1/012015

Impacts of 21st century climate changes on flora and vegetation in Denmark

Flemming Skov[1]*

Bettina Nygaard[1]

Peter Wind[1]

Finn Borchsenius[2]

Signe Normand[2]

Henrik Balslev[2]

Camilla Fløjgaard[1,2]

Jens-Christian Svenning[2]

[1]Department of Wildlife Ecology and Biodiversity, National Environmental Research Institute, Aarhus University, Grenaavej 14, DK-8410 Rønde, Denmark. [2]Department of Biological Sciences, Aarhus University, Ny Munkegade 114, DK-8000 Aarhus C, Denmark

*E-mail: fs@dmu.dk

Abstract. In this paper we examined the potential impacts of predicted climatic changes on the flora and vegetation in Denmark using data from a digital database on the natural vegetation of Europe. Climate scenarios A2 and B2 were used to find regions with present climatic conditions similar to Denmark's climate in the year 2100. The potential natural vegetation of Denmark today is predominantly deciduous forest that would cover more than 90% of the landscape. Swamps, bogs, and wet forest would be found under moist or wet conditions. Dwarf shrub heaths would be naturally occurring on poor soils along the coast together with dune systems and salt-marsh vegetation. When comparing the natural vegetation of Denmark to the vegetation of five future-climate analogue areas, the most obvious trend is a shift from deciduous to thermophilous broadleaved forest currently found in Southern and Eastern Europe. A total of 983 taxa were recorded for this study of which 539 were found in Denmark. The Sørensen index was used to measure the floristic similarity between Denmark and the five subregions. Deciduous forest, dwarf shrub heath, and coastal vegetation were treated in more detail, focusing on potential new immigrant species to Denmark. Finally, implications for management were discussed. The floristic similarity between Denmark and regions in Europe with a climate similar to what is expected for Denmark in year 2100 was found to vary between 48-78%, decreasing from North to South. Hence, it seems inevitable that climate changes of the magnitudes foreseen will alter the distribution of individual species and the composition of natural vegetation units. Changes, however, will not be immediate. Historic evidence shows a considerable lag in response to climatic change under natural conditions, but little is known about the effects of human land-use and pollution on this process. Facing such uncertainties we suggested that a dynamic strategy based on modeling, monitoring and adaptive management is adopted. Modeling techniques can be constantly improved, but will never be perfect and should therefore be linked to a fine-masked network of observatories to check model predictions and feed empirical data back into the models for calibration and further development.

1. Introduction

The looming threat of global climate change necessitates a reassessment of nature conservation goals and management strategies. Climate is a strong driver of biogeographical distribution patterns and many studies show that shifts in distribution of a large number of plants and animals have already taken place [1-5]. Given the magnitude of the expected 21st century climate change much larger range shifts are expected in the future. Nature conservation efforts traditionally focus on what is or – in nature restoration projects - what has been. Now, we are forced to include uncertain future species distribution patterns in conservation planning and not only manage what is, but also what might be. This is a demanding task that requires novel conservation strategies: in the face of an uncertain but rapidly changing future we will have to adopt a dynamic strategy based on modeling, monitoring and adaptive management. In this paper we try to estimate the possible consequences of climate change for a regional flora, using the Danish flora as an example.

Denmark is a small country (42.394 km^2, ~470 km in maximum extent) situated in the temperate climate zone characterized by cool summers with a mean summer temperature of around 16° C and winters with mean temperatures of around 0.5° C. The mean yearly temperature is 7.7° C. Precipitation occurs throughout the year, but the autumn months of September, October and November tend to be the wettest. Climatic differences in Denmark are not large, but it is generally warmer and drier in the South-eastern parts of the country (http://www.dmi.dk/). Reflecting the current macroclimate, the dominant natural vegetation (zonal habitats) would be deciduous broad-leaved forest, but the majority of species in the current flora are associated with anthropogenic or azonal habitats. The Danish landscape was to a large extent covered by glaciers during the Last Glacial Maximum, and the flora found in Denmark today is partly a result of natural postglacial immigration and subsequent ecological sorting and partly a result of human land-use and anthropogenic species introductions. However, many species suited for the Danish climate have probably not yet reached the country due to dispersal limitation (see, e.g., [6-10]). The exact number of higher plants in Denmark is difficult to establish. Most sources suggest 1,500 species of which 450 are introduced, leaving 1,050 indigenous species (http://www.denstoredanske.dk). Among the introduced species approximately 200 have become naturalized. The remaining group of exotics forms a sporadic and unstable element of the Danish flora, but many of these species originate from warmer climates and may play an increasingly important role as climate warms.

The major threats to the Danish flora today are all directly related to human land-use, primarily the usage of most of the terrestrial area for intensive agriculture, but also pollution with nutrients (notably N and P), cessation of traditional low-intensity land use practices (grazing and hay cutting), drainage of wetlands, high-intensity forestry, and the destruction and fragmentation of remnant patches of semi-natural vegetation play an important role. These pressures have resulted in a large number of higher plant species becoming rare or even threatened. The most recent Danish Red List from 1997 [11] includes a total of 220 species of which 23 have disappeared from the Danish flora, 36 are critically endangered, 66 are vulnerable and 95 species are considered rare. Species in the last two categories may become critically endangered in the near future unless some action is taken. The Nature Conservation Act provides Denmark's main legislative framework for nature conservation. It comprises four pillars: general protection of habitats against cultivation and settlement; coastal zone protection; land acquisition; and specific regulatory powers for the protection of nature. Furthermore, 246 Natura 2000 sites have been established according to the EU Habitats Directive, covering a land area of 3,590 ha, or 8.3 % of Denmark's terrestrial surface. However, the current biodiversity strategy does not take the expected climate change-driven shifts in species ranges into account. It is not an easy task for many reasons: little is known about the persistence of plants when they are exposed to an increasingly unfavourable climate. Will they disappear immediately or persist for an unknown time period (extinction delay) and what is the role of adaptation? What are the effects of increasing levels of carbon dioxide and how will that affect vegetation structure and species competition? Finally, what are the mixed effects of climate change and other well known pressures on plants and biodiversity (eutrophication, fragmentation, changing land-use patterns)?

In this paper we try to estimate the possible consequences of climatic change on the Danish flora by linking IPCC climate scenario predictions for Denmark to maps of the potential natural vegetation of Europe under the present climate. We look at the potential vegetation types for Denmark today and compare them to the potential vegetation types for parts of Europe where the climate today is comparable to the climate expected for Denmark by the year 2100. The comparisons make it possible to estimate the proportion of species that are likely to stay (present under both climatic scenarios), species likely to disappear (species present in Denmark today, but not growing under warmer conditions) and species that would potentially be able to immigrate under climate change.

2. Material and Methods

2.1. Distribution of the natural vegetation of Europe

The distribution of the natural vegetation of Europe (NVE) has been mapped in a large European project spanning 25 years and finalized in 2003 [12]. The aim of the NVE project was to map the potential vegetation in Europe at a scale of 1:2.500.000 according to the indigenous flora and the climatic and edaphic conditions at each site. We used the NVE data set rather than data sets on current land-cover (see, e.g., CORINE land cover project (http://www.eea.europa.eu/publications/COR0-landcover) to focus on the Danish and European native flora and vegetation likely to develop in the absence of human influence under a given climate. The mapping process was based on a phytosociological approach where the fundamental mapping units are characterized by a specific floristic composition. The 698 vegetation units described were linked in a hierarchical system consisting of 19 physiognomic-structural formations, of which 14 represent macroclimatic (zonal) formations and 5 represent azonal formations such as coastal vegetation, mires and swamps. The NVE data set includes a number of descriptors for each vegetation unit, including: total extent in Europe, structural characteristics, lists of typical or dominating species (divided into trees, shrubs, lianas, herbs and mosses), diagnostic characteristics, climatic and edaphic conditions where the unit is found, typical land-use, major threats, etc.

2.2. Climatic equivalents of Denmark's future climate in Europe today

The Intergovernmental Panel of Climate Change [13] has developed a number of scenarios that make different assumptions for future greenhouse gas pollution, land-use and other driving forces. We used the A2 and B2 future climate scenarios in the present study. They are commonly used to represent a severe, business-as-usual scenario (A2) and a more optimistic scenario where relatively smaller climatic changes are expected (B2). It is worth noting that emission so far has exceeded the A2 scenario ([13,14]).According to the A2 scenario, a mean annual increase of 3.1°C is predicted in year 2100 for Denmark and correspondingly 2.2°C for the B2 scenario (www.klimatilpasning.dk). Both scenarios predict the approximate same change in precipitation patterns with a yearly increase in precipitation of 8-9%, but with wetter winters and drier summers than today. Based on climatic data for the two scenarios (TYN_SC 1.0; [15]) we calculated a measure of available energy, extreme minimum temperature and water deficit (see [6] for details) and created climatic envelopes for Denmark in year 2100 using the method described in [6], (implemented in ArcGIS 9.3, ESRI,

Redlands, CA, USA, http://www.esri.com). The resulting envelope for Denmark's climate in year 2100 was used to find areas with similar present climate in Europe (future-climate analogue areas). Current climate data was based on the CRU_CL 2.0 data set. (http://www.cru.uea.ac.uk). Based on this map we created five broadly defined geographical subregions. This number of subregions was chosen to obtain a reasonable combination of spatial resolution and representativity and was drawn to include areas with a climate most similar to what is predicted for Denmark restricted to mainland Europe up to a distance of 1,500 km avoiding large mountain areas (e.g., the Alps and the Pyrenees). In turn, these regions were overlain the map of the natural vegetation of Europe. This makes it possible to extract and analyse the floristic composition of each subregion and make comparisons with the present Danish flora.

2.3. Analysis

Lists of present vegetation units were prepared for Denmark and each of the five future-climate analogue regions (see Figure 1), and all taxa belonging to these units were extracted from the NVE database. In the analysis we included taxa at the species and subspecies levels. The NVE database does not attempt to list all species from each unit, but focus on common, characteristic, or diagnostic species so our analysis does not quantify the potential changes for the whole flora, but concerns the native species that are typical for the different vegetation units registered for Denmark and future-climate analogue regions. The area of Denmark is relatively small compared to the other subregions and taxa from the Danish flora may therefore be underrepresented in the NVE database. To compensate for this we also created a complete list of taxa for all subregions and checked if taxa not recorded for Denmark in the NVE were known to be a part of the Danish flora.

Figure 1. Study area. Denmark (DK) and five sub-regions (future-climate analogue areas) used in the analysis. The sub-regions are based on the inserted map that shows where the climate today correspond to the expected climate in Denmark year 2100 under the IPCC scenarios, A2 and B2, respectively. I. Germany; II. Northern France; III. Southern France; IV: Northern Italy; and V: Bosnia/Serbia.

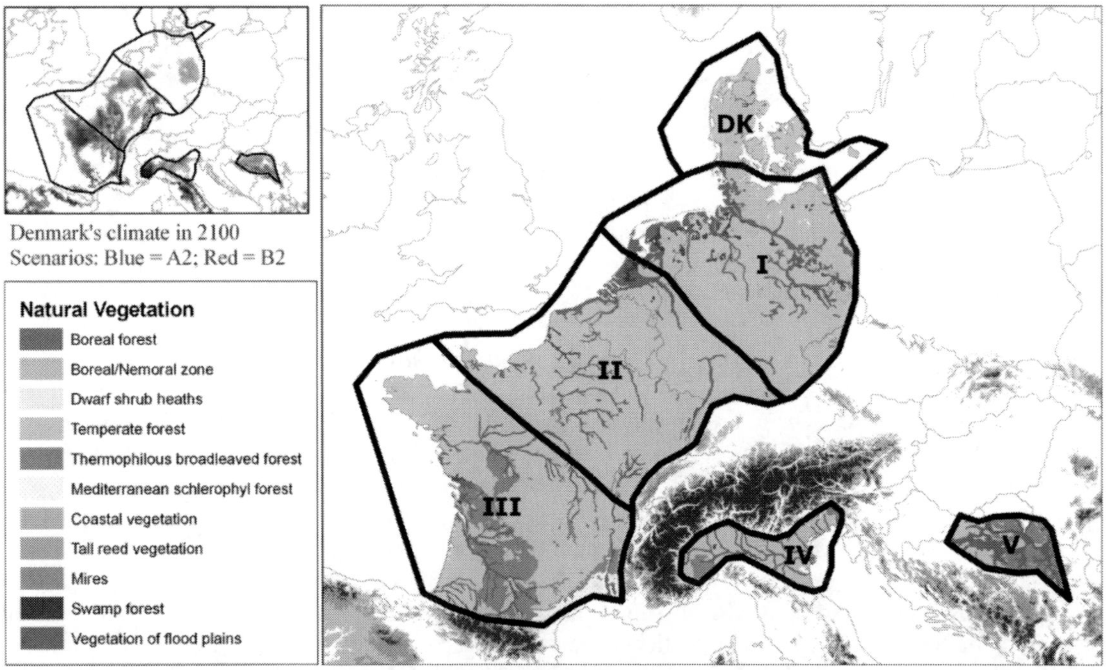

The *floristic similarity* of the sub-regions based on their species composition were analyzed using the Sørensen index (values between 0 and 1: from 0 to 100% similarity) [16]. The susceptibility to climatic warming of the current native flora of Denmark was appraised using each species overall distribution in Europe according to the NVE database.

An *Index of Resistance to warming (IoR)* was created by assigning each species a total score between 0 and 100 based on its current distribution. Scores are relative and reflect the span of a species' climatic niche. Species present in sub-region I were assigned a score of 5, species present in subregion II were given a score of 20 and the presence in subregion II, IV or V each added another 25 to the index. Species with a score of 100 are thus present in all of the subregions and considered to be least vulnerable to warming. Species present in Denmark today, but not listed as characteristic species in any of the future-climate analogue areas were considered most vulnerable (score = 0). It is important to understand, however, the limitations of the index. A low index of resistance may be a result of: (1) false negatives (i.e., a species not recorded for a vegetation unit in NVE, but present in the region anyway); and (2) rare cases where a species not present in any of the subregions, but occurring under warmer conditions elsewhere in Europe. The index of resistance should therefore be seen as a first, crude attempt to estimate the proportion of species at risk.

The immigration potential for a species not yet present in Denmark was estimated as a relative measure of the area they fill now and their nearness to Denmark. An *Index of Immigration (IoI)* was developed using the following scores: presence in (I): 70; (II): 15; (III): 5; (IV): 5; and (V): 5. A species, for example, not growing in Denmark today, but present in Northern Germany (subregion I) would be a more likely immigrant than a species much further away. Scores are additive and a species present in all subregions will thus have a score of 100.

Figure 2. Coastal dune landscape at the island Fanø in southwestern Denmark. The vegetation consists of "decalcified fixed dunes with *Empetrum nigrum*" (type 2140 according to Annex I of the Habitats Directive) and "fixed coastal dunes with herbaceous vegetation" ("grey dunes", type 2130). Both dune types are considered to be particularly vulnerable. Photo: Environmental Centre Ribe.

IOP Conf. Series: Earth and Environmental Science **8** (2009) 012015 doi:10.1088/1755-1315/8/1/012015

Table 1a. Zonal (depending mainly on climate) vegetation in Denmark and the percentage of the terrestrial area covered by each vegetation unit (summed values for formations) according to NVE [10].

Zonal and vegetation		93.5%
E - Atlantic dwarf shrub heaths		
Northwest European coastal heaths		
E10	Frisian-Danish coastal heaths (with *Calluna vulgaris, Empetrum nigrum, Erica tetralix*)	*1.5%*
F - Mesophytic deciduous broad-leaved and mixed coniferous-broad-leaved forests		
Species-poor acidophilous oak and mixed oak forests		
Lowland-colline (to submontane) types		
F8	Atlantic-subatlantic birch-pedunculate oak forests (*Quercus robur, Betula pendula, Populus tremula*)	*11.3%*
Mixed oak-hornbeam forests		
Lowland-colline (to submontane) pedunculate oak-hornbeam forests		
F37	South Scandinavian mixed hornbeam-pedunculate oak forests (*Quercus robur, Carpinus betulus*)	*1.7%*
Beech and mixed beech forests		
Species-poor oligotrophic to mesotrophic beech and mixed beech forests		
F77	(Atlantic-) subatlantic *Deschampsia flexuosa*-(oak-) beech forests (*Fagus sylvatica, Quercus robur, Q. petraea*)	*0.8%*
F79	South Scandinavian *Deschampsia flexuosa*-(oak-) beech forests (*Fagus sylvatica, Quercus robur*, partly *Q. petraea*)	*30.5%*
Species-rich eutrophic and eu-mesotrophic beech and mixed beech forests		
F108	South Scandinavian-north Central European *Galium odoratum*- and *Milium effusum*-beech forests	*47.5%*
F109	North subatlantic *Cephalanthera*-beech forests with *Hepatica nobilis*	*0.2%*

3. Results and discussion

3.1. The natural vegetation of Denmark

Table 1 lists the 15 vegetation units that comprise the natural vegetation of Denmark according to the NVE atlas. The main zonal vegetation formation is deciduous forest (mainly varieties dominated by *Fagus sylvatica* on richer soils and *Quercus robur/petraea* on poor, oligotrophic soils). Forest would cover an estimated 92% of the landscape under natural conditions (Table 1a).

The most common azonal vegetation types are sand-dunes and salt-marsh vegetation typical of exposed and sheltered coast zones, respectively. Swamps, bogs, and wet forest dominated by *Alnus glutinosa* or (rarely) *Quercus robur*, *Fraxinus excelsior*, and *Ulmus glabra* constitute the potential natural vegetation in moist habitats (Table 1b).

Table 1b. Azonal vegetation (depending mainly on soil and hydrological conditions) in Denmark and the percentage of the terrestrial area covered by each vegetation unit (summed values for formations) according to NVE [10].

Azonal vegetation		6.5%
P - Coastal vegetation and inland halophytic vegetation		
Vegetation of coastal sand dunes and sea shores		
Bothnian-Baltic sand-dune vegetation		
P4	Sundic-Baltic sand-dune vegetation complexes	*0.1%*
Atlantic sand-dune vegetation		
P5	Northwest European sand-dune vegetation complexes	*2.0%*
Halophytic vegetation		
Coastal halophytic vegetation, often in combination with sand-dune vegetation		
P19	West European salt-marsh vegetation	*0.2%*
P22	Sundic-west Baltic halophytic vegetation	*0.3%*
S - Mires		
Ombrotrophic mires (bogs)		
Sphagnum papillosum-blanket bog and raised bog complexes		
S8	Atlantic-subatlantic *Sphagnum magellanicum*-raised bogs	*0.6%*
T - Swamp and fen forests		
Alder carrs and swamp forests (Alnus glutinosa, A. barbata)		
T1	Alder carrs (*Alnus glutinosa*)	*0.9%*
U - Vegetation of flood-plains, estuaries and fresh-water polders		
Alluvial and wet lowland forests in the nemoral zone		
*Vegetation of estuaries and fresh water polders (*Fraxinus excelsior, Quercus robur, Ulmus glabra*)*		
U28	Pedunculate oak-ash forests *(Fraxinus excelsior, Quercus robur) and ash-elm forests (Ulmus glabra, Fraxinus excelsior)*	*2.2%*
R - Tall reed vegetation and tall sedge swamps, aquatic vegetation		
R1	Freshwater tall reed swamps *(Phragmites australis, Typha angustifolia, T. latifolia, Schoenoplectus lacustris)*	*0.1%*
Other		*0.2%*

As noted in the introduction the vegetation in Denmark has been strongly modified by human land-use during millennia, so the map of the natural vegetation does not reflect the current vegetation. In reality, forest covers about 12% of the area, but the majority is planted and managed intensively. Heathland and open grassland covered large areas in Denmark 100-200 years ago due to soil degradation and overgrazing, but most of this land is now under intensive agriculture. Nevertheless, vegetation similar to potential natural vegetation occurs patchily across Denmark.

3.2. Danish vegetation today compared to the vegetation of future-climate analogue areas

Table 2 shows the main vegetation formations found in Denmark and in the five future-climate analogue regions today. The most apparent pattern is the clear differences in the dominating major vegetation formations. In Denmark, temperate, deciduous forest is the absolute dominating major formation. This formation is gradually replaced by thermophilous broadleaved forest towards the

South and East in the future-climate analogue areas. Dwarf shrub heath is presently not common in Denmark, but very rare in future-climate analogue areas, or even virtually non-existent in Northern Italy and the Balkans. These differences give an indication that major vegetation changes might be expected given 21st century climate changes. Some dissimilarities are caused by special topographic or edaphic features, e.g., the occurrence of poor soils, high salinity, or occurrence of large rivers and deltas, and therefore unlikely to be representative for potential future vegetation in Denmark. Most of the other subregions, for example, have a high proportion of floodplain vegetation; a type that is very rare in Denmark that has no large river and delta systems. Another feature that separates Denmark from the other areas is the relative high proportion of sand dune and salt march vegetation found here.

Table 2. Distribution of major vegetation formations within the study areas. DK: Denmark; I: Northern Germany; II: Central France; III: Southern France; IV: Northern Italy; V: Serbia/Croatia. Letters for the formations correspond to those in tables 1.

Formation	DK	I	II	III	IV	V
C - Boreal forest	-	-	-	0.1%	-	-
D - Boreal/Nemoral zone	-	1.7%	0.1%	-	-	-
E - Dwarf shrub heaths	1.5%	-	0.0%	0.1%	-	-
F - Temperate forest	92.1%	83.5%	89.9%	71.1%	45.7%	19.2%
G - Thermophilous broadleaved forest	-	0.2%	0.4%	15.7%	44.1%	34.6%
J - Mediterranean schlerophyl forest	-	-	-	6.3%	0.0%	-
K - Xerophytic coniferous forest	-	-	-	-	0.8%	-
L – Forest steppes	-	-	-	-	-	3.8%
P - Coastal vegetation	2.5%	0.5%	0.2%	0.4%	0.7%	0.7%
R - Tall reed vegetation	0.1%	-	-	0.0%	0.2%	-
S – Mires	0.6%	0.2%	0.0%	-	-	-
T - Swamp forest	0.9%	2.6%	0.8%	0.2%	0.1%	-
U - Vegetation of floodplains	2.2%	10.5%	8.1%	6.1%	7.5%	41.6%

3.3. Richness – species numbers and floristic similarity between regions

The main results of the analyses of species richness are shown in Table 3. The total number of taxa recorded in this study is 983. The number of species recorded for Denmark includes both species listed in the descriptions of vegetation units mapped for Denmark and additional species which are known to occur in these vegetation units (Table 1) in the country (naturally or introduced). The similarity between different floras are described using the Sørensen similarity index [16]. The results are shown in the last column of Table 3. The number of species per future-climate analogue area was calculated as the number of taxa found in the description of each vegetation unit present in the region. These descriptions are not complete, but represent common, typical or diagnostic species of the particular vegetation unit in its total distribution area. The species numbers are therefore not total counts of the floristic diversity of the future-climate analogue areas.

Table 3. Natural vegetation plant richness: number of NVE species in Denmark and the five future-climate analogue areas as well as the number of shared species and the Sørensen similarity index between Denmark and each future-climate analogue area.

Region	# species	# shared species	Similarity
Denmark	[1]539	-	-
I – Germany	447	355	0.78
II – Northern France	509	371	0.76
III – Southern France	556	357	0.68
IV – Northern Italy	415	234	0.51
V – Balkan	441	226	0.48

1 This number includes all species recorded in this study that are known to be present in Denmark and not just the 270 species recorded for the NVE units in the country (see text).

3.4. Possible changes in species composition

3.4.1. Forest

Under the current climatic regime temperate forest would be the dominating natural vegetation type in Denmark and most of the study area without human disturbance. However, natural forest has probably had a certain degree of openness allowing many open to semi-open vegetation types to co-exist [17]. Many of these open vegetation types have become much more widespread in historical times due to traditional land-use, e.g., forest clearing and extensive grazing, and are frequently regarded as semi-natural secondary. Their flora, nevertheless, represents an integral part of the natural species pool. The temperate forest type formation includes forest on rich as well as very poor soil and is species rich, with 306 taxa recorded for the formation in Denmark (Table 4). Of these, 53 scores less than 100 points on the Index of Resistance and could thus be be considered threatened to some degree by the impending climate change. However, we checked the NVE distributions against an on-line, virtual flora published by The Swedish Museum of Natural History that shows the distribution of more than 2,000 taxa in the Nordic countries and on the Northern Hemisphere (http:/linnaeus.nrm.se/flora/). These maps showed wider distributions for all species but one (*Rubus saxatilis*) which indicates that the NVE approach may be too imprecise to provide a reliable measure of risk. A total of 179 species are found in warmer climates, but are not yet present in Denmark and might be considered potential immigrants (Index of Immigration > 0). As we have adjusted the list of species present in Denmark, this index is probably more useful than the IoR. Examples of potential immigrants are shown in Table 5.

One of the most prominent changes to be expected under a warmer climate is a shift from dominance of temperate deciduous trees and shrubs to a higher percentage of drought-resistant summer- or wintergreen broadleaved woody species characteristic of the thermophilous forest formation currently found in south-central Europe. An ongoing vegetation shift from indigenous deciduous to exotic evergreen species has already been observed in forests of southern Switzerland [3]. A conspicuous floristic element of the thermophilous broadleaved forests are species of the genus *Quercus*, e.g., *Q. pubescens, Q. cerris, Q. frainetto.* The thermophilous forest formation occupies a transitional position between temperate forest in the North and the Mediterranean sclerophyll forest and the xerophytic coniferous forest to the South and the forest steppes of the East. Floristically, the transitional position is shown by the number of shared species among the formations mentioned. The formation is not considered a naturally occurring formation for Denmark, but it is noteworthy that of 315 taxa recorded for this vegetation formation in the study area 167 are already present in Denmark and the remaining 148 taxa could be considered potential immigrants. Examples of potential immigrants typical of the thermophilous forest formation are shown in Table 5.

Table 4. Number of species recorded by NVE for the three major vegetation formations occurring in Denmark today and one potential vegetation type (thermophilous forest) that may become frequent in the future. The table also lists the number of species that are potentially threatened by global warming in Denmark (Index of Resistance (IoR) < 100) and the number of potential new species (Index of Immigration (IoI) > 0).

Formation	# today	# IoR < 100	# IoI > 0
Temperate forest	306	53	179
Dwarf shrub heath	52	18	8
Coastal vegetation	87	60	61
Thermophilous broadleaved forest	167	-	148

3.4.2. Dwarf shrub heath

Dwarf shrub heath is dominated by dwarf shrub species of the genera *Calluna, Empetrum, Erica* and *Genista*, but grasses, mosses and lichens are also common. Dwarf shrub heaths occur naturally on species poor, acid soils, but their distribution in Denmark and elsewhere increased through history as a result of specific forms of land-use (e.g. grazing, fire or turf-cutting). In the last century modern, intensive agriculture and afforestation has reduced the extent of dwarf shrub heaths dramatically. True climax heath vegetation occurs in coastal or mountain areas of the British Isles whereas its distribution in the study area is limited and mainly restricted to the Atlantic coast in Denmark and in Northern France. Nevertheless the remaining patches of dwarf shrub heath have a high priority in nature management in Denmark. The formation is relatively species poor and Table 4 shows that 52 taxa are recorded for Denmark. Of these, 18 have an Index of Resistance <100. The distribution of these taxa was checked againg the virtual flora as described above and only one species, *Empetrum nigrum*, can be considered seriously threatened by a warming climate. An additional eight species are found in warmer climates, but not in Denmark and might be regarded as potential immigrants. Examples of potential immigrants are shown in Table 5.

3.4.3. Coastal vegetation - dune and salt marsh vegetation

Coastal vegetation is formed by processes of sedimentation and removal of sediments along the coasts and by flooding as a result of tidal waters. The vegetation must be able to endure continuous erosion and deposition of sand and/or the impact of sea salt. Only a limited number of species have adapted to this very special habitat and climate is generally considered to be of secondary importance for the distribution of halophytic vegetation and the many species have similar, large and widespread distributions. The coastline of Denmark is very long compared to the size of the country (> 7,000 km) and the dune systems and salt marshes are well developed in different forms on exposed and sheltered coasts and the fraction of land with this formation is the highest in the study area (see Table 2) and more than five times the average value. Vegetation units belonging to the formation are distributed in a narrow fringe along the Atlantic coast from Denmark through Germany, Holland, Belgium and France to the Pyrenees. The formation has 83 taxa recorded for Denmark. Of these, 60 has an Index of Resistance <100 and can be considered threatened to some degree by climate change; this is the highest percentage threatened species for the formations considered. Also, when checking against the virtual flora (see above), 17 species remain in the list. Among the prominent taxa that may disappear from Danish coastal habitats are *Artemisia maritima, Astragalus danicus, Atriplex calotheca, A. glabriuscula, A. pedunculata, Cakile maritima subsp. baltica, Calammophila baltica, Centaurium littorale, Cochlearia officinalis, Crambe maritima, Eleocharis parvula, Elymus athericus, Empetrum nigrum, Festuca rubra subsp. arenaria, Helichrysum arenarium, Koeleria arenaria, Lathyrus maritimus, Leymus arenarius, Petasites spurius, Polygonum oxyspermum, Salix repens subsp. dunensis and Samolus valerandi.* However, 61 species are found in warmer climates and could be potential immigrants, which is also the highest number found in this study. As long-distance dispersal is likely to be much more effective in these coastal systems (because species are adapted to dispersal in a highly dynamic system and dispersal by sea currents), coastal vegetation may be expected to exhibit faster and stronger response to 21st climate change. This is particularly noteworthy as dune vegetation is a prioritized habitat within the European Natura 2000 network. Examples of potential immigrants are shown in Table 5. Likely immigrants include *Euphorbia paralias*, a potentially aggressively expanding species that is considered a noxious weed in Australia where it has been introduced. Another species that could benefit from global warming in Denmark is *Ulex europaeus*, a thorny shrub capable of forming large thickets in fixed dunes. This is an example of a species already introduced in Denmark which may naturalize rapidly when climate conditions become suitable.

Table 5. Examples of potential immigrants in Denmark as a response to global warming. The table includes the 20 species with the highest Index of Immigration scores. Species already introduced to Denmark and locally naturalized are marked with an *.

Temperate forest	Dwarf shrub heath	Coastal vegetation	Thermophilous forest
Acer opalus	*Anagallis tenella*	*Anthemis maritima*	*Acer monspessulanum*
Acer monspessulanum	*Cytisus scoparius subsp. maritimus*	*Arbutus unedo*	*Acer opalus*
Amelanchier ovalis	*Erica cinerea*	*Artemisia santonicum*	*Amelanchier ovalis*
Anthriscus nitida	*Erica vagans*	*Asparagus acutifolius*	*Carex halleriana*
Aposeris foetida	*Sanguisorba minor**	*Carex halleriana*	*Clinopodium vulgare*
Arabis pauciflora	*Scilla verna*	*Clematis flammula*	*Coronilla coronata*
Calamagrostis varia	*Thymus praecox*	*Ephedra distachya*	*Cytisophyllum sessilifolium*
Cardamine enneaphyllos	*Ulex europaeus subsp. latebracteatus*	*Erica cinerea*	*Dictamnus albus*
Cardamine trifolia	*Ulex gallii*	*Euphorbia paralias*	*Erythronium dens-canis*
Carex brizoides		*Euphorbia portlandica*	*Euonymus verrucosa*
Coronilla coronata		*Festuca juncifolia*	*Galium schultesii*
Cytisophyllum sessilifolium		*Frankenia laevis*	*Hierochloe australis*
Epipactis microphylla		*Helichrysum conglobatum*	*Lamium montanum*
Euonymus latifolia		*Juniperus oxycedrus*	*Lonicera caprifolium*
Euonymus verrucosa		*Limonium auriculae-ursifolium*	*Lonicera etrusca*
Lamium montanum		*Phillyrea latifolia*	*Luzula forsteri*
Lonicera nigra		*Pistacia lentiscus*	*Melica picta*
Lonicera etrusca		*Puccinellia limosa*	*Molinia arundinacea*
*Pinus nigra**		*Quercus ilex*	*Peucedanum cervaria*
Pyrus cordata		*Rubia peregrina*	*Primula acaulis*
Ranunculus nemorosus		*Ruscus aculeatus*	*Pyrus cordata*
Ranunculus platanifolius		*Sarcocornia fruticosa*	*Rubia peregrina*
Salix atrocinerea		*Sarcocornia perennis*	*Ruscus aculeatus*
Senecio nemorensis		*Smilax aspera*	*Sesleria albicans*
Sesleria albicans		*Spartina maritima**	*Sorbus domestica*

4. Implications for management

Most studies of the potential effects of global climate warming on plants and animals are based on information from atlas surveys (see, e.g., [18-20]). Here, we have explored a new approach using information extracted from a map of the potential natural vegetation of Europe (NVE). The map describes a set of vegetation units and the species that characterize them. The NVE maps are based on expert judgment and have a more qualitative nature than distribution data from an atlas survey. This study reveals some of the advantages and weaknesses of the approach. First, the benefits are: (1) The

NVE covers the whole flora of Europe in contrast to the Atlas Flora Europaea that is the traditional source of plant distribution data for Europe but covers only ca. 20% of the total flora. (2) The NVE uses a phytosociological approach and thus provides a good overview of vegetation assemblages (as seen today), the species that characterize them and their preferred environment. The shortcomings are: (1) the lists of species assigned to each vegetation unit are not complete which makes it difficult to compare units and to extract the precise distribution for individual taxa. (2) Some vegetation units cover relatively large areas which decrease the spatial resolution of the data.

In conclusion, plant distribution data from the NVE cannot replace traditional data sources, but it provides a wealth of suppplementary information and offers a valuable vegetation/ecosystem perspective to climate envelope modelling that traditionally focuses on individual taxa or measures of species richness and diversity. A number of general conclusions can be drawn from this study and from the literature. It seems reasonably clear that:

- Substantial changes will happen. Climate change of the magnitude foreseen will alter the distribution of the flora and, hence, the future species composition of the natural vegetation. Range shifts of individual species, however, are extremely difficult to accurately predict with the available model tools and the present knowledge of the climatic tolerances and dispersal abilities of the species.
- Natural change is slow. Analyses of biogeographic patterns of plant species and historic evidence show that many if not most plants will be unable to keep track of the expected climatic changes. This will probably cause time lags over perhaps hundreds of years before many of the new immigrants from south will appear in Denmark (and some are likely to simply fail to shift northwards). Nevertheless, potential immigrant species occurring close to Denmark, for instance in Germany, could arrive relatively fast, notably species adapted for long-distance dispersal and species that thrive in the current landscape, dominated by intensive agriculture. Fast immigration is also expected for coastal vegetation, with many species adapted for efficient long-distance dispersal by sea currents. In Denmark the invasive *Rosa rugosa* is a good example of this.
- The direction and speed of change in a human modified landscape is much more difficult to predict. Human land-use has transformed the landcover of Europe and most landscapes are characterized by fragmented patches of natural vegetation in a matrix of agricultural land. This will significantly reduce the natural dispersal capacity of many species (habitat specialists), but enhance the distribution of others (mainly generalist species). Translocation of species, e.g., the use of non-native, 'exotic' plants in gardens, parks or forests, is another important factor that may provide some species a head start, because they will already be present in the areas where the climate becomes suitable. This study may underestimate this threat to a certain degree as the NVE atlas exclude non-European species, e.g., *Robinia pseudacacia* and *Buddleya davidii* that are problematic invasive species in large parts of Europe south of Denmark. Both species are widely planted in Denmark and selfsown individuals already occur locally in urban areas.
- Changes affect species independently. The future composition of vegetation types may very well be different from what we see today. Historic evidence also shows that vegetation units vary through time [21]. There is, however, a large overlap of species present in Denmark now and in vegetation types occurring in the five subregions further south.
- Immigration of new species will not necessarily lead to extinction of native species. Recent evidence (Svenning et al., this volume) shows that many areas of Europe are not saturated with species. This indicates that most plant communities will be able to absorb new immigrants without losing native species. Native species already threatened by climate change are more vulnerable to competition with new immigrants and may therefore become extinct. Facing an uncertain, but rapidly changing future the best solution is probably to adopt a dynamic strategy based on modelling, monitoring and adaptive management. Modelling

techniques are constantly improved, but will never be perfect and should therefore be linked to a fine-masked network of observatories to check model predictions and feed empirical data back into the models for calibration and further development.

5. Acknowledgements

This is a paper resulting from the international conference "Beyond Kyoto: Addressing the Challenges of Climate Change - Science Meets Industry, Policy and the Public", held in Aarhus, Denmark, March 5th-7th 2009. We thank the Natural Vegetation of Europe project for access to the distribution data and gratefully acknowledge economic support from the Danish Natural Science Research Council (grant #272-07-0242 to JCS).

References

[1] Parmesan C *et al.* 1999 Nature **399** 579
[2] Parmesan C and Yohe G 2003 Nature **421** 37
[3] Walther GR *et al.* 2002 Nature **416** 389
[4] Lenoir J *et al.* 2008 Science **320** 1768
[5] Myers OB *et al.* 2009 Global Change Biol **15** 1434
[6] Skov F and Svenning JC 2004 Ecography **27** 366
[7] Svenning JC and Skov F 2004 Ecology Letters **7** 565
[8] Svenning JC and Skov F 2005 Journal Of Biogeography **32** 1019
[9] Svenning JC *et al.* 2008 Ecography **31** 316
[10] Svenning JC and Skov F 2007 Ecology Letters **10** 453
[11] Stoltze M and Pihl S, 1998 *Rødliste 1997 over dyr og planter i Danmark* (Miljø- og Energiministeriet, Skov- og Naturstyrelsen.
[12] Bohn U *et al.*, 2003 *Map of the Natural Vegetation of Europe* (Bundesamt für Naturschutz (BfN), Bonn, Germany, p. 1-655.
[13] IPCC, 2001 *Climate change 2001: Synthesis report - Summary for policy-makers.* (Intergovernmental Panel on Climatic Change, Cambridge University Press, Cambridge.
[14] IPCC, 2007 *Climate Change 2007: Synthesis Report. Contribution of Working Groups I, II and III to the Fourth Assessment Report of the Intergovernmental Panel on Climate Change* (IPCC, Geneva, Switzerland, p. - 104.
[15] Mitchell TD *et al.* 2003 Journal of Climate
[16] Legendre P and Legendre L, 1998 *Numerical ecology* (Elsevier Science BV, Amsterdam, p. - 853.
[17] Svenning JC 2002 Biological Conservation **104** 133
[18] Guisan A and Zimmermann NE 2000 Ecological Modelling **135** 147
[19] Araujo MB *et al.* 2006 Journal Of Biogeography **33** 1712
[20] Normand S *et al.* 2007 Journal for Nature Conservation **15** 41
[21] Markova AK *et al.* 2009 Quaternary International **201** 53

Potential 21st century changes to the mammal fauna of Denmark – implications of climate change, land-use, and invasive species

Camilla Fløjgaard[1,2*]

Naia Morueta-Holme[1]

Flemming Skov[2]

Aksel Bo Madsen[2]

Jens-Christian Svenning[1]

[1]The Ecoinformatics and Biodiversity Group, Department of Biological Sciences, Aarhus University, Ny Munkegade 114, DK-8000 Aarhus C, Denmark.
[2]Department of Wildlife Ecology and Biodiversity, National Environmental Research Institute, Aarhus University, Grenaavej 14, DK-8410 Rønde, Denmark

*E-mail: camilla.flojgaard@biology.au.dk

Abstract. The moderate temperature increase of 0.74 °C in the 20th century has caused latitudinal and altitudinal range shifts in many species including mammals. Therefore, given the more dramatic temperature increase predicted for the 21st century, we can therefore expect even stronger range shifts as well. However, European mammals are already faced with other anthropogenic pressures, notably habitat loss, pollution, overexploitation, and invasive species, and will have to face the combined challenge posed by climate change in a landscape highly influenced by human activities. As an example of the possible consequences of land use, invasive species, and climate change for the regional-scale mammal species composition, we here focus on the potential 21st century changes to the mammal fauna of Denmark. Supported by species distribution modelling, we present a discussion of the possible changes to the Danish mammal fauna: Which species are likely to become locally extinct? Which new species are most likely to immigrate? And, what is the potential threat from invasive species? We find that future climate change is likely to cause a general enrichment of the Danish mammal fauna by the potential immigration of seventeen new species. Only the northern birch mouse (*Sicista betulina*) is at risk of extinction from climate change predicted. The European native mammals are not anticipated to contribute to the invasive-species problem as they coexist with most Danish species in other parts of Europe. However, non-European invasive species are also likely to enter the Danish fauna and may negatively impact the native species.

1. Introduction

During the past century the global average temperature has increased by ~0.74 °C [1], causing many species to shift their geographic ranges. For example, in Yosemite National Park in western North America, the altitudinal range limit of many small mammals have shifted upwards [2], while in the Northern Great Lakes Region the small mammal community is becoming increasingly dominated by expanding southern species, while northern species are declining in numbers[3]. In Europe, species range shifts have principally been reported for other taxa, e.g., an upwards altitudinal range shift for forest plants and a latitudinal range shift for butterflies [4,5]. Considering the climate changes predicted for the future [1], we can expect dramatic spatial changes in species composition as climate-sensitive species capable of tracking the changing climate will shift their ranges and climate-sensitive poor dispersers will contract their ranges or go extinct [6].

Finds of mammal remains from the Last Ice Age, phylogeographic studies, and species distribution modelling studies indicate that at least the widespread European mammals have the capacity to respond to climate change by large geographic range shifts [7-10]. However, the effect of the 21st century climate change may be more severe than Pleistocene climate changes, given that many mammal species are already negatively affected by other anthropogenic pressures: A recent report on the global status of mammals shows that a fourth of all mammals are threatened by extinction [11] and it has been assessed that 15% of the European mammals are threatened (vulnerable, endangered or critically endangered), another 9% are close to being threatened, and 27% have declining populations [12]. The assessment of the status of mammals does not evaluate the possible risk from future climate change. The status is attributed alone to anthropogenic threats such as habitat loss, disturbance, pollution, overexploitation, and invasive species; threats that are also often the focus in conservation strategies [12]. In this paper we discuss the possible consequences of future climate change scenarios and its interplay with other anthropogenic factors to assess the likely 21st century changes in the Danish mammal fauna.

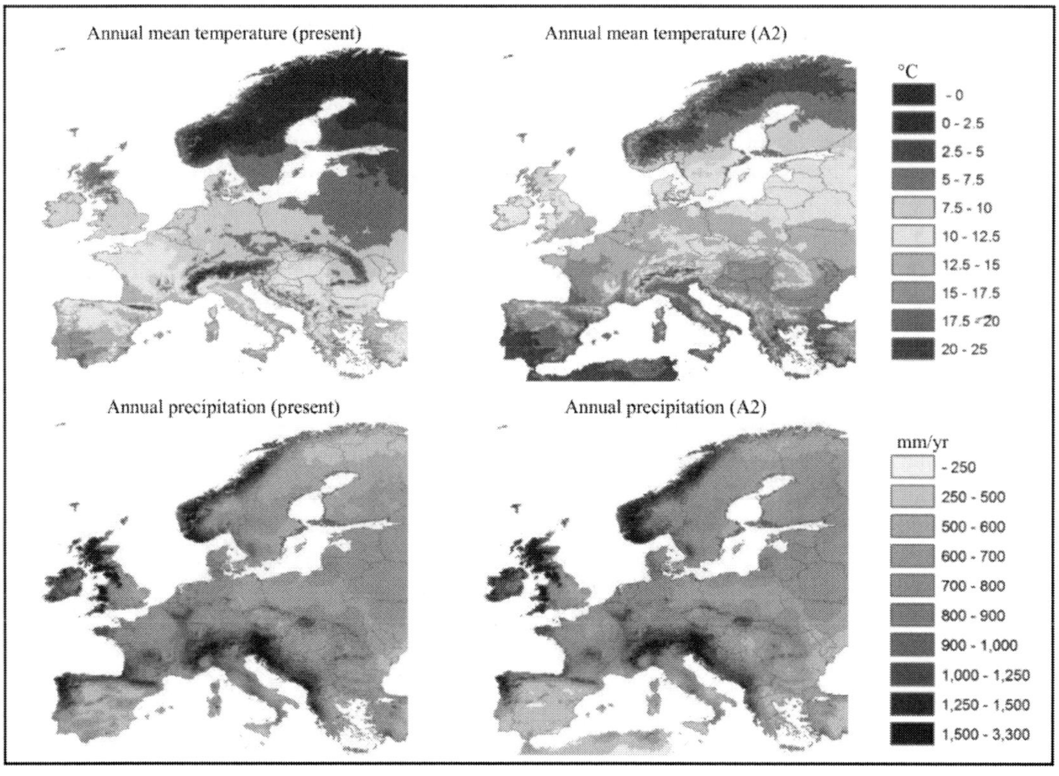

Figure 1. Annual mean temperature and annual precipitation in Europe at the present (1961-1990) and as predicted under the A2 future climate scenario (2070-2099).

Under the IPCC A2 scenario, that is often described as a business-as-usual scenario and assumes a moderate to high increase of atmospheric CO_2, a global temperature increase of 3-4 °C by 2090-2099 relative to 1980-1999 is expected [1] (Fig. 1). In European scale, climate conditions are expected to shift in a north-eastern direction under the A2 scenario [1], mainly reflecting a temperature shift by an increase in annual mean temperature by 3-4 °C (figure 1 and 2). The pattern for precipitation is more complex: Southern Europe and in particular the Iberian Peninsula becomes remarkably drier, Central Europe becomes slightly drier, whereas the precipitation in Scandinavia is predicted to increase relative to the present. Metzger et al. [13] modelled the shift in environmental zones in Europe under the A2 scenario. They predicted a trend towards a general northward shift of especially the Southern and Central European environmental zones, but also that the Atlantic environments in Denmark and the British Isles would be relatively stable over the 21st century.

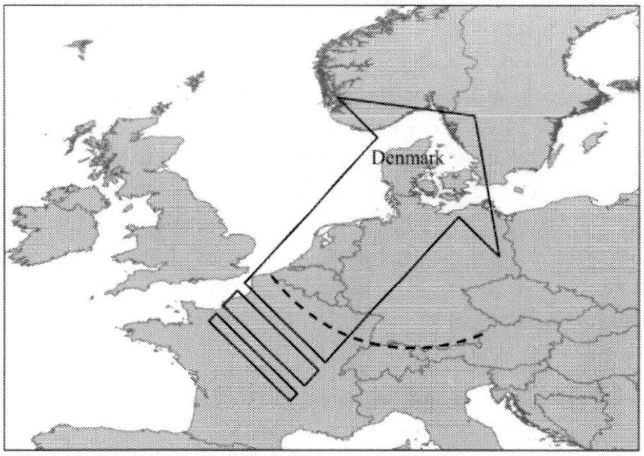

Figure 2. The scenarios for the climate in Europe at the end of the 21st century generally show a climatic shift in a north-eastern direction (indicated by the arrow). The dashed line marks an approximate 'immigration' limit, with species occurring north of this line being within migration distance of Denmark in the next 100 years, given a maximum mammal migration velocity of ~7.9 km/yr (see Methods) and the direction of the climatic shift.

Today, Denmark has a highly fragmented landscape in terms of natural habitats that persist in small areas in a matrix of intensively managed agricultural land and urban areas. A number of especially large species have been lost over the last centuries or millennia due to hunting and persecution, notably moose (*Alces alces*), aurochs (*Bos primigenius*), lynx (*Lynx lynx*), wildcat (*Felis sylvestris*), and brown bear (*Ursus arctos*) more than a thousand years ago, and wild boar (*Sus scrofa*) and wolf (*Canis lupus*) in the last 2-300 years [14]. The extant mammal fauna of Denmark is also highly affected by the intense past and present human utilization of the landscape. Species such as bank vole (*Clethrionomys glareolus*), European polecat (*Mustela putorius*), pine marten (*Martes martes*) and badger (*Meles meles*) have experienced population declines and contingent losses of genetic diversity over the last 100 years due to habitat loss and fragmentation driven by an increased demand for cultivated land and human-induced barriers (e.g., [15-18]).

In this paper, we outline the most likely changes in the Danish mammal fauna over the 21st century, assuming climate changes as predicted under the A2 scenario [1]. We discuss (i) the likely impacts of climate change and other potential drivers of changes in population size or geographic range on the resident Danish native fauna, as well as (ii) which species are likely new immigrants to the Danish fauna taking climate, land-use, and biotic interactions into account. We supplement the discussion with species distribution modelling for selected species.

2. Methods

2.1. Species and climate data

Data on the distributions of European mammals came from the Atlas of European Mammals [19] and the Danish Mammal Atlas [20]. In this paper we focused on the terrestrial non-volant mammals, excluding bats and marine species such as seals. We used the A2 future climate scenario as modelled in the TYN_SC 1.0 data set: ([21], http://www.cru.uea.ac.uk/~timm/data/index-table.html) that is commonly used in predictions of species potential future distributions (e.g.,

[22]). It represents a moderate to high increase in global temperature of 3-4 °C by 2090-2099 relative to 1980-1999. The corresponding climate data for the present was also available at http://www.cru.uea.ac.uk/~timm/data/index-table.html (CRU_CL 2.0, [23]). Annual mean temperature (AMT) and annual precipitation (PANN) were calculated according to Ohlemüller *et al.* [24]. All GIS operations were performed using ArcGIS 9.3 (ESRI, Redlands, CA, USA).

2.2. Estimation of migration speed and direction of mammals

We used two studies on the observed spread of introduced mammals in Europe to estimate a likely maximum migration rate for mammals across modern European landscapes. The grey squirrel (*Sciurus carolinensis*) spread at an average speed of 7.7 km yr^{-1} in Britain from 1965 to 1980 ([25] calculated from data in [26]), while the muskrat (*Ondatra zibethicus*) spread at an average speed of 10.9 km yr^{-1} during 1900-1930, expanding from its introduction site near Prague to most of Central and Eastern Europe [27]. In the period 1930-1960 the spread of the muskrat slowed down to an average speed of 5.1 km yr^{-1}, which is attributed to the trapping programmes initiated around 1925-30 [27]. Averaging across these three migration rates of spread we use ~7.9 km yr^{-1} as the likely maximum migration rate. We note that there is some uncertainty in the estimate due to the limited sample size on migration rates and that many species are likely to spread much more slowly than these fast-spreading invasive species. It is therefore a liberal estimate that species occurring within 800 km of Denmark are likely to immigrate to Denmark in the future.

To further reduce the number of likely new immigrants to Denmark we also estimate the direction of the spread of species. The direction that mammal species are likely to migrate in if they track the 21st century climate changes is probably towards the north-east, reflecting the general direction of the expected climate shift, especially noticeable in the north-east shift in temperature clines (figure 1 and 2), as also predicted by previous studies [22,28]. Hence, mammals with a northern range limit within 800 km south-west of the Danish border are considered relatively likely 21st century immigrants to Denmark.

2.3. Species distribution modelling

For the selected relevant species such as likely 21st century immigrants to Denmark and species judged most at risk of extinction from climate change given their current geographic range, we used species distribution modelling to quantify their potential distributions under present and future climatic conditions. This was done using maximum entropy species distribution modelling, a machine-learning technique based on the principle of maximum entropy [29], as implemented in Maxent 3.2.1 (http://www.cs.princeton.edu/~schapire/maxent/). Maxent has performed very well in comparative studies of species distribution modelling algorithms [30,31]. First, we modelled the present distribution of suitable climatic conditions for each species using AMT and PANN, two climatic variables that have been successfully used in previous studies for predicting mammal species distributions in the present and under such a different climate as the Last Glacial Maximum [7,32]. Thirty percent of the samples were randomly selected as test data. The Area Under the receiver operating characteristic Curve (AUC) when applying the fitted models to the test data was ≥0.731 for all species, indicating the models to have moderate to good predictive ability [30]. Models calibrated on the full dataset were then projected onto the A2 scenario predictions for AMT and PANN for 2070-99 (figure 1). We used Maxent's logistic output format, as it can be interpreted as an estimate of a species' probability of occurrence. For native species we used the threshold that maximises both sensitivity and specificity, i.e., the threshold that maximizes the number of correctly predicted presences and absences. To reduce the risk of underestimating the introduced species' climatic niche we therefore used Maxent's "10 percentile training presence"-threshold, which is a more liberal threshold and only allows wrongful prediction of 10% of the occurrences. It was not possible to create a realistic model of the climatic niche for four introduced and invasive species, *Muntiacus reevesi, Tamias sibiricus, Procyon lotor* and *Sciurus carolinensis*, based on their currently highly limited European distributions.

IOP Conf. Series: Earth and Environmental Science **8** (2009) 012016 doi:10.1088/1755-1315/8/1/012016

Table 1. The Danish mammal fauna under 21st century climate change: The 44 extant terrestrial mammal species in Denmark and the most likely new 21st century immigrant species [12,19,20], their current conservation status, and the expected main 21st century threats in Denmark.

Common name	Scientific name	Status in DK		Possible 21st century threats
Hedgehog	*Erinaceus europaeus*	N	C	
Common shrew	*Sorex araneus*	N	C	
Pygmy shrew	*Sorex minutus*	N	C	
Water shrew	*Neomys fodiens*	N	C, D	Habitat destruction
Mole	*Talpa europaea*	N	C	
Brown hare	*Lepus europaeus*	N	C, D	Intensive agriculture
Rabbit	*Oryctolagus cuniculus*	Ni	R	Intensive agriculture
Red squirrel	*Sciurus vulgaris*	N	C	Grey squirrel, forest fragmentation
*Siberian chipmunk	*Tamias sibiricus*	I	R, I	Eradication
European beaver	*Castor fiber*	N	R, I	Habitat destruction
Bank vole	*Clethrionomys glareolus*	N	C	Habitat destruction
Water vole	*Arvicola terrestris*	N	C	American mink
*Muskrat	*Ondatra zibethicus*	I	R, I	Eradication
Field vole	*Microtus agrestis*	N	C, D	Habitat destruction
Common vole	*Microtus arvalis*	N	R	
Harvest mouse	*Micromys minutus*	N	R	Habitat destruction
Striped field mouse	*Apodemus agarius*	N	R	
Yellow-necked mouse	*Apodemus flavicollis*	N	C	
Woodmouse	*Apodemus sylvaticus*	N	C	
Brown rat	*Rattus norvegicus*	N	C	Management
Black rat	*Rattus rattus*	Ni	R	Management
House mouse	*Mus musculus*	N	C	
Common dormouse	*Muscardinus avellanarius*	N	R, T	Habitat destruction
*Garden dormouse	*Eliomys quercinus*	N	R	
*Northern birch mouse	*Sicista betulina*	N	R, T	Climate change
*Coypu	*Myocaster coypus*	I	R	Eradication
Red fox	*Vulpes vulpes*	N	C	
*Racoon dog	*Nyctereutes procyonoides*	I	R	Eradication
*Racoon	*Procyon lotor*	I	R	Eradication
Stoat	*Mustela erminea*	N	C, D	Habitat destruction
Weasel	*Mustela nivalis*	N	C, D	Habitat destruction
Polecat	*Mustela putorius*	N	C	American mink
American mink	*Mustela vison*	I	C	Eradication
Stone marten	*Martes foina*	N	C	
Pine marten	*Martes martes*	N	R	Habitat destruction
Badger	*Meles meles*	N	C	Habitat destruction
Otter	*Lutra lutra*	N	R	Habitat destruction
Wild boar	*Sus scrofa*	N	R	Eradication
Fallow deer	*Dama dama*	Ni	C	
Red deer	*Cervus elaphus*	N	C	
Sika	*Cervus nippon*	I	R	
Roe deer	*Capreolus capreolus*	N	C	
Mouflon	*Ovis ammon*	Ni	R	

Table 1 continued.

Potential immigrant species		Status in EU		
*Millet's shrew	*Sorex coronatus*	N	C/ LC	Intensive agriculture
*Miller's water shrew	*Neomys anomalus*	N	C/ LC	Habitat destruction
*Bi-coloured white-toothed shrew	*Crocidura leucodon*	N	C/ LC	
*Greater white-toothed shrew	*Crocidura russula*	N	C/ LC	
*Lesser white-toothed shrew	*Crocidura suaveolens*	N	C/ LC	
*Grey squirrel	*Sciurus carolinensis*	I	R	Eradication
*Common hamster	*Cricetus cricetus*	N	C/LC	
*Southern water vole	*Arvicola sapidus*	N	R/NT	American mink
*Common pine vole	*Microtus subterraneus*	N	C/LC	
*Fat dormouse	*Glis glis*	N	C/LC	
*Reeves' muntjac	*Muntiacus reevesi*	I	R	Mortality on roads

N, native to Europe; Ni, native to Europe, but introduced to Denmark; I, introduced in Europe. Status: C, common; D, declining; R, rare; T, threatened; I, increasing/spreading. European Red List status: LC, least concern. *Climatic suitability models and current distributions are shown in figure 4-6.

Because these species are widely distributed in their native regions and exist under a broad span of climates it is likely that the current climate in Europe is suitable for them from the Mediterranean region in the south as far north as the northern limit of the temperate zone, and for *Tamias sibiricus* even into the boreal zone.

↑ **Bison (*Bison bonasus*)**
Photo: Camilla Fløjgaard
↓ **Wild boar (*Sus scrofa*)**
Photo: http://www.biopix.dk

↑ **Lynx (*Lynx lynx*)**
Photo: Hans Thomsen
↓ **Wolf (*Canis lupus*)**
Photo: Hans Thomsen

Figure 3. Large mammals that given major changes in land-use and/or public attitudes could occur in Denmark under the current climate as well as under the climate predicted for Denmark by year 2090-2099 under the A2 scenario.

3. Results and Discussion

We assembled information on the conservation status of mammal species currently found in Denmark as well as the species that could potentially migrate to Denmark during the 21st century (Table 1) [12,19,20]. According to the estimated migration distance, many species are within migration distance of the Danish border. However, we did not include species with habitat requirements not found in Denmark now and unlikely to be found there during the 21st century, e.g., montane and alpine species such as Alpine shrew (*Sorex alpinus*). Furthermore, species such as wolf (*Canis lupus*), lynx (*Lynx lynx*), brown bear (*Ursus arctos*), and wildcat (*Felis silvestris*) would clearly find suitable climatic conditions in Denmark today, but are extinct from Denmark due to historic hunting and persecution [14]. Although the climate in the future is likely to continue to be suitable for these species as well as other large species such as the European bison (*Bison bonasus*), the establishment of these mammals in Denmark in the future would require major changes in land-use and/or changes in public attitudes towards these species. We therefore do not discuss these species further in the present study, although we do not imply to rule out that they could become re-established as part of the wild mammal fauna in Denmark during the 21st century. Similarly, the climate is also suitable for wild boar (*Sus scrofa*) in Denmark today and is likely to continue to be suitable in the future. However, as long as wild boar is considered a threat to domestic pigs and persecuted when crossing the border into Denmark, it is unlikely that this species will establish a significant wild population in Denmark in the future (figure 3).

Potentially invasive species, species at risk of going extinct in Denmark, and native European species that are relatively likely immigrants to the Danish fauna during the 21st century were selected for species distribution modelling to quantify their potential future geographic distribution (figure 4-6). The modelling generally predicts a north-eastern shift in the species ranges over the 21st century. We recognize that the modelling does not directly take non-climatic factors such as habitat and food availability into account although such factors can be important range determinants. However, most of the species discussed here are relatively generalized in terms of feeding preferences and associated with widely available habitats in Denmark and the neighbouring regions, and we therefore think that habitat and food availability are less likely to be important constraints on the range dynamics of these species in relation to Denmark.

3.1. Mammal species at risk of extinction from Denmark

There are no endemic mammals in Denmark and there are very few small-ranged species, that is, the kind of species usually considered at high risk of extinction due to climate change [6]. Most of the Danish mammal species are widespread and found in much of Europe or even widely across Eurasia. In relation to the climate change perspective, it is particularly important to note that most of the species are widely distributed today in Central and Southern Europe under climates considerably warmer than the current Danish climate and often at least as warm as the A2 scenario predictions for the end of the 21st century.

The only clear exception is the Northern birch mouse (*Sicista betulina*), which despite having a wide Eurasian distribution and considered a Least Concern species on the IUCN Red List [12] is one of the rarest mammals in Denmark with two isolated populations [33]. Predictions of the species' future distribution indicate that the current climate in Jutland is suitable, but the biology, status, and distribution of the species in Denmark require further investigation. However, climate change might be one of the biggest threats to this species' survival in Denmark. Given the A2 scenario, the climate is predicted to become unsuitable for this species by 2070-2099 (figure 4).

There are a few species that can be considered at risk due to competition with introduced species. Competition between introduced American mink (*Mustela vison*) and native European polecat (*Mustela putorius*) does occur, but it is not considered to be a threat to the existence of the latter in Denmark [34]. Populations of water vole (*Arvicola terrestris*) have declined precipitously in the British Isles due to predation by American mink, but it is not considered a threat to water voles in Denmark since the populations here are more catholic in their habitat selection than the British populations [35].

Northern birch mouse (*Sicista betulina*)

Figure 4. Only one Danish mammal species, Northern birch mouse (*Sicista betulina*) is predicted to go extinct in Denmark by 2100 due to climate change under the A2 scenario (right). Left, the present distribution and current predicted climatic suitability. Photo: Westy Esbensen.

Common dormouse (*Muscardinus avellanarius*) is considered threatened and at risk of extinction [36] in Denmark. The preferred habitat is mixed-deciduous forest with a high plant diversity [36], which is rarely found in Denmark. In addition, common dormouse has declined due to intensive forest management, clear-cutting, and pesticide use at forest edges. Common dormouse is widely distributed in the warmer Central and Southern Europe and is considered a species of Least Concern on the IUCN Red List [12]. Habitat destruction must therefore be considered the most immediate and main threat to the continued occurrence of common dormouse in Denmark.

The effects of agricultural practises on mammals in the future are difficult to predict because they depend on development of new technologies and policies, e.g., with respect to set-aside areas, biofuel crops, or new tall-growing crops such as maize and sunflower.

3.2. New mammals in Denmark

Species that are within migration distance of Denmark and therefore reasonably likely to immigrate during the 21st century include Millet's shrew (*Sorex coronatus*), Miller's water shrew (*Neomys anomalus*), bi-coloured (*Crocidura leucodon*), lesser (*Crocidura suaveolens*) and greater white-toothed shrew (*Crocidura russula*), common hamster (*Cricetus cricetus*), common pine vole (*Microtus subterraneus*), southern water vole (*Arvicola sapidus*), and fat dormouse (*Glis glis*). The climatic niche projections for these species indicate that most are also likely to find suitable climatic conditions in Denmark at the end of the 21st century (figure 5). However, for common hamster future climatic conditions are predicted to be generally unsuitable.

An additional likely immigrant species may already now be establishing itself: The garden dormouse (*Eliomys quercinus*) is predicted to find highly suitable climate in the future in Denmark, especially in Jutland (figure 5). As this species thrives in gardens and parks it seems likely that it will find widely suitable conditions in Denmark in the future. Two individuals have been caught and a few more have been observed in Southern Jutland over the last 10 years [37]. Although these were probably introduced or escaped pets, it is therefore speculated that a small population might already have established itself in Jutland and could facilitate a quick future colonization of a larger part of Denmark.

The species mentioned here are all native to Europe and we consider it unlikely that they will provide problems akin to invasive species if they establish populations in Denmark, hereunder out-competing native Danish species. Notably, although there is some interspecific competition at local scales, e.g., between common shrew, Millet's shrew, and greater white-toothed shrew for specific habitats, at larger scales these species all coexist today in other parts of Europe [19]. While competition between the water vole and the southern water vole is likely to occur, we do not expect competitive exclusion of either species on a larger scale. In Denmark, we find fossorial, aquatic and semi-aquatic ecological forms of the water vole [35]. Given the aquatic life style of the southern water vole it is unlikely to interact much with the less aquatic forms of the water vole. Furthermore, the distributions of water vole and southern water vole overlap at larger scales in southwestern Europe and sympatry between the southern water vole and even the aquatic ecological form of water vole occurs at local scales [19].

3.3. Introduced and invasive species

Of the introduced mammals found in Europe only the American mink (*Mustela vison*) has become established and common in Denmark so far. It is found in most of the country except for a few minor islands [38]. Although the catastrophic decline of water vole in the British Isles is strongly linked to predation by mink [39], it has not had a major negative impact on any native mammals in Denmark except for possible minor competition with polecat for food and habitat. Given the wide distribution of mink in Europe and in its native range it is unlikely that it will disappear from Denmark because of climate change. Eradication efforts have been quite unsuccessful at large scales, but experimental re-introductions of otter (*Lutra lutra*) in the UK have been associated with rapid mink declines [40]. Since otter is rapidly re-expanding in Denmark after a strong 20th century decline [41], otter may limit the distribution and abundance of American mink in Denmark in the future.

Regular muskrat (*Ondatra zibethicus*) activity has been observed in Southern Jutland since 2000 [42]. It is already carefully watched by the authorities because it is known to cause damage to dikes and vegetation around lakes and streams elsewhere [42]. The species distribution modelling and its large North American range (from southern USA to Alaska), indicates that the climate in Denmark is predicted to continue being suitable for this species in the future (figure 6). We can consequently expect that muskrat will continue its spread to all of Denmark over the course of the 21st century.

The relatively few individuals of coypu (*Myocastor coypus*) (figure 7), a species native to South America, found in Denmark are presumed to be escaped animals from fur farms [43]. In Central Europe, Italy and the Balkans coypu is considered an invasive species that can reach high population densities and cause damage to crops and natural vegetation [19]. So far coypu has not been successful at establishing populations in the wild in Denmark probably because the winters are too cold [43]. We can expect the establishment and spread of coypu in Denmark as the winters get warmer. Already now coypu lives outside and successfully breeds in a Danish zoo in Jutland (Aqua Freshwater Aquarium; http://www.ferskvandscentret.dk/aqua/nyheder/).

Racoon (*Procyon lotor*) and Racoon dog (*Nyctereutes procyonoides*) (figure 7) are also occasionally found in the wild in Denmark, but are generally considered escaped from fur farms [44,45] although their invasion from Germany into Jutland has been expected for several decades. The model for racoon dog likely underestimates the species' climatic niche because the species' European distribution is not yet in equilibrium with climate (figure 6). Racoon dog occurs from southern China to south-eastern Russia and we can therefore assume that the climate in large parts of Europe today is suitable. The racoon occurs from Canada to northern South America and should therefore also find suitable climatic conditions in most of Europe today, but because of the limited European distribution (figure 6) it was not possible to make any realistic prediction of suitable climatic conditions for this species based on its European records. However, from the subtropical-tropical southern limits of these species' native ranges, we infer that the climate in large parts of Europe including Denmark is likely to continue being suitable for them in the future.

Southern water vole (*Arvicola sapidus*)

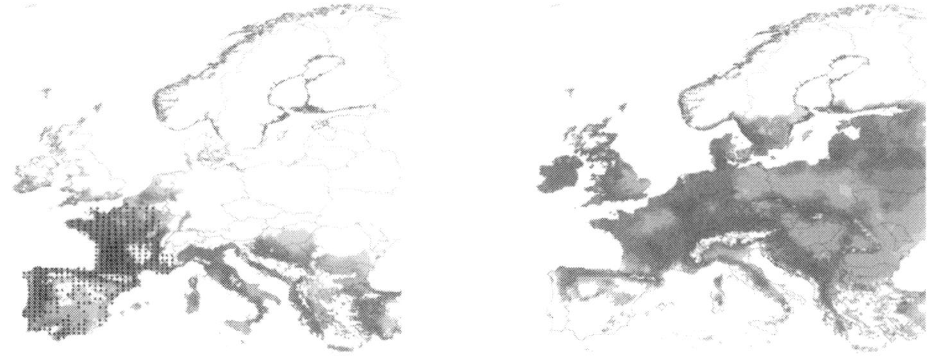

Common hamster (*Cricetus cricetus*)

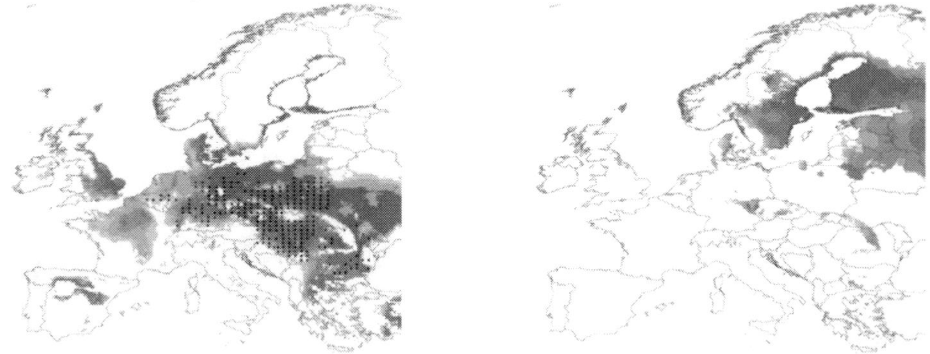

Bi-coloured white-toothed shrew (*Crocidura leucodon*)

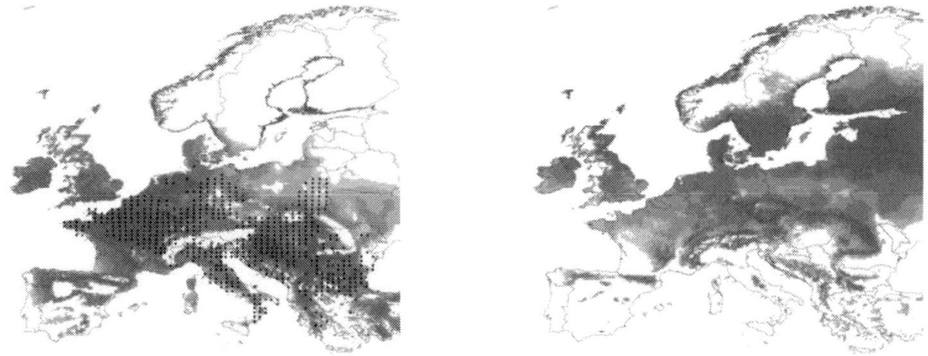

Greater white-toothed shrew (*Crocidura russula*)

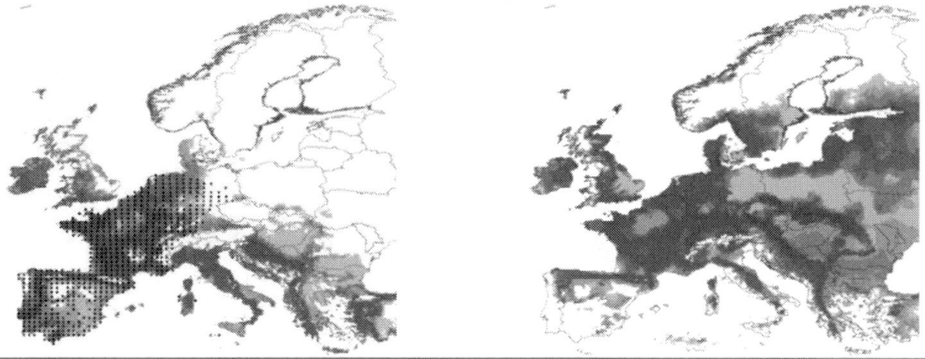

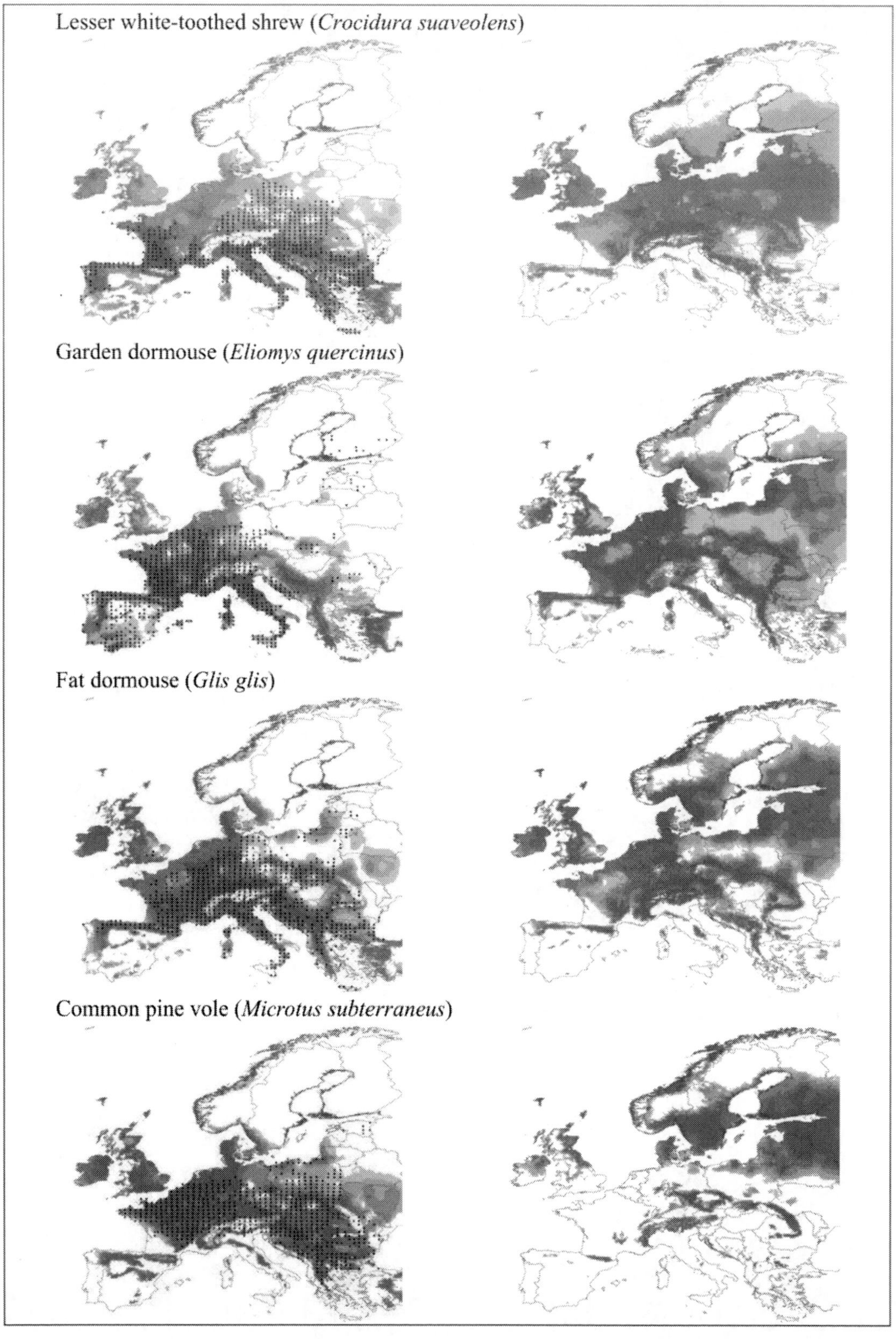
Lesser white-toothed shrew (*Crocidura suaveolens*)
Garden dormouse (*Eliomys quercinus*)
Fat dormouse (*Glis glis*)
Common pine vole (*Microtus subterraneus*)

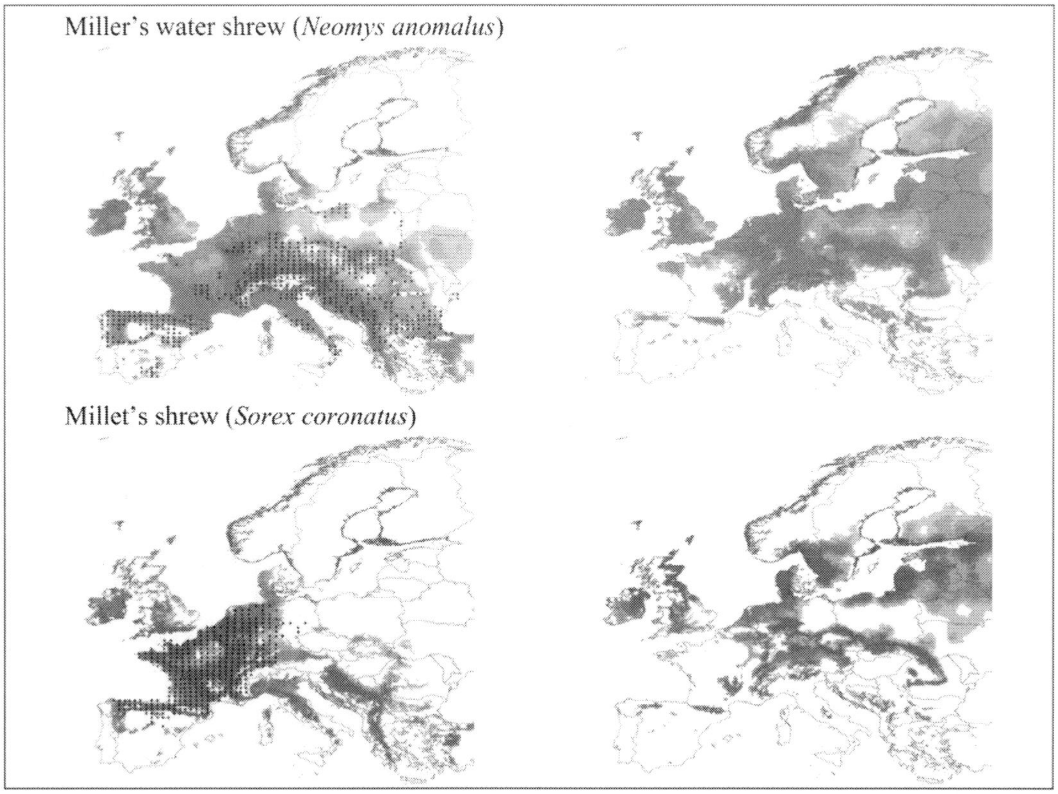

Figure 5. Ten species of mammals native to Europe south of Denmark that are reasonable likely to immigrate into Denmark over the 21st century given that the Danish climate will become highly suitable for these species under the A2 scenario (right) and that their current range (left) is considered within migration distance of Denmark (figure 2). *Cricetus cricetus* is a marginal case.

Similarly, the European distributions of grey squirrel (*Sciurus carolinensis*) (figure 6) and Siberian chipmunk (*Tamias sibiricus*) are too limited to realistically represent the species climatic niche. The grey squirrel is native in eastern North America from north of the Great Lakes to Florida in the south and from this it is likely that the current distribution in Europe does not at all fully represent its climatic niche. The species has spread rapidly on the British Isles as well as recently also in northern Italy replacing the native red squirrel (*Sciurus vulgaris*) as it moves along [46,47]. The population of grey squirrel on the British Isles is isolated from continental Europe and the population in northern Italy is so distant that there is not any immediate threat of grey squirrel spreading to Denmark in the near future. However, since it is very likely that grey squirrel will find suitable climatic conditions in Denmark both today and in the future, it is also very likely that a population will establish itself in Denmark should the species be introduced by humans or from escaped pets. In addition, given the expanding Italian populations, a larger expansion of grey squirrel across non-boreal parts of continental Europe is to be expected. The native range of Siberian chipmunk stretches from northern European Russia to China and Japan, while the sporadic populations in Central Europe all stem from released or escaped pets [19] (figure 6). In Denmark the species has survived in the wild in northern Zealand since 1990 and the small population is assessed to be increasing [48]. Again, from the wide native distribution it seems likely that Denmark will continue to be climatically suitable in the future. Reeves' muntjac (*Muntiacus reevesi*) (figure 7) was introduced to Britain and has been found in the wild since the 1920s where it is a considerable pest to undergrowth in forests and native and garden flora and competes with roe deer (*Capreolus capreolus*) and another exotic deer species, the Chinese water deer

IOP Conf. Series: Earth and Environmental Science **8** (2009) 012016 doi:10.1088/1755-1315/8/1/012016

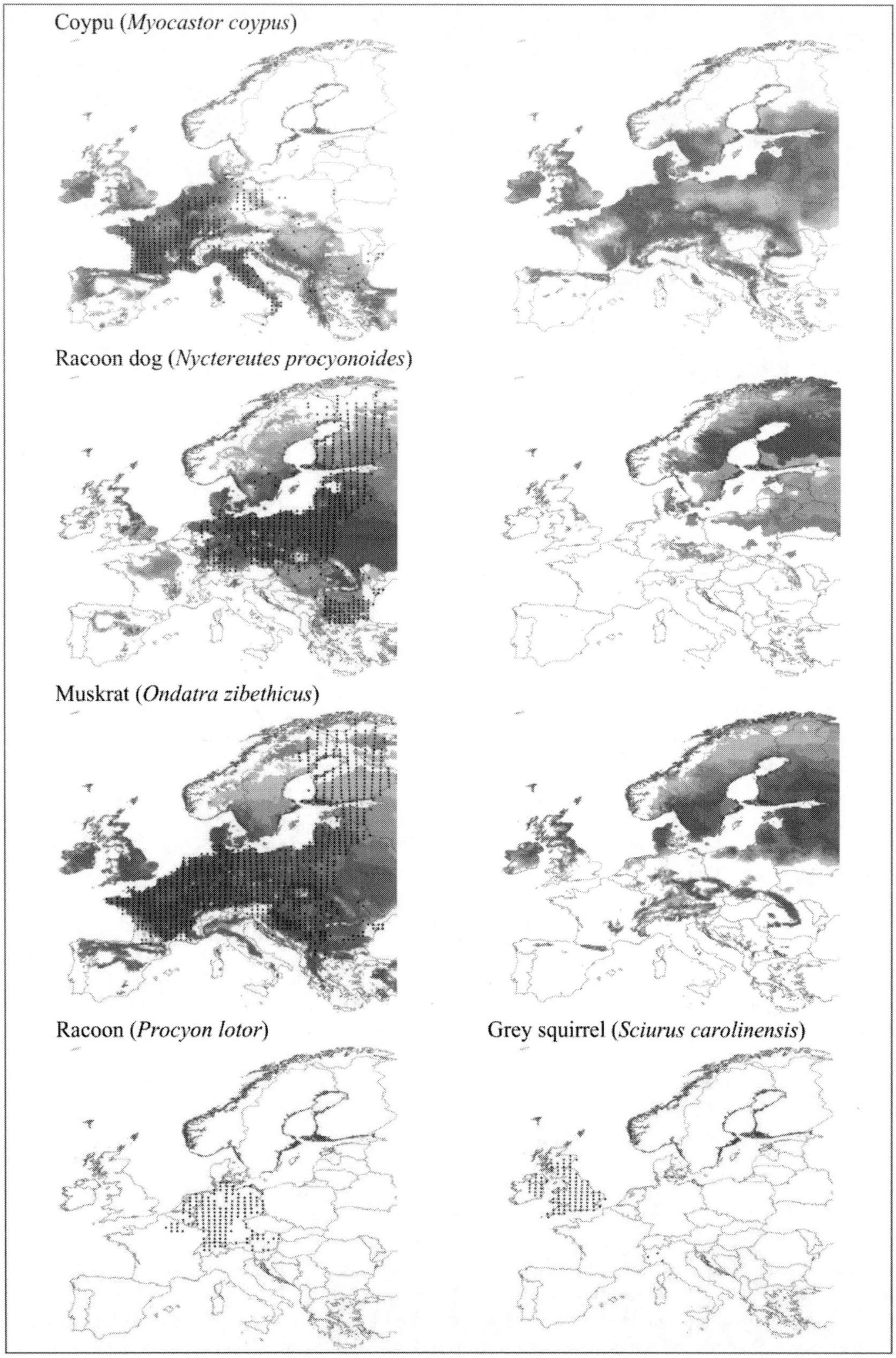

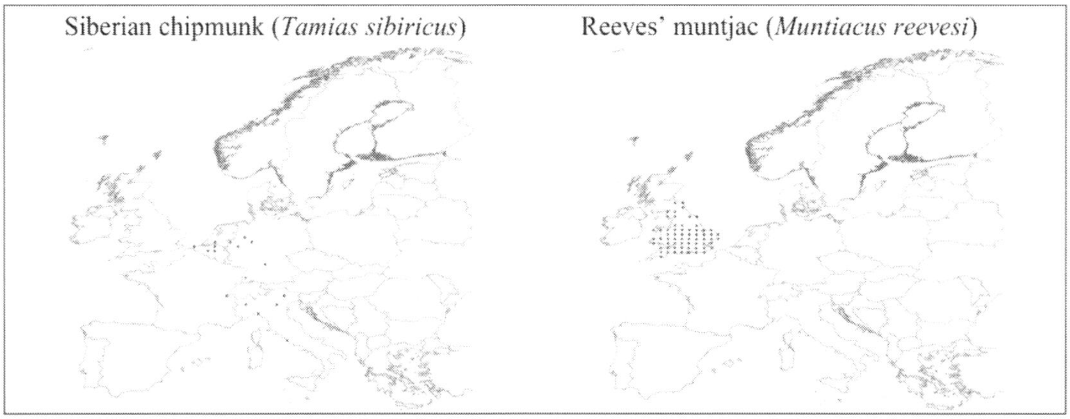

Figure 6. Six species of introduced mammals that are invasive in Europe outside Denmark at present and likely to expand in or spread to Denmark during the 21st century. They all tolerate or benefit from the climate change expected under the A2 scenario and have present distributions not too distant from Denmark (somewhat questionably for *Sciurus carolinensis*) and some of them are occasionally found in Denmark as escapes from fur farms. It was not possible to make a meaningful model of the climatic suitablitly of *Muntiacus reevesi, Procyon lotor*, *Sciurus carolinensis*, and *Tamias sibiricus*, but considering the climates covered by their native and current European ranges we think that both the present and expected future Danish climate is suitable for these species (see Discussion).

(*Hydropotes inermis*) at high densities [19]. Given the species native distribution in eastern China and the current distribution in the British Isles [19], it is likely that the climate in Denmark is suitable both today and in the future. There is no immediate threat of muntjac spreading from the British Isles to Denmark, but since it does have hunting value, introduction for sport shooting could be a motivation for introducing the species in Denmark as well.

4. Conclusion

The conclusions of the present study are based on results from species distribution modelling and a review of the literature on the species' ecology. This infers that the conclusions partly hinge on the extent to which we can trust the predictions of the species' future distributions. Besides the uncertainty in the future climate scenario and the uncertainty of how individual species will respond to climate change, there are also uncertainties in the modelling procedure, that is, to what extent species distributions can be fully determined by these models. Given these uncertainties, we conclude the following.

There are a number of species already threatened and at risk of extinction in Denmark today. However, reflecting the wide geographic distribution of Danish mammal species, only the Northern birchmouse (*Sicista betulina*) is likely at risk of extinction in Denmark over the 21st century due to climate change per se. However, the species most immediately at risk of extinction in Denmark is the common dormouse (*Muscardinus avellanarius*), which is now very rare in Denmark, probably due to loss of suitable forest habitats. The future persistence of this species will depend on habitat restoration, while it is unlikely to be threatened by the warming climate.

Invasive species may become an increasing problem for the native Danish fauna over the 21st century. American mink (*Mustela vison*) already now constitutes a competitor to the native polecat (*Mustela putorius*) and has elsewhere caused precipitous declines of the water vole (*Arvicola terrestris*). Additional introduced species are likely to spread to Denmark in the future or expand from recently established populations, perhaps especially muskrat (*Ondatra zibethicus*), coypu (*Myocaster coypu*), racoon (*Procyon lotor),* and racoon dog (*Nyctereutes procyonoides*). However, the scientific basis for evaluating whether these species will have a negative impact on the native

IOP Conf. Series: Earth and Environmental Science **8** (2009) 012016 doi:10.1088/1755-1315/8/1/012016

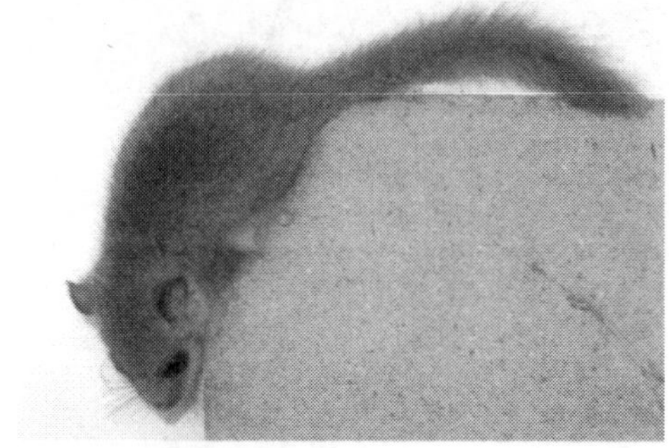

↑ **Reeves' muntjac (*Muntiacus reevesi*)**
Photo: Jens-Christian Svenning

↓ **Grey squirrel (*Sciurus carolinensis*)**
Photo: http://www.biopix.dk

↑ **Fat dormouse (*Glis glis*)**
Photo: Wikipedia Commons, public domain

↓ **Coypu (*Myocastor coypus*)**
Photo: http://www.biopix.dk

↑ **American mink (*Mustela vison*)**
Photo: http://www.biopix.dk

↓ **Racoon (*Procyon lotor*)**
Photo: http://www.biopix.dk

↑ **Muskrat (*Ondatra zibethicus*),** Photo: Dave Menke, Wikipedia Commons, public domain

↓ **Racoon dog (*Nyctereutes procyonoides*)**
Photo: http://www.biopix.dk

Figure 7. Some of the more likely 21st century immigrant species to the Danish mammal fauna.

fauna is at best weak. Clearly less problematic is the immigration of species native to Europe south of Denmark, as these species already now coexist with the Danish species in wide areas of Europe. These species include garden dormouse (*Eliomys quercinus*), Millet's shrew (*Sorex coronatus*), Miller's water shrew (*Neomys anomalus*), bi-coloured (*Crocidura leucodon*), lesser (*Crocidura suaveolens*) and greater white-toothed shrew (*Crocidura russula*), common pine vole (*Microtus subterraneus*), southern water vole (*Arvicola sapidus*), and fat dormouse (*Glis glis*).

In conclusion, future climate change per se is likely to cause a general enrichment of the Danish mammal fauna by the potential immigration of 17 new species, while only threatening one species with extinction in Denmark. Future management and conservation plans should include control strategies for non-European invasive species with negative impacts on the native biota and include monitoring programs to continually assess faunistic changes, hereunder interactions between the species currently found in Denmark and new immigrants.

5. Acknowledgements

This paper is a result of the international conference "Beyond Kyoto: Addressing the Challenges of Climate Change - Science Meets Industry, Policy and the Public", held in Aarhus, Denmark, March 5th-7th 2009. We are grateful to Signe Normand for preparation of the climate data and thank the Danish Natural Science Research Council for economic support (grant #272-07-0242 to JCS). We also thank Niels Sloth for making pictures from biopix.dk available.

6. References

[1] IPCC 2007 Summary for Policymakers in Solomon S *et al* 2007 *Climate Change 2007: The Physical Science Basis. Contribution of Working Group I to the Fourth Assessment Report of the Intergovernmental Panel on Climate Change* (Cambridge: Cambridge University Press) p 18
[2] Moritz C *et al* 2008 *Science* **322** 261
[3] Myers P *et al* 2009 *Global Change Biology* **15** 1434
[4] Lenoir J *et al* 2008 *Science* **320** 1768
[5] Parmesan C *et al* 1999 *Nature* **399** 579
[6] Thomas CD *et al* 2004 *Nature* **427** 145
[7] Fløjgaard C *et al* 2009 *Journal of Biogeography* **36** 1152
[8] Sommer RS and Nadachowski A 2006 *Mammal Review* **36** 251
[9] Sommer RS *et al* 2009 *Mammal Review* **39** 1
[10] Banks WE *et al* 2008 *Quaternary Science Reviews* **27** 2568
[11] Schipper J *et al* 2008 *Science* **322** 225
[12] Temple HJ and Terry A (compilers) 2007 *The status and distribution of European mammals* (Luxembourg: Office for Official Publications of the European Communities) p 45
[13] Metzger MJ *et al* 2008 *Environmental Conservation* **35** 64
[14] Erslev E 1871 *Historisk Tidsskrift* **4** 841
[15] Pertoldi C *et al* 2008 *Biological Journal of the Linnean Society* **93** 457
[16] Redeker S *et al* 2006 *Mammalian Biology - Zeitschrift fur Saugetierkunde* **71** 144
[17] Pertoldi C *et al* 2006 *Journal of Zoology* **270** 102
[18] Pertoldi C *et al* 2005 *Journal of Zoology* **265** 387
[19] Mitchell-Jones AJ *et al* 1999 *The Atlas of European Mammals* (London: T & AD Poyser Natural History) p 484
[20] Baagøe HJ and Jensen TS ed 2007 *Dansk Pattedyratlas* (Copenhagen: Gyldendal) p 392
[21] Mitchell TD *et al* 2003 *Tyndall Centre Working Paper 55* (http://www.tyndall.ac.uk/publications/working_papers/wp55.pdf)
[22] Skov F and Svenning JC 2004 *Ecography* **27** 366
[23] New M *et al* 2002 *Climate Research* **21** 1

[24] Ohlemüller R *et al* 2006 *Global Change Biology* **12** 1788
[25] Okubo A *et al* 1989 *Proceedings of the Royal Society of London. Series B, Biological Sciences* **238** 113
[26] Williamson MH *et al* 1986 *Philosophical Transactions of the Royal Society of London B, Biological Sciences* **314** 505
[27] van den Bosch F *et al* 1992 *Journal of Biogeography* **19** 135
[28] Levinsky I *et al* 2007 *Biodiversity and Conservation* **16** 3803
[29] Phillips S *et al* 2006 *Ecological Modelling* **190** 231
[30] Elith J *et al* 2006 *Ecography* **29** 129
[31] Phillips SJ and Dudik M 2008 *Ecography* **31** 161
[32] Martínez-Meyer E *et al* 2004 *Global Ecology and Biogeography* **13** 305
[33] Jensen TS and Møller JD 2007 Birkemus *Sicista betulina* (Pallas, 1779) *Dansk Pattedyratlas* ed Baagøe HJ and Jensen TS (Copenhagen: Gyldendal) pp 170-173
[34] Hammershøj M *et al* 2007 Ilder *Mustela putorius (*Linnaeus, 1758) *Dansk Pattedyratlas* ed Baagøe HJ and Jensen TS (Copenhagen: Gyldendal) pp 194-197
[35] Lodal J 2007 Mosegris *Arvicola terrestris* (Linnaeus, 1758) *Dansk Pattedyratlas* ed Baagøe HJ and Jensen TS (Copenhagen: Gyldendal) pp 122-125
[36] Vilhelmsen H 2007 Hasselmus *Muscardinus avellanarius* (Linnaeus, 1758) *Dansk Pattedyratlas* ed Baagøe HJ and Jensen TS (Copenhagen: Gyldendal) pp 164-167
[37] Jensen B and Erritzøe J 2007 Havesyvsover *Eliomys quercinus* (Linnaeus, 1766) *Dansk Pattedyratlas* ed Baagøe HJ and Jensen TS (Copenhagen: Gyldendal) pp 168-169
[38] Hammershøj M *et al* 2007 Mink *Mustela vision* (Schreber, 1777) *Dansk Pattedyratlas* ed Baagøe HJ and Jensen TS (Copenhagen: Gyldendal) pp 198-201
[39] Jeffries DJ *et al* 1989 *Mammal Review* **19** 111
[40] Bonesi L and Macdonald DW 2004 *Oikos* **106** 9
[41] Madsen AB *et al* 2007 Odder *Lutra lutra* (Linnaeus, 1758) *Dansk Pattedyratlas* ed Baagøe HJ and Jensen TS (Copenhagen: Gyldendal) pp 214-219
[42] Ramsgaard NR 2007 Bisamrotte *Ondatra zibethicus* (Linnaeus, 1766) in Baagøe HJ and Jensen TS. Dansk Pattedyratlas Gyldendal pp 126-127
[43] Baagøe HJ 2007 Sumpbæver *Myocastor coypus* (Molina, 1782) *Dansk Pattedyratlas* ed Baagøe HJ and Jensen TS (Copenhagen: Gyldendal) pp 174-175
[44] Baagøe HJ and Ujvári M 2007 Mårhund *Nyctereutes procyonoides* (Gray, 1834) *Dansk Pattedyratlas* ed Baagøe HJ and Jensen TS (Copenhagen: Gyldendal) pp 182-183
[45] Baagøe HJ and Ujvári M 2007 Vaskebjørn *Procyon lotor* (Linnaeus, 1758) *Dansk Pattedyratlas* ed Baagøe HJ and Jensen TS (Copenhagen: Gyldendal) pp 184-185
[46] Okubo A *et al* 1989 *Proceedings of the Royal Society of London. Series B, Biological Sciences* **238** 113
[47] Bertolino S *et al* 2008 *Biological Conservation* **141** 2564
[48] Ujvári M 2007 Sibirisk jordegern *Tamias sibiricus* (Laxmann, 1769) *Dansk Pattedyratlas* ed Baagøe HJ and Jensen TS (Copenhagen: Gyldendal) pp 112-113

IOP Conf. Series: Earth and Environmental Science **8** (2009) 012017 doi:10.1088/1755-1315/8/1/012017

Big moving day for biodiversity? A macroecological assessment of the scope for assisted colonization as a conservation strategy under global warming

Jens-Christian Svenning[1]*

Camilla Fløjgaard[1,2]

Naia Morueta-Holme[1]

Jonathan Lenoir[1]

Signe Normand[1]

Flemming Skov[2]

[1]The Ecoinformatics & Biodiversity Group, Department of Biological Sciences, Aarhus University, Ny Munkegade 114, DK-8000 Aarhus C, Denmark. [2]Department of Wildlife Ecology and Biodiversity, National Environmental Research Institute, Aarhus University, Grenaavej 14, DK-8410 Rønde, Denmark

*E-mail: svenning@biology.au.dk

Abstract. Future climate change constitutes a major threat to Earth's biodiversity. If anthropogenic greenhouse gas emissions continue unabated, 21st century climate change is likely to exceed the natural adaptive capacity of many natural ecosystems and a large proportion of species may risk extinction. A recurrent finding is that the degree of negative impact depends strongly on the dispersal potential of the species. However, there is a growing realization that many, if not most species would be unlikely to disperse as fast and far as required. As a consequence, it has been proposed that species at risk should be actively translocated into unoccupied, but environmentally suitable areas that are likely to stay suitable over the next 100 or more years (assisted colonization or assisted migration). This solution is controversial, though, reflecting negative experiences with introduced exotics and probably also the traditional emphasis in conservation management on preserving a certain local, often historical situation with a static species composition, and a tendency among ecologists to think of biological communities as generally saturated with species. Using the European flora as a case study, we here estimate the main environmental controls of plant species richness, assess how the maximum observed species richness depends on these environmental controls, and based here on estimate how many species could at least be added to an area before further species additions would perhaps inevitably lead to corresponding losses locally. Our results suggest that there is substantial room for additional plant species across most areas of Europe, indicating that there is considerable scope for implementing assisted colonization as a proactive conservation strategy under global warming without

IOP Conf. Series: Earth and Environmental Science 8 (2009) 012017 doi:10.1088/1755-1315/8/1/012017

necessarily implicating negative effects on the native flora in the areas targeted for establishment of translocated populations. Notably, our results suggest that 50% of the cells in Northern Europe, the likely target area for many translocations, could harbor at least 1/3 as many additional species as they have native species. However, we also emphasize that other, more traditional conservation strategies should also be strengthened, notably providing more space for nature and reducing nitrogen deposition to increase population resilience and facilitate unassisted colonization. Furthermore, any implementation of assisted colonization should be done cautiously, with a careful analysis on a species-by-species case.

1. Introduction

Anthropogenic climate change constitutes a major threat to Earth's biodiversity. If anthropogenic greenhouse gas emissions continue unabated, 21st century climate change is likely to exceed the natural adaptive capacity of many natural ecosystems and a large proportion of species may risk extinction [1]. A large number of studies have assessed the potential consequences of 21st century climate change for species ranges, hereunder the proportion of species risking extinction (reviewed in [1]). These studies vary in the estimated climate sensitivities of the species studied, probably reflecting not only organismic and geographic differences, but also differences in modeling methodologies. Nevertheless, a recurrent finding is that the degree of negative impact, in terms of predicted range reduction, depends strongly on the dispersal potential of the species [2-6]. Typically, dispersal is modeled under two contrasting scenarios for the time period considered: "no dispersal", where species are assumed to be unable to migrate beyond their current range, and "full dispersal", where species are assumed to be able to migrate from their current range to any area that becomes environmentally suitable. The "no-dispersal" scenario generally results in predictions of much stronger range contractions and many more species at risk from extinction [2-6], also cf. [7].

The sensitivity of the climate change risk estimates to species' dispersal potential makes the latter a key concern both for basic ecology, to provide a better understanding of how fast and far species are likely to disperse, and for conservation management, to facilitate the dispersal of species to track climate change [8-12]. It is clear that many species are already responding to the current climate changes by migrating north- or upwards [13-15], just as many species have responded to major past climate changes by shifting their range [16]. Nevertheless, it is also becoming clear that many species are dispersing at a slower pace than the climate is shifting [14]. Furthermore, there is increasingly strong evidence that many species have strongly dispersal-limited ranges, e.g., often still being constrained by dispersal in their postglacial range expansion from their ice age refugia [9-11,17]. In particular, species with small ranges (often called 'endemic' species) are often constrained to areas that have functioned as long-term refugia due to consistently favorable climate conditions [18,19]. In addition, for organism groups like trees and forest herbs empirical and model-based estimates of migration rates are also much slower than the rates needed for efficient tracking of the expected climatic shifts over the 21st century [4,10]. The strong fragmentation of natural habitats in many regions will additionally strengthen dispersal limitation for many species [20,21]. In sum, there is a growing realization that (1) biodiversity losses due to climate change will be strongly exacerbated if species fail to efficiently track the shifting climate by dispersal, and (2) that many species would be highly unlikely to disperse as fast and far as required.

In response to this, several strategies for the conservation of species diversity under climate change have been proposed. The more traditional and less controversial strategies involve landscape management to facilitate species dispersal [22] and a focus on regions, notably mountainous areas, where dispersal is likely to be better able to keep track with climate change (due to strong climate gradients over short distances) [4]. However, as already discussed many species may not be able to efficiently track climate change even if such measures are extensively implemented. As a consequence, a more radical solution is increasingly being advocated, i.e., assisted colonization (or assisted migration), where species are actively translocated into unoccupied, but environmentally suitable areas that are likely to stay suitable over the next 100 or more years [4,22-24]. This solution is controversial, though, reflecting (1) the many negative experiences with introduced exotics [24-26], but probably also (2) the traditional emphasis in conservation management on preserving a certain local, often historical situation with a static species composition, and (3) a tendency among ecologists to think of biological communities as generally saturated with species (discussed in [27,28]), i.e., meaning that introducing new species would inevitably entail losses among the species already present. In addition, assisted colonization is also likely to be a rather expensive conservation strategy.

The recent study by Hoegh-Guldberg et al. [24] outlines a decision framework for assessing possible species translocations on a species-by-species basis. In contrast, here we aim to elucidate the potential for assisted colonization to be implemented already under the current climatic conditions as a proactive conservation strategy in the face of the impending dramatic climate changes. The degree to which biological communities are saturated with species is crucial for the feasibility of assisted colonization as a conservation strategy. However, this is a long debated, but still scientifically unresolved issue, and we therefore base our assessment on a conservative treatment of this question. We use the European flora as a case: European plant species distributions are well documented to be influenced by climate as well as by dispersal limitation [9,11,29]. More specifically, we:

(1) estimate the main environmental controls of plant species richness at a 50 × 50 km scale using spatial regression modeling;

(2) assess how the maximum observed species richness (as a proxy for the potential species richness, i.e., the number of species that could coexist at this scale) depend on these environmental predictors using quantile regression [30], i.e., conservatively accepting that environmental conditions may limit the maximum number of species that can coexist in an area, without assuming that all communities are saturated with species;

(3) assess the minimum number of species that could be added to an area before further species additions might inevitably lead to corresponding losses locally. The latter is achieved by comparing the predicted maximum richness of each grid cell to the realized richness.

2. Materials and Methods

2.1. Species and environmental data

Species data came from Atlas Florae Europaeae (AFE; volumes 1-13), which covers ~20% of the European flora and maps species distributions on an equal-area grid with cells of ~50 × 50 km (AFE cells; [31] and http://www.fmnh.helsinki.fi/english/botany/afe/). The study area (n=2312 AFE cells) included most of Europe within 34°N–71°N and 11°W–32°E, excluding the former Soviet Union due to a relatively low and variable sampling effort in that region. Species richness across Europe was computed from the native distribution of 3069 plant species as the total number of species per AFE cell. Only species-level information was used, i.e., subspecies and varieties were merged.

As potential environmental controls we focused on climatic variables and topographic heterogeneity for two reasons: (1) these variables have been shown by previous studies to be highly important at the large scale of this study [32-34], and (2) these variables are largely unmodifiable by conservation management, while other potential controls such as land-use and, to some extent at least, soil conditions (e.g., soil fertility or hydrology) are not and could therefore be changed to accommodate the translocated species if there was a need for it. The climatic and topographic variables considered were standard variables considered in studies

IOP Conf. Series: Earth and Environmental Science **8** (2009) 012017 doi:10.1088/1755-1315/8/1/012017

of large-scale patterns of plant species richness [29,33-35] (Table 1): Data for AET, PET, and WD, came from the 30'-resolution United Nations Environment Programme GNV183 data set (http://www.grid.unep.ch), which covers the approximate time period of 1920-1980 [36,37], data for AMT, GDD, TMIN, PANN, and WBAL were obtained from the 10'-resolution climate grid CRU CL 2.0 [38]; http://www.cru.uea.ac.uk/cru/data/hrg.htm), and the altitude data came from the 30''- resolution Worldclim grid [39]; http://www.worldclim.org/). The climatic and topographic variables were recomputed as means (or range, in the case of topography) for each AFE cell. See [4,35] for further explanation.

2.2. Statistical modeling

We first estimated the main environmental controls of plant species richness in Europe using regression modeling. Multicollinearity can strongly compromise parameter and significance estimates for the individual predictors in multiple regression modeling [40]. To avoid this issue, we first assessed the main patterns of inter-correlation among the potential environmental controls by a Principal Components Analysis (PCA) on their correlation matrix [41]. Variables were then grouped according to whether they loaded strongly on PCA axes 1, 2 or 3 (Table 1). To avoid multicollinearity problems, only variables from different PCA groups were allowed into the same model in the subsequent regression modeling.

Table 1. Environmental variables and their loadings on the first three axes of a Principal Components Analysis of their inter-correlations

Variable	*Acronym*	*Unit*	*PCA axis 1*[a]	*PCA axis 2*[a]	*PCA axis 2*[a]
Annual mean temperature	AMT	°C	**0.952**	0.095	0.249
Growing-degree-days	GDD	°C	**0.972**	-0.061	0.066
Minimum temperature	TMIN	°C	**0.853**	0.276	0.244
Potential evapotranspiration	PET	mm $year^{-1}$	**0.961**	0.102	-0.146
Actual evapotranspiration	AET	mm $year^{-1}$	**0.680**	0.233	0.292
Annual precipitation	PANN	mm $year^{-1}$	-0.168	**0.936**	0.218
Water deficit	WD	mm $year^{-1}$	**0.834**	-0.021	-0.397
Water balance	WBAL	mm $year^{-1}$	-0.643	**0.745**	0.104
Ln (topographic range)	TOPO	log_e m	0.201	0.647	**-0.664**

[a]The PCA axis 1,2, and 3 accounted for 57.1%, 22.3%, and 9.9% of the variation, respectively. Bold face indicates to which PCA axis group each variable was assigned (see Materials and Methods)

The spatial regression modeling was implemented using ordinary least-squares (OLS) multiple regression both without and with spatial filtering. Given the above constraint, all the 12 possible three-variable models were developed, and then compared using information-theoretic model selection using Akaike's Information Criterion [42]. However, as significant spatial autocorrelation remained in the regression residuals of the best model, we repeated the regression modeling using eigenvector-based spatial filtering [43,44] to ensure that the model selection was not biased by spatial autocorrelation [45]. The spatial filters were generated as the eigenvectors of a Principal Coordinate Analysis of a pairwise matrix of geographic distances between all AFE cells, following the approach of Diniz-Filho and Bini [43]. Geographic distances were truncated at 1000 km [43]. The selection of spatial filters was done for the best unfiltered model, successively adding (at $P<0.05$) higher-order filters until spatial autocorrelation was negligible in the regression residuals (Moran's $I<0.100$ for the 22 default distance classes), avoiding the inclusion of filters that were highly correlated with the environmental predictors. The selected spatial filters were added to all 12 regression models, and the model selection was then re-run. In all regression models, the variance inflation factor (VIF) was <<10 for all predictors and filters, e.g., 1.6 or less for the best model with spatial filters.

After having established the best predictive model for plant species richness, we then re-ran the model using quantile regression [30,46] to establish the upper limit of observed richness given the predictor variables in the best spatial regression model. Quantile regression allows the analysis of the upper or lower limits of a variable rather than focusing on the mean as an OLS regression does. This is appropriate here as a conservative assessment of how many species could potentially coexist in an AFE cell given its environmental conditions and taking into account that many cells are likely to harbor much fewer species than they could due to historical constraints, notably postglacial dispersal limitation [9-11,29]. Following a similar logic, Brown and Peet [46] used quantile regression to analyze controls of exotic species richness. We note that species richness may also be influenced by other factors than those modeled and long-term historical effects, notably human land use and soil, but as discussed above these factors could be modified to accommodate translocated species if necessary. Here, we fitted the 95% quantile regression model and then projected it back onto the study area to estimate the potential maximum richness per AFE cell and by subtracting the number of native species present computed the potential additional richness, i.e., the number of additional species that would be expected to be able to co-occur with the native species in a given AFE

Table 2. Plant species richness and its potential environmental determinants: Information-theoretic model selection based on multiple linear regression (a) without and (b) with spatial filters[a,b]. The adjusted R^2, the Akaike's Information Criterion difference (ΔAIC)[c],and the Akaike weight (w)[d] are given.

Model			*(a)*	R^2_{adj}	ΔAIC	w	*(b)*	R^2_{adj}	ΔAIC	w
AET	PANN	TOPO		0.380	121	0.00		0.555	3	0.19
AMT	PANN	TOPO		0.148	857	0.00		0.450	494	0.00
GDD	PANN	TOPO		0.122	927	0.00		0.411	653	0.00
PET	PANN	TOPO		0.141	875	0.00		0.482	354	0.00
TMIN	PANN	TOPO		0.129	907	0.00		0.435	556	0.00
WD	PANN	TOPO		0.148	857	0.00		0.364	829	0.00
AET	**WBAL**	**TOPO**		**0.446**	**0**	**1.00**[e]		**0.556**	**0**	**0.81**[f]
AMT	WBAL	TOPO		0.157	833	0.00		0.453	481	0.00
GDD	WBAL	TOPO		0.121	930	0.00		0.409	660	0.00
PET	WBAL	TOPO		0.149	854	0.00		0.490	322	0.00
TMIN	WBAL	TOPO		0.130	906	0.00		0.435	555	0.00
WD	WBAL	TOPO		0.175	782	0.00		0.366	824	0.00

[a]Five spatial filters were used (no. 2-6). [b]The best model is shown in bold face. [c]$\Delta AIC=AIC$ for a given model minus the minimum AIC value in the set of candidate models; ΔAIC=0 for the best model, while models with $\Delta AIC \leq 2$ have substantial support and models with ΔAIC >10 have essentially no support. [d]Interpreted as the probability that a given model is the best model in the candidate set. [e]Standardized regression coefficients and their P-values: 0.596 AET (P<0.001) + 0.204 WBAL (P<0.001) + 0.158 TOPO (P<0.001). [f]Standardized regression coefficients and their P-values (not given for the spatial filters): 0.508 AET (P<0.001) + 0.033 WBAL (P=0.062) + 0.445 TOPO (P<0.001).

cell given its climatic and topographic environment. The main reason why we consider this assessment to be conservative is that we constrain the potential richness of a grid cell to close to the maximum observed within Europe for a given combination of climate and topography, while we are not aware of any strong empirical evidence that plant communities are indeed saturated with species anywhere in Europe, and especially not at a 50 × 50 km scale. As a further conservative aspect of our analysis, we modeled the 95% quantile rather than e.g. the 99% quantile.

The OLS regressions and associated analyses including the PCA were all implemented in SAM 3.0 [47], while quantile regression was implemented in R 2.8.1 [48] using the library quantreg [49]. Maps were produced in ArcGIS 9.3.

3. Results

The spatial regression modeling and associated model selection clearly showed that plant species richness (number of species) in Europe was mainly related to actual evapotranspiration (AET), water balance (WBAL), and topographic range (TOPO; Table 2). This result was robust to whether or not the modeling was done with spatial filtering, although adding the spatial filters clearly improved the regression models in terms of explanatory power (Table 2). According to the standardized regression coefficients, the strongest predictor was clearly AET (Table 2). When spatial filters were included, the strength of TOPO increased to approximate that of AET, while WBAL became very weak (Table 2).

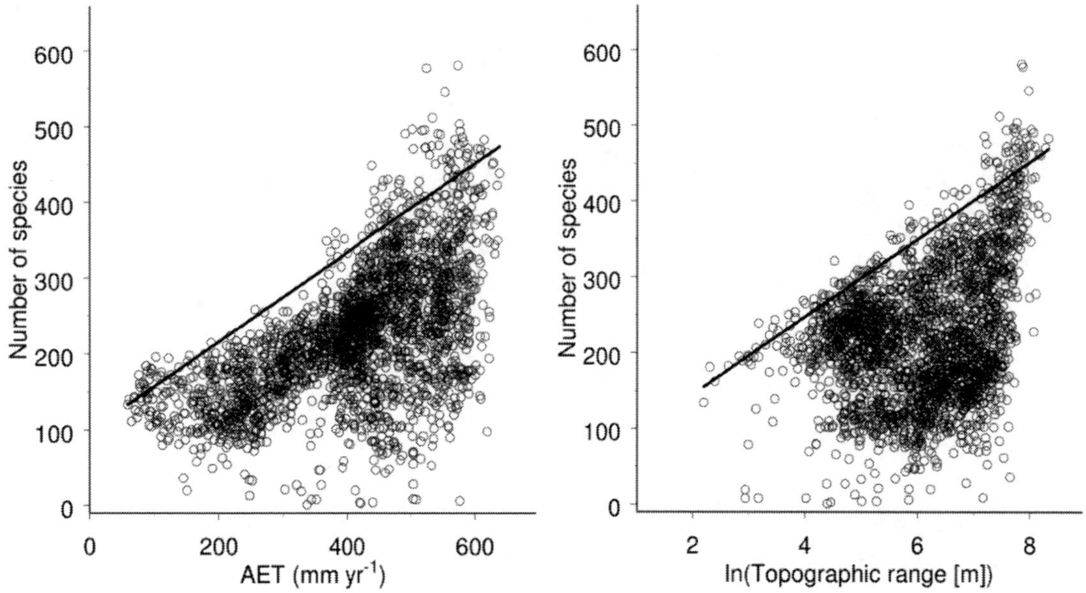

Figure 1. Number of native species per ~50 × 50 km grid cell as a function of annual evapotranspiration (AET) and topographic range (TOPO). Lines show univariate 95%-quantile regression fits.

The 95% quantile regression modeling was done using only AET and TOPO as predictors, as WBAL turned out to be completely nonsignificant ($P=0.53$) in the AET+WBAL+TOPO quantile regression model.

Furthermore, the fits of the two models were not different (t-test, df=1, F-value=1.20, P=0.274). The resulting 95% quantile regression model equation for species richness was: -94.4 + 0.548 AET (P<0.00001) + 31.9 TOPO (P<0.00001). The 95% quantile species richness relationships to AET and TOPO were both approximately linear (Fig. 1).

Projecting the quantile regression back onto Europe showed that the potential additional richness was on average 99.3 species (±76 SD; median=82.3 species, 25%-quartile=44.8 species) for the whole region (Fig. 2), i.e., most cells could harbor at least an additional 45 species and many could harbor 100 or more additional species. Potential additional richness tended to be higher in Southern Europe than in Northern Europe (Fig. 2). Nevertheless, considering only Northern Europe (≥50° latitude) potential additional richness was on average 63.2 species (±41.1 SD; median=63.7 species, 25%-quartile=33.9 species). Furthermore, if these values were recalculated in percent of the realized native richness for a given AFE cell, the median and 25%-quartile were 39.1% and 17.9% for the whole region and 33.3% and 15.5% for Northern Europe (means are not given since the percentages were highly skewed) (Fig. 2). Hence, half the cells in Northern Europe could harbor ≥1/3 as many additional species as they have native species.

4. Discussion

4.1. Potential for assisted colonization

Our quantile regression results suggest that there is substantial room for additional plant species across most areas of Europe at the scale of 50 × 50 km grid cells. Hereby, our results indicate that there may be ample scope for implementing assisted colonization as an element in a proactive conservation strategy under global warming without necessarily implicating negative effects on the native flora in the areas targeted for establishment of translocated populations. As global warming is likely to force suitable climatic conditions for many species >500 km or even >1000 km towards the north or northeast, it is especially relevant that our results suggest that 50% of the cells in Northern Europe could already now harbor at least 1/3 as many additional species as they have native species. Furthermore, given the increase in maximum richness with AET (Fig. 1) we can predict that as global warming proceeds these numbers will even increase. Of course, this interpretation hinges on the interpretation of the quantile regression results:

How much certainty can we place in our interpretation that the quantile regression predictions represent the environmentally determined potential richness of a grid cell? We do not claim that our analysis unambiguously resolves the debate on community saturation [28,33,34,50-53]. However, we note that the approach is already conservative in actually assuming that climate and topography sometimes do limit species richness, as theoretically it is an unresolved, controversial issue whether such limits really exist, i.e., whether communities are ever saturated with species [28,50-53]. In addition, the geographic pattern in our results provides at least some support for this interpretation. There are two clear geographic patterns in potential additional richness (Fig. 2). (1) Southern Europe appears to have many areas that are particularly poor in native species relative to their climate and topography when compared to Northern Europe. Several factors may be responsible for this pattern: (i) Human land-use effects are likely to be particularly strong in the south, where intensive land-use has the longest history and resulted in extensive environmental degradation, e.g., erosion. (ii) Southern Europe has a complex environmental history on a Pleistocene time-scale, with numerous relatively small ice age refuge areas and interspersed non-refuge areas [54]. This is reflected in the current flora by strong concentrations of localized endemics in the vicinity of former refuges [54] and may cause strong richness variation beyond what can be explained by the current environment, especially since the climate would generally be conducive for higher species richness than in the north (cf. Fig. 1). (iii) Dispersal limitation is likely to be particularly strong in Southern Europe due to the geographic fragmentation of this region by the Mediterranean Sea and numerous high mountain ranges. (2) Furthermore,

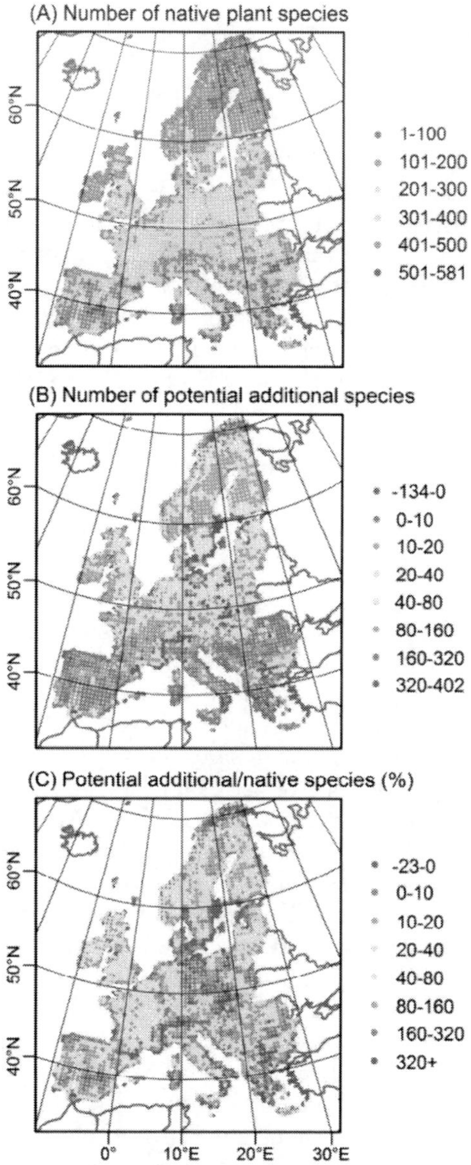

Figure 2. (A) Number of native species from the Atlas Flora Europaeae (~20% of the whole flora) per ~50 × 50 km grid cell. (B) Maximum potential species richness per grid cell given its climate and topography, as estimated by the 95%-quantile regression model on actual evapotranspiration (AET) and topographic range (TOPO). (C) The potential additional species richness, computed as (B) minus (A), i.e., the minimum number of additional species that a grid cell would be expected to contain without loss of native species, relative to its number of native species.

within Northern Europe, potential additional species richness is highest towards the north and west (Fig. 2). This follows nicely the expectation if species distributions and species richness are still limited by postglacial immigration from ice age refugia in southern, south-central, and south-eastern Europe [10,11]. Hence, these geographic patterns lend support to interpreting the quantile regression estimates of potential maximum richness as largely reflecting the number of species a given grid cell would contain given its climate and topography if historical dispersal constraints were absent (and limitations by land-use or other unrepresented environmental factors were only of minor importance).

4.2. Implementing assisted colonization as a conservation strategy under global warming

The climatic changes over the 21st century are likely to require plant species to migrate northwards or northeastwards at ~4 km yr^{-1}, or under highly optimistic circumstances (the IPCC B1 scenario) ~2 km yr^{-1}, migration rates far exceeding the empirically observed migration rates for many plant species [4]. As a consequence, dispersal limitation is likely to become an increasingly strong constraint on the northward expansion of plant species as global warming proceeds [3,5]. Already, there is evidence that even in mountain systems where the temperature gradient occurs over much shorter distances than in lowlands, forest plant migrations are so slow that even the moderate 20th century warming has only been incompletely tracked [14]. As a consequence, the need and the potential for assisted colonization as part of the nature conservation strategy for Europe, and elsewhere, is likely to steadily increase as global warming proceeds. Relieving other negative pressures [24,55], notably by providing more space for nature and reducing nitrogen deposition, would help populations to cope with the negative effects of the changing climate. However, these strategies are unlikely to be sufficient in themselves in the face of climate change, given its expected magnitude [3,5]. Furthermore, increasing temperature and/or drought are already now forcing local population declines in Southern Europe [14,56-60]. Hence, it would be timely to now begin a concerted research effort into assisted colonization as part of a European-wide conservation strategy, especially as a proactive approach would be prudent given that the establishment of new populations may be slow and even require repeated attempts to succeed. Controversially, in North America, a volunteer organization is already now implementing assisted colonization to ensure the survival of *Torreya taxifolia*, a highly threatened conifer from southeastern United States that is declining dramatically in its native range, possible due to a failure to migrate to a more suitable, cooler climate in more northern areas [23]. In addition, in Europe and most likely elsewhere nurseries, gardeners and foresters are already unintentionally facilitating assisted colonization for numerous species, with commercial northern limits often exceeding natural northern limits by 500-1000 km [61]. Numerous of these species have already escaped and established naturalized populations far north of their native range as a result, e.g., *Abies alba, Acer pseudoplatanus, Asarum europaeum, Eranthis hyemalis, Fagus sylvatica, Helleborus foetidus*, and *Juglans regia* [4,9; J.-C. Svenning, pers. obs.]. However, it is important to note that these unintentional assisted colonizations may entail reduced genetic diversity and atypical genetic composition [61] and are being made without considering the potential for negative effects on the native biota.

Assisted colonization is controversial as a conservation strategy because of the numerous problems associated with invasive exotic species [25]. However, this risk can be minimized by only moving species within the same broad biogeographic region [24] as well as cautious implementation [25,62]. Notably, a key risk for translocations within a biogeographic region is that even within such regions there are groups with allopatric ecologically vicariant species. In such cases, translocations could threaten locally native species, e.g., by hybridization and introgression as in the well-known case of *Hyacinthoides non-scripta* [63]. Hence, the situation for each candidate species needs to be carefully studied from both ecological and biogeographic perspectives. Nevertheless, as indicated by the present study as well as additional substantial evidence against community saturation [28,50-53] there is little reason to think that translocations in general should inevitably result in losses of native species.

4.3. Conclusion

The results of the present study suggest substantial scope for assisted colonization as a proactive conservation strategy under global warming. Notably, our results suggest that 50% of the cells in Northern Europe could harbor at least 1/3 as many additional species as they have native species. However, we also emphasize that other, more traditional conservation strategies should also be strengthened, notably providing more space for nature and reducing nitrogen deposition to increase population resilience and facilitate unassisted colonization. In fact, the success of assisted colonization as a conservation strategy is likely to also depend on reducing these non-climatic negative anthropogenic pressures on nature. Furthermore, any implementation of assisted colonization should be done cautiously, with a careful analysis on a species-by-species case.

5. Acknowledgements

This is a paper resulting from the international conference "Beyond Kyoto: Addressing the Challenges of Climate Change - Science Meets Industry, Policy and the Public", held in Aarhus, Denmark, March 5th-7th 2009. We thank the Atlas Florae Europaeae project for access to the distribution data, and gratefully acknowledge economic support from the Danish Natural Science Research Council (grant #272-07-0242 to JCS).

References

[1] Fischlin A *et al.* 2007 in *Climate change 2007: impacts, adaptation and vulnerability. Contribution of Working Group II to the Fourth Assessment Report of the Intergovernmental Panel on Climate Change* ed ML.Parry *et al.* (Cambridge, Cambridge University Press), 211
[2] Thomas CD *et al.* 2004 *Nature* **427** 145
[3] Thuiller WD *et al.* 2005 *PNAS* **102** 8245
[4] Skov F and Svenning JC 2004 *Ecography* **27** 366
[5] Svenning JC and Skov F 2006 *Biodiversity and Conservation* **15** 3341
[6] Normand S *et al.* 2007 *Journal for Nature Conservation* **15** 41
[7] Williams JW *et al.* 2007 *Proceedings of the National Academy of Sciences USA* **104** 5738
[8] Nathan R 2006 *Science* **313,** 786
[9] Skov F and Svenning JC 2004 *Ecology Letters* **7** 565
[10] Skov F and Svenning JC 2007 *Ecology Letters* **10** 453
[11] Svenning JC *et al.* 2008 *Ecography* **31** 316
[12] Nathan R *et al.* 2008 *Trends in Ecology & Evolution* **23** 638
[13] Parmesan C and Yohe G 2003 *Nature* **421,** 37
[14] Lenoir J *et al.* 2008 *Science* **320** 1768
[15] Moritz C *et al.* 2008 *Science* **322** 261
[16] Huntley B and Webb III T 1989 *Journal of Biogeography* **16** 5
[17] Van der Veken S *et al.* 2007 *Journal of Ecology* **95** 343
[18] Skov F and Svenning JC 2007 *Global Ecology and Biogeography* **16** 234
[19] Araújo MB *et al.* 2008 *Ecography* **31** 8
[20] Honnay O *et al.* 2002 *Ecology Letters* **5** 525
[21] Collingham YC and Huntley B 2000 *Ecological Applications* **10** 131
[22] Davis MB 1989 *Climatic Change* **15** 75

[23] Barlow C and Martin PS 2007 *Wild Earth Forum* [Fall/Winter] 52-56
[24] Hoegh-Guldberg O *et al.* 2008 *Science* **321** 345
[25] McLachlan JS *et al.* 2007 *Conservation Biology* **21** 297
[26] Van der Veken S *et al.* 2008 *Frontiers in Ecology and the Environment* **6** 212
[27] Davis MA *et al.* 2005 *Ecography* **28** 696
[28] Ricklefs RE 2006 *Ecology* **87** S3
[29] Skov F and Svenning JC 2005 *Journal of Biogeography* **32** 1019
[30] Cade BS and Noon BR 2003 *Frontiers in Ecology and the Environment* **1** 412
[31] Jalas J and Suominen J ed 1972-1994. *Atlas Florae Europaeae: Distribution of vascular plants in Europe, Vol. 1-10.* (Helsinki, The Committee for Mapping the Flora of Europe and Societas Biologica Fennica Vanamo)
[32] Pearson RG and Dawson TP 2003 *Global Ecology and Biogeography* **12** 361
[33] O'Brien EM *et al.* 2000 *Oikos* **89** 588
[34] Francis AP and Currie DJ 2003 *American Naturalist* **161** 523
[35] Svenning JC et al. 2009 *Ecography* **32** 13
[36] Ahn CH and Tateishi R 1994 *Journal of the Japan Society of Photogrammetry and Remote Sensing* **33** 12
[37] Ahn CH and Tateishi R 1994 *Journal of the Japan Society of Photogrammetry and Remote Sensing* **33** 48
[38] New M *et al.* 2002 *Climate Research* **21** 1
[39] Hijmans RJ *et al.* 2005 *International Journal of Climatology* **25** 1965
[40] Graham MH 2003 *Ecology* **84** 2809
[41] Legendre P and Legendre L 1998 *Numerical ecology* 2nd ed (Amsterdam, Elsevier)
[42] Burnham KP and Anderson DR 2002 *Model selection and multi-model inference: a practical information-theoretic approach* (New York, Springer-Verlag)
[43] Diniz-Filho JAF and Bini LM 2005 *Global Ecology and Biogeography* **14** 177
[44] Griffith DA and Peres-Neto PR 2006 *Ecology* **87** 2603
[45] Diniz-Filho JAF *et al.* 2008 *Global Ecology and Biogeography* **17** 479
[46] Brown RL and Peet RK 2003 *Ecology* **84** 32
[47] Rangel TF et al. 2006 *Global Ecology and Biogeography* **15** 321
[48] R Development Core Team 2008 R: a language and environment for statistical computing (Vienna, R Foundation for Statistical Computing), http://www.r-project.org
[49] Koenker R 2008 *quantreg: Quantile regression. R package version 4.26*, http://www.r-project.org
[50] Cornell HV 1999 *Ecoscience* **6** 303
[51] Hillebrand H 2005 *Oikos* **110** 195
[52] Stohlgren TJ *et al.* 2008 *Ecology Letters* **11** 313
[53] Loreau M 2000 *Ecology Letters* **3** 73
[54] Médail F and Diadema K 2009 *Journal of Biogeography* doi:10.1111/j.1365-2699.2008.02051.x
[55] Sala OE *et al.* 2000 *Science* **287** 1770
[56] García D *et al.* 1999 *Biological Conservation* **87** 215
[57] Mejías JA *et al.* 2002 *Botanical Journal of the Linnean Society* **140** 297
[58] Castro J *et al.* 2004 *Journal of Ecology* **92** 266
[59] Arrieta S and Suarez F 2006 *Flora* **201** 152
[60] Jump AS *et al.* 2006 *Global Change Biology* **12** 2163
[61] Van der Veken S *et al.* 2008 *Frontiers in Ecology and the Environment* **6** 212

[62] Hunter ML 2007 *Conservation Biology* **21** 1356
[63] Bleeker W *et al.* 2007 *Biological Conservation* **137** 248

How will climate change affect spatial planning in agricultural and natural environments? Examples from three Dutch case study regions

Margaretha Blom-Zandstra[1,*], Maurice Paulissen[2], Herman Agricola[2] & Ben Schaap[1]

[1] Plant Research International, Wageningen UR, P.O. Box 616, 6700 AP Wageningen, the Netherlands

[2] Alterra, Wageningen UR, P.O. Box 47, 6700 AA, Wageningen, the Netherlands

E-mail: greet.blom@wur.nl

Abstract. Climate change will place increasing pressure on the functioning of agricultural and natural areas in the Netherlands. Strategies to adapt these areas to stress are likely to require changes in landscape structure and management. In densely populated countries such as the Netherlands, the increased pressure of climate change on agricultural and natural areas will inevitably lead, through the necessity of spatial adaptation measures, to spatial conflicts between the sectors of agriculture and nature. An integrated approach to climate change adaptation may therefore be beneficial in limiting such sectoral conflicts. We explored the conflicting and synergistic properties of different climate adaptation strategies for agricultural and natural environments in the Netherlands. To estimate the feasibility and effectiveness of the strategies, we focussed on three case study regions with contrasting landscape structural, natural and agricultural characteristics. For each region, we estimated the expected climate-related threats and associated trade-offs for arable farming and natural areas for 2040. We describe a number of spatial and integrated adaptation strategies to mitigate these threats. Formulating adaptation strategies requires consultation of different stakeholders and deliberation between different interests. We discuss some trade-offs involved in this decision-making.

1. Introduction

The Netherlands is a very densely populated country, in which strongly intermixed agricultural and natural areas (Figure 1) are fragmented by urbanized areas and infrastructural networks. This results in competition for the actual use of land, which will further intensify [1,2] as urbanization continues. Requirements for reconstruction of the countryside will increase and pressure will be put on the introduction of more environmentally-friendly management in agriculture, inevitably leading to spatial conflicts.

* To whom any correspondence should be addressed

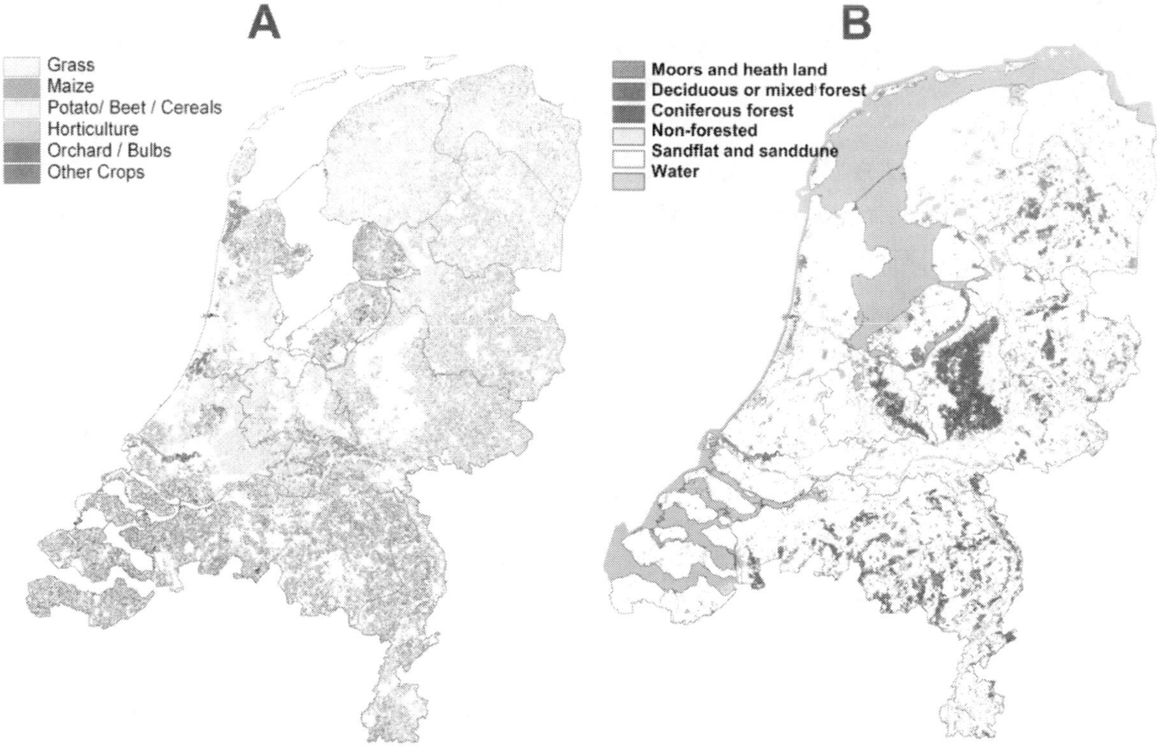

Figure 1. Main crops in agriculture (A) and main natural units (B) in the Netherlands. Map derived from [10].

Climate change in the Netherlands will increase these spatial conflicts [3], due to the likely changes in the functioning of agriculture, ecosystems and the urban environment. Effects of climate change will be both direct, through changes in CO_2, temperature and precipitation, as well as indirect, mainly through changes in water-related habitat factors like salinity, acidity and the availability of moisture, oxygen and nutrients. This will introduce many stressors to Dutch agricultural and natural systems such as flooding, drought, heat stress, salinization [4,5], an increase of pests and diseases in agricultural crops and changes in species composition in natural ecosystems [6,7]. Thus, it is expected that climate change will have a negative impact on Dutch economy and ecology. As a result, a reduction of the negative climatic effects on the functioning of agriculture, ecosystems and the urban environment is high on the Dutch political and research agenda [8].

An innovative and integrated adaptation policy is required in order to anticipate and respond to the effects of climate change. This requires the visualization of conflicting and synergistic properties of agricultural and natural systems, followed by the formulation of integrated adaptation strategies suitable for resolving spatial bottlenecks in various sectors. To ensure an effective implementation of these adaptation strategies, they should be carefully discussed with the stakeholders involved in spatial adaptations [9].

The damage to agricultural crops or natural ecosystems from climate change depends on the local situation (altitude, soil and hydrological characteristics), the season (crop growth stage and sensitivity), and the resilience of agricultural crops and natural ecosystems [10,11,12,13]. Thus, the effectiveness of adaptation strategies is also region-specific. This requires an approach on local level and the use of the specific expertise of stakeholders, taking the local interests and requirements into consideration.

In this paper we describe a semi-quantitative approach for the evaluation of climate change-related risks and integrated spatial regional adaptation measures within the Netherlands for a time horizon of 2040. Based on both expert judgement [10] and published evidence [14,15], we considered the main threats of climate change to Dutch agricultural and natural systems. We then explored the conflicting and synergistic properties of different climate adaptation strategies for agricultural and natural environments. To evaluate the urgency of the climate change-related risks on the one hand, and the feasibility and effectiveness of the adaptation strategies to mitigate these threats on the other, we focussed on three regional case studies [16] (see Figure 2). The regional level in the Netherlands is very important in legislation on climate adaptation, because it is the 'work floor' on which climate change-related risks and adaptation measures become tangible. In addition, the role of regional stakeholders and policy-makers has become increasingly important due to administrative decentralization. The chosen case study regions have contrasting soil characteristics (predominantly clayey, sandy and peaty soils, respectively), hydrology, landscape structure, agricultural and natural environments. They have also experienced different opportunities for changes in land-use such as the introduction of new crops or the development of new natural units. We will discuss some trade-offs involved in the regional decision-making process.

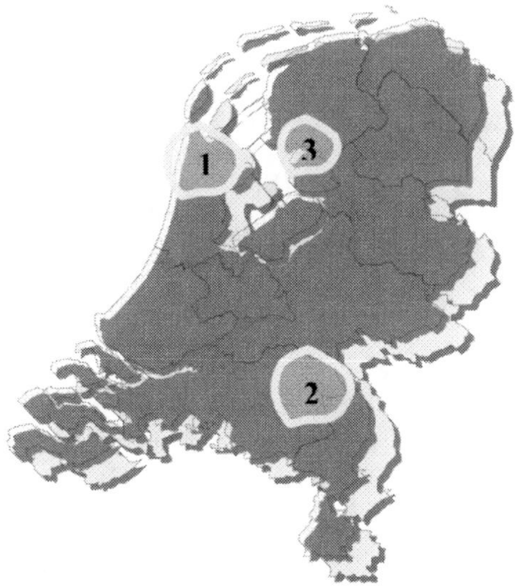

Figure 2. Case study regions. 1. Northern part of 'Noord-Holland' with mainly clayey soil. 2. Eastern 'Noord-Brabant' with mainly sandy soil. 3. 'Zuidwest-Friesland' & 'Noordwest-Overijssel' with mainly peaty soil.

2. The Netherlands: a low-lying country

A large part of the Dutch coastal plain is situated below sea level (Figure 3), characterized by polders, i.e. areas with various and controlled water levels. In order to facilitate agriculture, polder water levels in (dry) summers are often maintained at a higher level than during the winter [17], by supplying large amounts of water from the Rhine and Meuse rivers, [18,19]. It is expected that climate change will increase the occurrence of mild winters and warm summers in the Netherlands [20]. On average, winters will likely become wetter with higher extreme precipitation sums. The amount of rain is expected to decrease although the intensity of extreme showers will increase. Finally, the calculated changes in wind speed and storm activity are small compared to their natural variability. Thus, both water shortages in the summer and water excess in the winter (but also during extreme summer precipitation events) are expected to occur more frequently.

Apart from changes in water characteristics, salinity stress may increase considerably as well [21]. At present, the soil in the western and northern part of the country already contains extended volumes of underground salt glands and the shallow groundwater shows a gradual transition from fresh water in the east towards brackish and salt water in the western parts of the country. Due to an ongoing rise in sea level at about 2 mm/year [21], an eastward shift of the brackish waterfront may be expected in the nearby future.

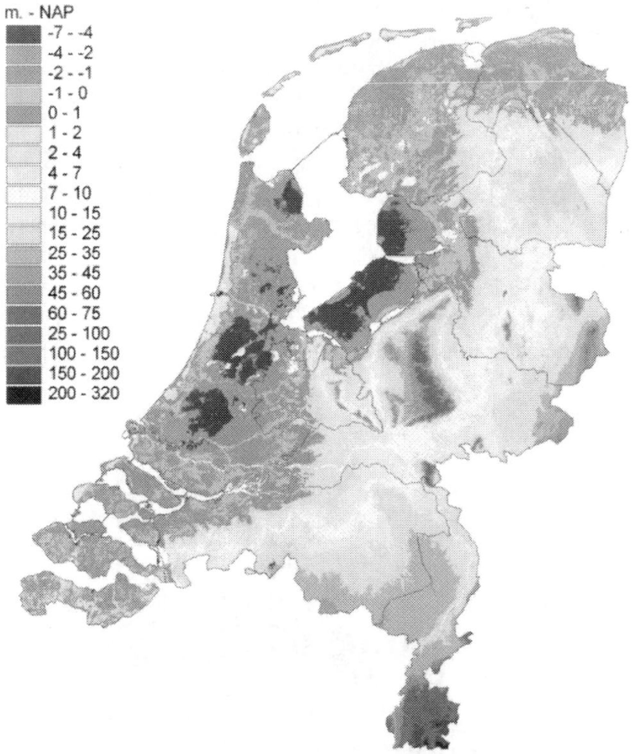

Figure 3. Contour map of the Netherlands (legend: elevation in m below or above sea level). Map derived from [10].

The increase in precipitation extremes as well as salinity stress may have an enormous impact on the functioning of agricultural and natural systems. In many areas the adaptation to these stresses will require changes in the landscape and its management.

3. Main climate-related threats to agricultural and natural areas of the Netherlands

The sensitivity of agricultural crops to drought in the Netherlands as estimated in the expert-judgement study [10] is shown in Figure 4. Especially the sandy soils in the more elevated areas of the central, eastern and southern Netherlands (i.e. soils with a low water retaining capacity) are vulnerable to dry conditions. Prolonged periods of drought are expected to frequently occur in 2040, especially from March until October [20]. In spring this may negatively affect seedling growth and survival, especially in sugar beet, potato and maize, while growth will be retarded during the summer and grass turf quality will deteriorate.

In the western Netherlands, agricultural crops growing on mainly clayey soils are sensitive to flooding (see Figure 5). Due to an increase or even possibly a doubling [22] of heavy rainfall (especially in spring and midsummer) and difficulties with runoff in the clayey soils, bulbs may be drowned, potatoes may become vulnerable to rot and pastures may become unsuitable for mowing and grazing.

Salinity problems may threaten agricultural crops within a broad strip along the coast as soon as freshwater shortages occur during prolonged dry periods (data not shown). The effect on crop growth depends on the crops' sensitivity to salinity. Especially bulbs and ornamental plants (very important export products) are extremely sensitive to saline water. Climate change will therefore increase the pressure on the cost-effectiveness of these sectors.

The incidence of heat extremes may double by 2040, especially in the central and eastern parts of the country, with a negative effect on yield or unwanted development of potato tubers [23].

In general, the occurrence of pests and diseases will increase [7] due to relatively warm winters (without frost) and the increase of warm, wet weather in the summer, although the extent of the increase can not be forecasted.

Figure 4. The estimated sensitivity of the agricultural regions in the Netherlands to drought (green: low sensitivity; red: high sensitivity; white: no agriculture). Map derived from [10].

The sensitivity of natural systems to climate change-related threats was estimated by a similar semi-quantitative expert-judgement approach [10]. Figure 6 shows the results of sensitivity to drought and heat stress. Coinciding with the generally higher sensitivity of agricultural areas in the elevated, sandy parts of the Netherlands, many natural systems on these sandy soils (such as pine forests and heathlands) are considered intermediately sensitive. Wetlands, including fen marshes on clay and fen peat soils on the one hand and raised bog (remnants) on the elevated sandy soils on the other, are considered to be the most sensitive systems to heat and drought stress [10].

IOP Conf. Series: Earth and Environmental Science 8 (2009) 012018 doi:10.1088/1755-1315/8/1/012018

Figure 5. The estimated sensitivity of the agricultural regions in the Netherlands to flooding (green: low sensitivity; red: high sensitivity; white: no agriculture). Map derived from [10].

Figure 6. The estimated sensitivity of natural areas in the Netherlands to drought and heat stress. Map derived from [10].

Figure 7 depicts the percentage of contracting target species (for nature conservation) in the natural areas of the Netherlands. For these mainly cold-adapted species, the Netherlands is at the southern (or low elevation) edge of their distribution range. As a result of climate change, the risk that Dutch habitats may become unsuitable for these species is high. Again, the high risk areas include pine forests and the moors and heath lands of the sandy soils (with many boreal elements). Here, wetlands appear to be less at risk, while e.g. dune areas, which contain relatively few cold-adapted species, are even less threatened by climate change.

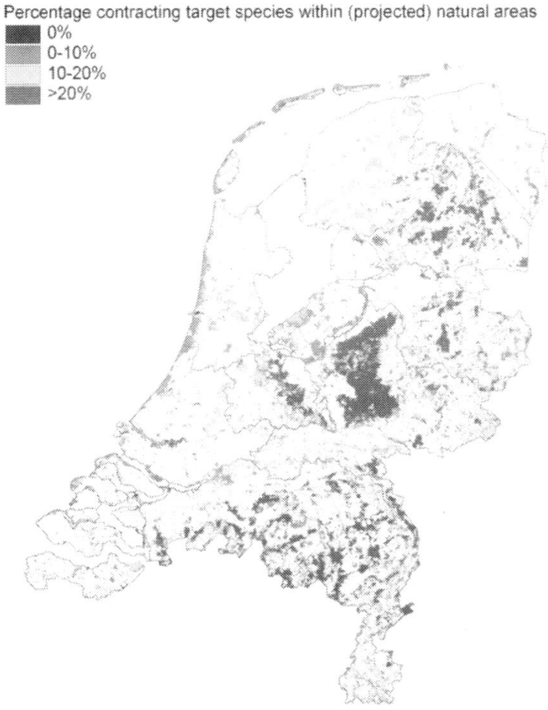

Figure 7. The estimated sensitivity of (projected) natural areas in the Netherlands to drought and heat stress. Map derived from [10].

4. Adaptation strategies

To overcome the bottlenecks of climate change, various adaptation strategies have been formulated for the different categories and scales present. Table 1 presents a rough overview of adaptation strategies for three different categories in agriculture. Measurements at the farm level can be performed by the farmers themselves. As far as innovations are concerned, research is responsible for their development and the industry for their funding and implementation. For the adaptation of water management at the farm and regional level, it is especially the provincial authorities and water boards that are responsible. They are also responsible for reconstruction of the countryside. So it is important to involve the various stakeholders in the development of adaptation strategies in order to discuss their feasibility and to gain support for their implementation [9].

Spatial adaptation strategies may interfere with the neighbouring agricultural and natural areas. Especially water management strategies will affect different functions within the region. Strategies that aim to relieve crops from water excess may cause drought to ecosystems in the same area. Irrigation of crops with water extracted from the groundwater may also cause drought stress in natural systems in the same area. However, an increase of the groundwater level to provide water to the root zone may cause undesired water excess in the neighbouring natural areas. Some water management strategies for agriculture were developed in more detail (Table 2) and their positive, neutral or negative effects on natural systems were evaluated. In order to realise spatial strategies, risks, effectiveness and priorities must be properly weighed [16]. To elaborate on this, we will focus on the three case study areas mentioned earlier. One effect of climate change on each of these areas will be described in detail.

Table 1. Adaptation strategies for agriculture at different scales. Information derived from [16].

Category	Adaptation strategies
Measurements at farm level	Adaptation schedules (rotation, sowing and harvest dates, etc)
	Irrigation
	Switch to other (resistant) cultivars or other crops (halophytes, new crops)
	Field border management
	Use of natural resources for production (biofuel crops)
	Introduction hydroponics (i.e. containers for ornamental hedges; horticulture)
	Investment in high-tech solutions (especially for horticulture: aquifers, floating greenhouses, energy conservation plants, etc.)
Reconstruction of the countryside	Multifunctional farming
	Introduction of climate coats
	Agricultural extensification by introduction of other functions (agro tourism, recreation, social farming, house sale, sea culture parks, etc)
Adaptation water management at farm or regional level	Intensive water level management (by pumping)
	Water conservation or retention (during flooding, storage for periods of drought)
	Subsoil drainage
	Broadening of ditches
	Rinsing brooks and ditches (to resist salinity)
	Deregulation of water level, possibly inlet saline water (in combination with agriculture extensification)
	Blue green veining
	Increase subsoil organic material to increase water absorbing capacity of the soil (to resist drought)

Table 2. Effects of intensive water management strategies for agriculture on ecological systems. Information derived from expert judgement [16].

Water management strategies	Effects on species diversity in natural areas
Regional water conservation - water retention areas - increase amount of brooks - increase runoff capacity	Negative when water level is dictated by agricultural interests and strictly managed ↓ Positive when both agriculture and natural ecosystems are taken into account (integrated approach)
Water conservation on the field - boggy fields allowance - water inlet	Positive
Subsoil drainage - deepening drains - intensification drainage - increase water level ditches	Negative (due to rapid water discharge) ↓ Neutral (when water level ditches is increased)
Increase water absorbing properties soil - ploughing organic matter	Neutral
Rinsing brooks and ditches	Negative

5. Considering different adaptation strategies in case study regions

5.1. The first case study region (Northern part of 'Noord-Holland') is situated below sea level. In the western and northern part it borders on the North Sea. In the eastern part it borders on the lake IJsselmeer. The soil mainly consists of clay and is hereafter referred to as 'clay region'. The region has an intensive and highly cost-effective agriculture with bulbs, arable crops and greenhouses. The natural systems are situated near the coast and consist of dunes, grassland, and fens. As agriculture has a high economical value in this region, provincial policy focuses on maintenance or further expansion of agriculture, mainly horticulture.

Salinity is an important problem in the 'clay region'. Various salt glands already contaminate the groundwater. Climate change will enhance this contamination as the inward flow of water from the sea will increase due to a rise in sea level. Figure 8 shows the areas in which salinity stress may become a serious risk for agriculture and natural systems in 2040, if the present land use will be continued. Highly sensitive bulbs are presently grown in areas 1-4, and subsoil salt glands exist in the areas indicated by the arrows

To cope with the future salinity problems, various adaptation strategies are possible varying from intensive water management to a more extensive approach. The characteristics of the two diverging strategies are described in Table 3.

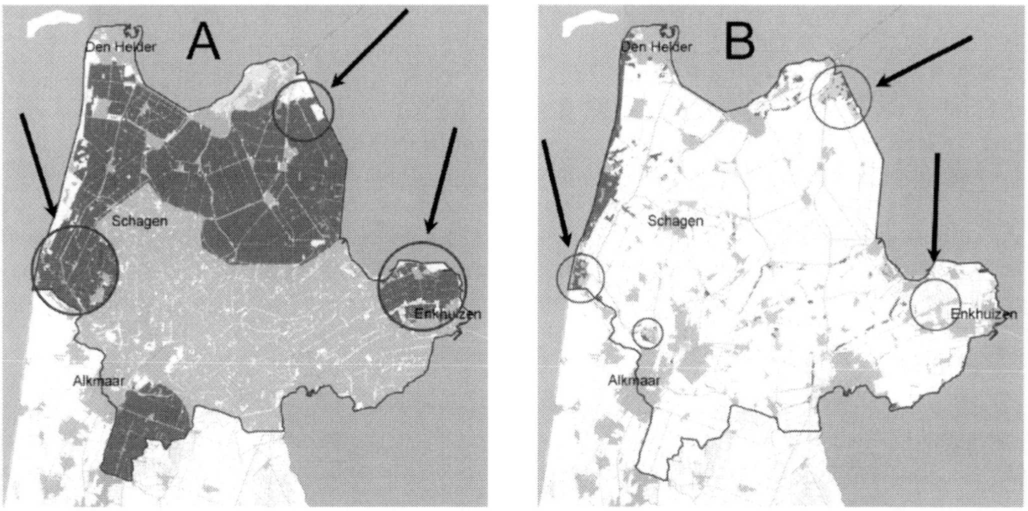

Figure 8. Estimated increase of salinity problems for agriculture (A: intense red: high increase; soft red: hardly any increase; green: no increase) and natural areas (B; red: high increase; yellow: hardly any increase; green: no increase) by climate change in the 'clay region' in 2040. Arrows indicate the presence of subsoil salt glands. Map derived from [16].

An intensive water management strategy was chosen when a sectoral approach was needed. This approach focuses on the interest of agriculture. Especially in the north-eastern part of the region with a strong and high cost-effective agriculture, this approach might be preferred, applying it to small plots as boundary condition. The choice for extensive water management approach is more attractive in the south-western part of the region where a polder is situated. This area is relatively far below sea level, contains a subsoil salt gland and requires large (economical) effort to withdraw the brackish groundwater. Introduction of salt tolerant crops and a combination with recreation might be interesting alternative sources of income.

Table 3. Possible adaptation strategies in 'the 'clay region'. Data derived from [16].

Characteristics	Intensive water management	Supply of saline water
Implication	Contrasting water regimes for agricultural and natural areas, intensive pumping needed	Extensive water management
Approach	Sectoral	Integrated
Application	For small plots	Growth of salt tolerant crops; combination with recreation
Realization time	Short-term	Long-term
Costs	Low	High
Feasibility	Only when used for cost-effective and low situated agricultural regions	For extensive agriculture combined with the realization of nature values

5.2. The second case study region (Eastern 'Noord-Brabant') is situated above sea level, is hilly and has a sandy soil. Therefore, the region is hereafter referred to as 'sand region'. Agricultural and natural areas are strongly intermixed. Agriculture mainly comprises arable crops, maize, and intensive cattle breeding. The natural areas consist of forested land, heath lands and drifting sand. Agricultural and natural areas are fragmented by several major cities at the edges of the region and a rather dense network of roads.

Drought is already a problem in the 'sand region' and agriculture requires intensive irrigation during the summer. Periods of serious water shortage regularly lead to competing claims from different water-using sectors and force the local authorities to prohibit farmers from irrigating their crops for certain periods. Climate change will probably enhance this threat, causing a need for robust interventions in water management. Figure 9 shows both the agricultural and natural areas in which drought may become a risk by 2040, and the need for an integrated approach. The conventional adaptation strategy 'irrigation' is compared to a more sophisticated approach 'increase of water-absorbing soil properties in the upstream valley areas' in Table 4. The creation of wetter conditions upstream will increase the potential for the development of high-quality nature there, while downstream both nature and agriculture will benefit. Extreme runoff peaks will occur less frequently. Moreover, this measure will provide a water buffer during wet conditions, while more water becomes available during dry periods. It must be clear that this adaptation strategy is only applicable in hilly areas with stream valleys and will take a great deal of time to be realized.

Figure 9. Increased drought risks (indicated by intense red) due to climate change for agricultural (A) and natural areas (B) in the 'sand region' in 2040. From [16].

Table 4. Possible adaptation strategies in the 'sand region'. Data derived from [16].

Characteristics	Irrigation systems	Increase of water absorbing properties soil
Implication	High (competing) claims for water availability	Enriching soils with organic material)
Approach	Sectoral, focus on agriculture	Integrated
Application	High cost-effective farm systems	At regional scale, governmental driven
Realization time	Short term	Long-term
Costs	High costs	High costs
Feasibility	If sufficient water available	Effective in hilly areas and regions with ditches

5.3. The third case study region ('Zuidwest-Friesland' & 'Noordwest-Overijssel') is situated near the Lake IJsselmeer, at or slightly below sea level. The soil is primarily peaty, but is intersected by low sandy ridges (largely coinciding with the green areas in Figure 10). Therefore, the region is thereafter referred to as 'peat region'. Wetland areas including shallow lakes are the most distinct natural features in this region. Agriculture consists mainly of dairy farming. In a part of the region the soil gradually sinks due to oxidation processes in the peat soils. The 'peat region' is scarcely populated and recreation is the main source of income. Therefore, provincial policy focuses on the attraction for aquatic sports. Tension between the agricultural and natural sectors may occur in the southern part of the 'peat region' if climate change increases the water excess (see Figure 10). Measurements to prevent damage to agricultural crops (e.g., lowering the groundwater level) may cause drought stress to natural areas, and thus requires an integrated approach. As recreation, the main source of income dominates in this region, two adaptation strategies are conceivable to cope with these threats: 1. Agriculture shifts towards a more multifunctional approach by introducing a stronger emphasis on recreation services and house sale or 2. Deregulation of the water level towards a more natural course in combination with an increase in the use of natural products (biomass for biofuel). Table 5 shows the characteristics of both adaptation strategies.

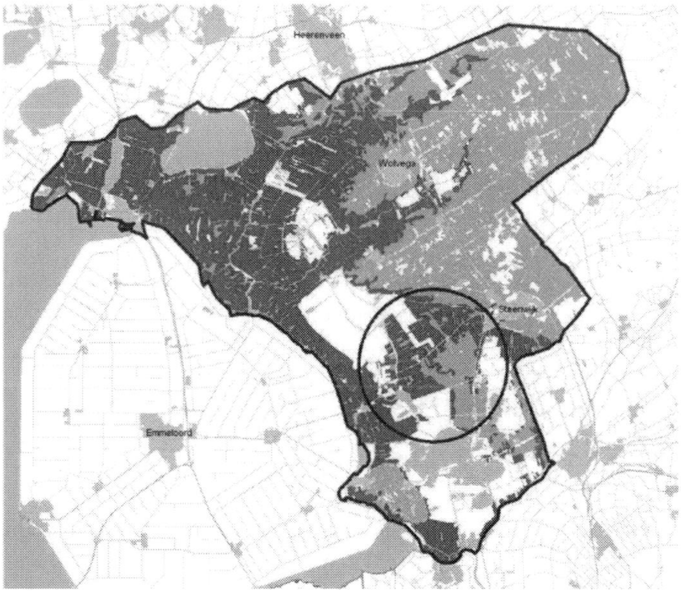

Figure 10. Estimated risks of water excess for agriculture by climate change (red: increase; green: no increase) in the 'peat region' in 2040 [16]. The circle indicates an area which lies extremely low below sea level (approximately -7 m NAP).

Both strategies enhance the environmental value for recreation. However, the choice for one or the other strategy will depend on the cost-benefit balance for this region and the willingness of farmers to adopt it.

Table 5. Possible adaptation strategies in the 'peat region'. Data derived from [16].

Characteristics	Multifunctional agriculture	Deregulation of the water level
Implication	Introduction green-blue services, recreation and house sale	Increased use of natural products, requires strong entrepreneurship
Approach	Integrated approach	Integrated approach
Application	Small scale approach	Large scale
Realization time	Short term	Long term
Costs	Low cost	High investment
Feasibility	Requires human-directed attitude farmer and sufficient consumer demand	Requires economical profit

6. Trade-offs

The choice of an adaptation strategy very will often require weighing the interests of the agriculture and nature sectors. Figure 11 presents a quadrant in which 10 adaptation strategies, applicable to the three case study regions, are situated. Along the x-axis the strategies may vary between a sectoral approach and an integrated approach. Along the y-axis the strategies are divided between an advantageous strategy for the agriculture sector and a strategy advantageous for the nature sector. High-tech adaptation strategies such as the introduction of hydroponics is principally a sectoral approach in which the agriculture sector has the largest advantage, although it will not negatively affect the nature sector.

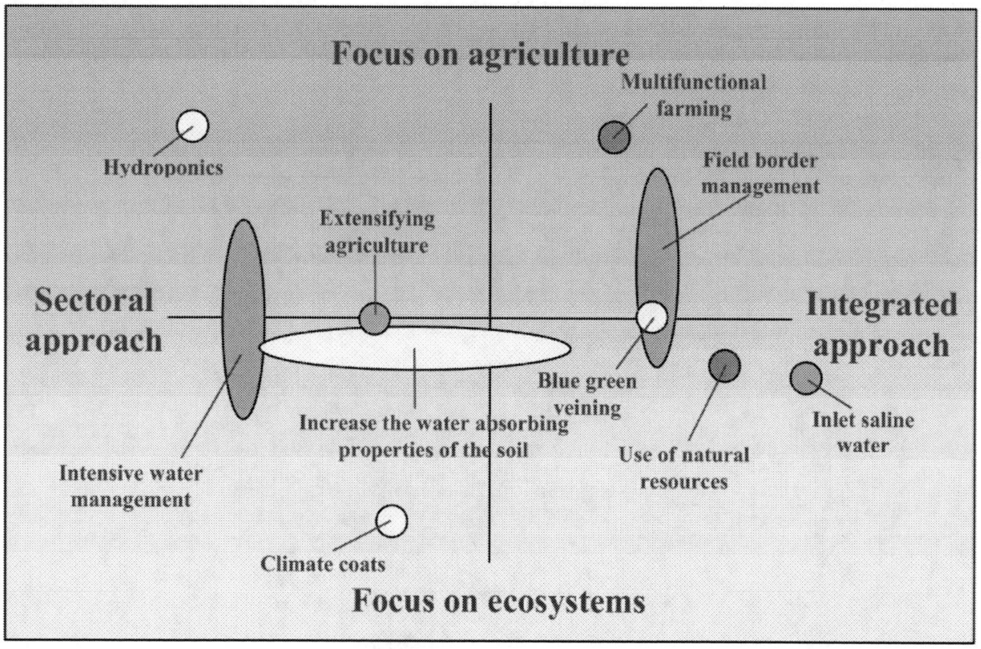

Figure 11. Some adaptation strategies, applicable to the case study regions (blue = 'clay region', yellow = 'sand region', red = 'peat region'), situated in a quadrant in which the x-axis represents a sectoral or integral approach and the y-axis whether the focus is on agriculture or on the natural systems. Adapted from [16].

When more space becomes subject to salinization, however, the natural sector will gain more (in terms of the potential for high-quality saline nature) than the agricultural sector. Here, agriculture will have to lose ground, unless the farmer is willing and able to switch to salt-tolerant crops.

Often different stakeholders are involved in decision-making on spatial interventions and even so, many conflicting interests may have to be considered [2]. For the formulation of feasible adaptation strategies and making them appropriate for implementation, both research and stakeholders should participate in their development. Therefore brainstorm groups were organized with farmer organizations, representatives from government, provinces and municipalities and ngo's to discuss the criteria for the choices to be made.

Table 6. Themes and criteria for the formulation of adaptation strategies derived from brainstorm meetings with stakeholders.

Themes	**Criteria**
Characteristic of the measurement	Sense of urgency
	No regret
	Period of realization
	Availability of innovative technology
	Desired scale level
Stakeholders involved	Affinity and willingness to participate in the implementation
	Image building
	Complexity of interaction between stakeholders, network
Economical consequences	Costs and effectiveness
	Who has to pay, investments
	Who will benefit from it
	Risks
Side effects	Impact to other parties (infra structure, urban live, ..)
	Win-win for other initiatives

Table 6 presents the themes and criteria that emerged and were discussed during these sessions. As policy makers felt the urgency of climate change threats in this process, a basis was formed for the successful implementation of adaptation strategies. A basis was also formed for cooperation between the parties involved in the challenges by climate change, working towards a resilient country in the future. Thus, the 'climate change' theme was placed on the agenda of the regional politicians and the perspectives were recognised described earlier by Koomen etal. [2].

7. Concluding remarks

The focussing on case study regions allowed us to the risks of climate change, address the feasibility and effectiveness of potential adaptation strategies. We found that the suitability of the adaptation strategies clearly differed between regions. In regions of a predominant agricultural character, a sectoral approach is possible, while in regions in which agricultural and natural areas are strongly intermixed, an integrated approach is required. Important criteria for choosing one adaptation strategy or another are the scale of implementation, availability of innovative technology, commitment by different stakeholders involved in the implementation, and the economical perspectives.

Acknowledgements

The authors would like to thank Harold Kuipers (Alterra) for providing the natural areas map (figure 1B) and Tom Dueck (Plant Research International) for improving the English.

References

[1] Van der Most, H., Marchand, M., Bucx, T., Nauta, T. & Van Staveren, M., 2009, Towards sustainable development of deltas, estuaries and coastal zones. Description of eight selected deltas. Deltares, Delft.

[2] Koomen, E., Dekkers, J. & van Dijk, T., 2008, Open-space preservation in the Netherlands: Planning, practice and prospects. Land Use Policy 25, 361-377

[3] Carton L.J. & Thissen W.A.H. , 2009, Emerging conflict in collaborative mapping: Towards a deeper understanding? Journal of Environmental Management 90, 1991 – 2001.

[4] Howden, S.M., Soussana, J.F., Tubiello, F.N., Chhetri, N., Dunlop, M. & Meinke, H., 2007, Adapting agriculture to climate change, Proceedings of the national Academy of sciences of the United States of America 104, 19691-19696.

[5] Thuiller, W., Lavorel, S., Araújo, M.B., Sykes, M.T., Prentice, I.C. & Mooney, H.A., 2005, Climate Change Threats to Plant Diversity in Europe , Proceedings of the National Academy of Sciences of the United States of America 102, 8245-8250 .

[6] Tubiello, F.N., Soussana, J.F., Howden, S.M., 2007, Crop and pasture response to climate change, Proceedings of the National Academy of Sciences of the United States of America 104, 19686-19690.

[7] Fuhrer J, 2003, Agroecosystem responses to combinations of elevated CO2, ozone, and global climate change. Agriculture Ecosystems & Environment 97 (1-3), 1-20

[8] Raadgever, G.T., Mostert, E. & van de Giesen, N.C., 2008, Identification of stakeholder perspectives on future flood management in the Rhine basin using Q methodology, Hydrology and Earth System Sciences 12, 1097-1109.

[9] Caniels, M.C.J., and H.A. Romijn. 2008. Strategic niche management: towards a policy tool for sustainable development. Technology Analysis & Strategic Management. 245-266.

[10] Blom, M., Paulissen, M., Vos, C. & Agricola, H., 2008, Effecten van klimaatverandering op landbouw en natuur, Nationale Knelpuntenkaart en Adaptatiestrategieën, PRI Rapport 182, pp 38 + attachments.

[11] Lavalle C., Micale, F., Houston, T.D., Camia, A., Hiederer, R., Lazar, C., Conte, C., Amatulli, G. & Genovese, G., 2009, Climate change in Europe. 3. Impact on agriculture and forestry. A review. Agronomy for sustainable Development 29, 433 – 446.

[12] Nonhebel, S., 1996, Effects of temperature rise and increase in CO2 concentration on simulated wheat yield in Europe, Climate change 34, 73 – 90.

[13] Witte, J.P.M., Runhaar, J. & Van Ek, R., 2009, Ecohydrologische effecten van klimaatverandering op de vegetatie van Nederland. KWR Watercycle Research Institute, by order of PBL and Deltares. KWR, Nieuwegein.

[14] IPCC, 2007. Climate Change 2007 – The Physical Science Basis. Contribution of Working Group I to the Fourth Assessment Report of the IPCC. Cambridge University Press. 2007. (ISBN 978 0521 88009-1 Hardback; 978 0521 70596-7 Paperback); Climate Change 2007 – Impacts, Adaptation and Vulnerability. Contribution of Working Group II to the Fourth Assessment Report of the IPCC, Cambridge University Press. 2007. (978 0521 88010-7 Hardback; 978 0521 70597-4 Paperback); Climate Change 2007 – Mitigation of Climate Change. Contribution of Working Group III to the Fourth Assessment Report of the IPCC. Cambridge University Press. 2007. (978 0521 88011-4 Hardback; 978 0521 70598-1 Paperback).

[15] IPCC web site, http://www.ipcc.ch/as

[16] Blom, M., Paulissen, M., Geertsema, W. & Agricola, H., 2009, Klimaatverandering in 3 casestudiegebieden. Integratie van adaptatie strategieën voor landbouw en natuur. Wageningen UR Rapport 255, Pp 74.

[17] Graveland, J. & Coops, H., 1997, Verdwijnen van rietgordels in Nederland. Oorzaken, gevolgen en strategie voor herstel. Landschap 14, 67-86

[18] Roelofs, J.G.M., 1991, Inlet of alkaline river water into peaty lowlands: effects on water quality

and *Stratiotes aloides* L. stands. Aquatic Botany 39, 267-293.

[19] RIZA, 2005, Droogtestudie Nederland. Aard, ernst en omvang van watertekorten in Nederland. Eindrapport. RIZA-rapport 2005.016, 104 p.

[20] Van den Hurk, B., Klein Tank, A., Lenderink, G., Van Ulden, A., Van Oldenborgh, G.J., Katsman, C., Van den Brink, H., Keller, F., Bessembinder, J., Burgers, G., Komen, G., Hazeleger, W. & Drijfhout, S,. 2006, KNMI Climate Change Scenarios 2006 for the Netherlands. KNMI Scientific Report WR 2006-01. KNMI, De Bilt.

[21] Safecoast, 2008, Coastal Flood Risk and Trends for the Future in the North Sea Region, synthesis report. Safecoast project team. The Hague, pp. 136.

[22] Goosen, H., Stuyt, L., de Groot, M., den Braber, M., Bessembinder, J., Steingröver, E., van Rooij, S., Witte, F. & Blom-Zandstra, G., 2009, Klimaateffect Atlas 1.0 Eindrapportage fase 2: Van Schetsboek naar klimaateffectatlas, in press.

[23] Schaap, B., Blom-Zandstra, G., Geijzendorfer, I., Hermans, T., Smidt, R. & Verhagen, J., 2009, Project Klimaat en Landbouw Noord-Nederland Fase 2, Nota Plant Research International en Alterra, pp 105.

Emissions of greenhouse gases (methane and nitrous oxide) from cattle slurry storage in Northern Europe

L Rodhe[1, 2], J Ascue[1] and Å Nordberg[1]

[1]JTI – Swedish Institute of Agricultural and Environmental Engineering, P.O. Box 7033, SE-750 07 Uppsala, Sweden

E-mail: lena.rodhe@jti.se

Abstract. Total greenhouse gas (GHG) emissions from stored manure corresponded to 14% of overall GHG emissions from Swedish agriculture in 2006 according to calculations using standard values for a cool climate. The present study identified storage conditions for cattle slurry in different regions of Sweden, developed methodology for measuring GHGs from slurry stored under similar conditions to full-scale storage, and determined annual GHG emissions (methane and nitrous oxide) from stored cattle slurry under Swedish conditions. Temperature measurements in full-scale storage of cattle slurry on farms showed a mean annual slurry temperature of 9.7°C in south-west Sweden and 5.6°C in the north. The closed chamber methodology and equipment developed for measuring GHG emissions were implemented for one year in a pilot-scale plant with similar conditions to full-scale storage as regards slurry temperature, climate and filling/emptying routines. During winter (Oct-April), methane emissions from stored cattle slurry were 3.6 g CH_4-C per kg VS, while during summer (May-Sept) they were 6.5 g CH_4-C per kg VS. This corresponded to an annual methane conversion factor (MCF) of 2.7%. Losses of nitrous oxide were close to zero.

1. Introduction

Greenhouse gases (GHG) contribute to global warming. The GHG methane (CH_4) is emitted from stored liquid manure, while nitrous oxide (N_2O) is the main GHG emitted from stored solid manure. The main agents in producing GHG are microorganisms, which participate in mineralization of organic carbon and nitrogen. Methanogens produce CH_4 and methanotrophs consume it, while N_2O is produced by both nitrifying and denitrifying bacteria, but only the latter group possesses the capacity to consume this form of nitrogen [1]. The rates of the various processes in the ecosystem (*e.g.* slurry container or soil) are regulated by the amount of substrate and by physical and chemical variables such as water, oxygen and temperature. These factors may vary in a diurnal and seasonal pattern and thus, depending on the conditions, the system can act as a source or sink of GHG. In order to quantify GHG emissions and suggest mitigation measures, factors influencing the microbial community must be known and taken into account.

Knowledge about greenhouse gas (GHG) emissions from the Swedish agricultural sector is lacking, as data based on actual measurements are limited [2]. This is especially true for the emissions from stored slurry and from field applications of slurry. Today, the main fraction of the manure produced in Sweden is handled as slurry, and the trend is for an increasing proportion of slurry and a decreasing proportion of solid manure [3]. In the absence of measured data, the annual Swedish National Inventory Report (NIR) on national emissions to the United

[2] To whom any correspondence should be addressed.

Nations Framework Convention on Climate Change (UNFCCC) is currently based on calculations using default values [4]. According to the Swedish inventory in 2006, about 13% of GHG emissions originate from the agricultural sector and take the form of nitrous oxide (N_2O) or methane (CH_4) [4]. The main sources are livestock (enteric fermentation), agricultural soils (arable land and grazing areas) and slurry management. These calculations are based mainly on standard values according to the Guidelines [5] from the Intergovernmental Panel on Climate Change (IPCC) . The Guidelines [5] state that at mean air temperatures below 10°C, the default methane conversion factor (MCF) is 17% for slurry handling (housing and storage) when there is no natural crust during storage, and 10% when a natural crust cover is formed. These factors are the judgement of the IPCC Expert Group in combination with [6] and [7]. In addition to temperature, factors such as feed and animal characteristics at the measurement site [8], manure characteristics [9], length of storage, time between indoor and outdoor storage, amount left in the storage facility (methanogenic inoculum) [9] and timing of storage/application also affect CH_4 emissions from slurry handling [5], while agitation of slurry seems to have negligible impact [10]. These affecting factors can be individual for regions but also for farms. Swedish conditions such as daily cleaning of animal houses (no slurry storage in houses) and average air temperatures lower than 10°C indicate that a smaller MCF may be more appropriate than the IPCC Guidelines default values. This is in accordance with IPCC intentions, as they encourage countries to use factors based on local studies. The default values for emissions from slurry management originate from regions with warmer climates and often with different slurry management systems than Sweden [11, 12]. Hence, there is an obvious risk of overestimation in the reported Swedish GHG emissions from slurry management.

GHG emissions from stored slurry have been studied in full-scale storage [13-15], pilot-scale storage [16-18] and at laboratory scale [9, 18]. It was found by [13] that CH_4 rates increased significantly with storage temperature and that the presence of a rather thick natural surface crust (up to 1 m thick) reduced CH_4 emissions from slurry by a factor of 11 to 12. The CH_4 emissions from cattle slurry were reported to be in the range 0-34.5 g CH_4 m^{-3} d^{-1} during a year, with lower emissions in winter (0.02-1.4 g CH_4 m^{-3} d^{-1}) and higher in summer (17.5-34.5 g CH_4 m^{-3} d^{-1}). For measuring, an open chamber technique with two chambers (area 0.2 by 0.7 m^2) was used in full-scale storage. As the spatial variations in emissions were large, larger sampling areas and large numbers of sampling plots may be required. The difference in CH_4 emissions from stored cattle slurry originated from two different dairy farms was large [9]. Slurry originated from cows with a high-energy diet, rich in concentrate and mainly composed of alfalfa silage and corn silage generated much lower CH_4 emissions than slurry from cows on a diet based on hay. One explanation was that the high-energy diet generated more concentrated slurry, which may have contained compounds, such as ammonia or volatile fatty acids (VFA) in concentrations that would inhibit methanogenic activity. Methane production was simulated from slurry storage structures for a one-year period, and the results indicated that the depth of residual manure remaining in the tank after removal for land application had a considerable impact on methane emissions [9].

The aims of the present study were to define storage conditions for cattle slurry in different regions of Sweden; to develop methodology for measuring GHGs from slurry stored under similar conditions to full-scale storage; and to determine annual GHG emissions (methane and nitrous oxide) from stored cattle slurry under Swedish conditions.

2. Materials and methods

The studies were performed over a two-year period (2006 to 2008). In year 1, full-scale storage conditions were studied on farms in three different climate regions of Sweden and methodology and equipment for measuring GHG emissions with a closed chamber technique were developed and tested. In year 2, a pilot-scale plant with nine containers was constructed to create similar conditions to full-scale storage in terms of slurry temperature, climate, filling/emptying routines, *etc*. Greenhouse gas emissions from cattle slurry in this outdoor pilot-scale plant were measured for a full year using the closed chamber technique.

2.1. Conditions in full-scale slurry storage in different regions of Sweden

2.1.1. Farm descriptions

Slurry storage on three dairy farms was studied for 1.5 years. The farms were located in Halland county in south-west Sweden (57°1.3′N, 13°3.3′E), Uppland county in eastern Sweden (60°0.9′N, 17°59.3′E) and Jämtland county at the northern boundary of agriculture in Sweden (62°57.6′N, 14°23.4′E). The farm data collected are presented in Table 1.

Table 1. Description of animal production, manure handling and storage on the study farms.

Farm, measuring period	Production, number of animals	Type of bedding and amount	Storage volume, m^3	Diameter, m	Depth of storage, m	Filling method	Emptying strategy
Halland, 20/12/05-26/06/06	Loose housing system, 60 dairy cows, 20 heifers	Sawdust, 6.5 m^3 per month	1700	23.3	4 (below soil surface)	Filling at bottom	Mainly in spring, also autumn
Uppland, 2/12/05-04/09/07; 10/11/06-11/04/07	Loose housing system, 60 dairy cows	Sparse straw, approx. 20 kg per week	1500	21.8	4 (half below soil surface)	Filling at bottom	Spring, summer and autumn
Jämtland, 7/11/05-29/04/06; 14/08/07-12/09/08	Tied system, 58-70 cows	Sawdust	1200	22.6	3 (half below soil surface)	Filling at bottom, scrapers	Spring and autumn

2.1.2 Measurements

Temperatures loggers encased in watertight boxes and attached by a chain to a buoy were placed at 0.5 m and 1.5 m below the surface of the stored slurry. Temperature data were logged every 3 hours. The farmers were also given pre-printed forms and asked to record slurry level in the tank, activities such as mixing and emptying of the tank and their visual evaluation of the slurry crust. Air temperature was taken from nearest meteorological station [19].

2.2. Development of pilot-scale methodology for closed chamber measurements of gas emissions from stored slurry

A pre-study was conducted during year 1 in order to develop materials and methodology for gas measurements from stored slurry in pilot-scale storage using a closed chamber technique. Measuring GHG emissions from stored animal slurry at pilot-scale is a compromise between the real conditions in full-scale storage (volume >1000 m^3), and the fully controlled conditions at laboratory scale (~0.010 to 0.100 m^3). The following critical conditions to be maintained in the pilot-scale storage were identified: the temperature of the slurry and its fluctuations should be similar to those in full-scale storage; the slurry should be exposed to changing climate conditions (wind, rain, snow, sun radiation); the tank should be filled and emptied with the same regularity as on a working farm; and the amounts of slurry entering and leaving the tank and its properties must be known. Test containers and a flexible roof were constructed for the pilot-scale gas measurement installation. The aim was to identify the appropriate enclosed gas volume (headspace) and sampling times after closure of the container when using the closed chamber technique for measuring emissions of CH_4 and N_2O from stored animal slurry. The target values for headspace and sample timing were studied for both winter and summer conditions.

2.2.1.Test containers

Three containers (height 1 m, diameter 1.92 m), figures 1 and 2, were placed on the ground to study slurry temperature variations during winter and summer conditions. Container 1 was placed indoors with a temperature above 10°C. Containers 2 and 3, both of which had an insulated base (0.2 m mineral wool) and walls (0.2 m mineral wool, plasticized and then compressed to 0.1 m thickness) were placed outdoors. In order to avoid freezing of the whole slurry volume in cold winters, container 2 had an insulated lid that could be temporarily placed on top, whereas container 3 had a low-intensity heating system in the lower half of its walls that could be switched on during extreme cold periods. The outdoor containers were placed on concrete stands. Temperature in the slurry in each container was measured with thermocouples at 0.1 m above the bottom and 0.1 m beneath the surface, and registered every hour (Intab PC-logger AAC-2f).

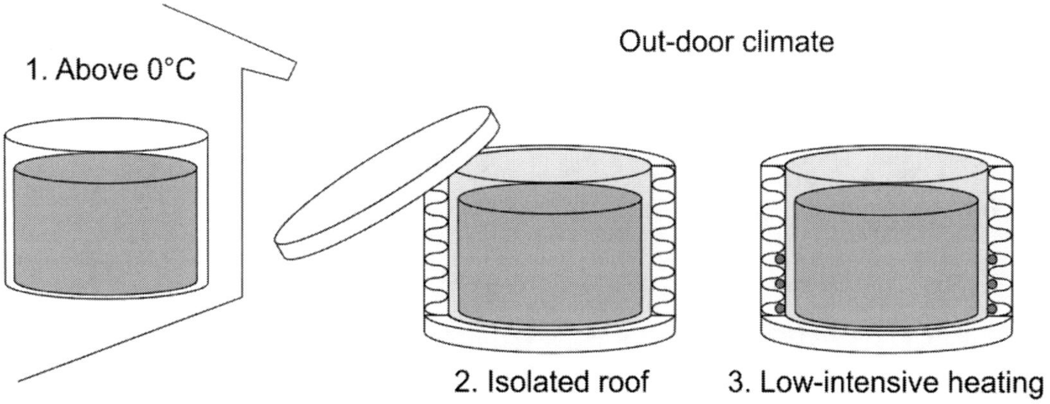

Figure 1. Test containers used in pilot-scale closed chamber method for gas measurements from stored slurry.

2.2.2. Flexible roof construction

A roof that could enclose a variable volume above the slurry surface stored in a container was constructed, figure 2. This roof consisted of a L-beam (50 mm by 50 mm) with the vertical side against the inside of the container, bent in a circle with diameter 1.875 m year 1 and 1.587 m year 2. A steel sheet with four valves for gas sampling was welded fast inside the L-beam. Sealant was used to make the roof airtight. An inflatable rubber tube (Pneuma-seal®, Type 2, PRS 423; maximal pressure 210 kPa) was placed around the periphery of the roof for sealing when the roof had been placed at the desired height above the slurry surface. Four spacer rods with measuring tapes were mounted in the roof in order to suspend the roof on the edges of the containers at the desired height before sealing with the rubber tube. Before measurements in summer, an axial fan (40 mm x 40 mm, 12V DC, 10.9 m^3 h^{-1}) was mounted on the lower side of the roof to create uniform gas distribution in the closed chamber. During measurements, a temperature logger (Tiny View Plus, fabricate INTAB AB) was placed underneath the roof so that gas emissions (Nm^3) could be calculated.

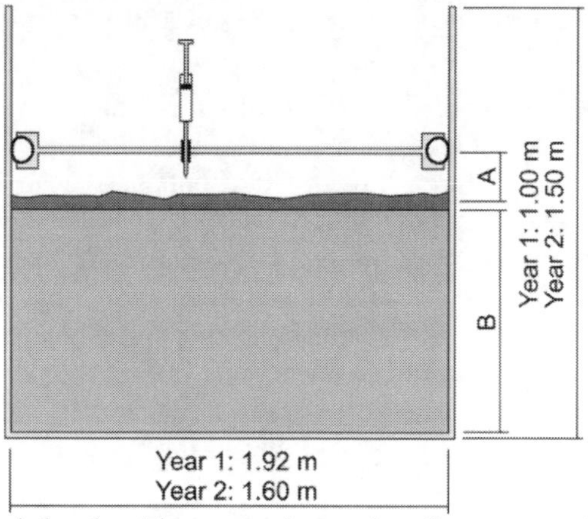

Figure 2. Drawing of closed chamber (slurry container and roof) at gas sampling. A = distance between slurry surface and roof (headspace); B = level of slurry in container. The dimensions of the container and roof are shown for the test containers in year 1 and for the containers in the experimental plant constructed in year 2.

2.2.3. Gas sampling

Table 2 shows the schedule used for finding the appropriate enclosed gas volume (headspace) and sampling times after roof closure when using closed chamber technique for measuring CH_4 and N_2O emissions from stored animal slurry. Gas samples were taken through two valves, placed opposite each other on the roof.

Table 2. Schedule for gas sampling during winter and summer conditions.

Season and date	Container	Slurry level, m	Distance between roof and slurry surface, m	Time of sampling, minutes after closing
Winter 23-24/02/06	1. Indoor	0.6	0.11, 0.21, 0.31	0, 15, 30, 45, 60, 75 and 90
Winter 08-09/03/06	2. Outdoor	0.6	0.30 (incl. 0.20 of snow)	0, 15, 30, 45, 60, 75 and 90
Winter 08-09/03/06	2. Outdoor	0.6	0.10 and 0.30 (snow removed)	0, 15, 30, 45, 60, 75 and 90
Summer 20/06/06	2. Outdoor	0.6	0.10 and 0.30	0, 15, 30 and 60
Summer 20/06/06	3. Outdoor	0.6	0.10 and 0.30	0, 15, 30 and 60
Summer 21/06/06	2. Outdoor	0.3	0.10 and 0.30	0, 15, 30 and 60
Summer 21/06/06	3. Outdoor	0.3	0.10 and 0.30	0, 15, 30 and 60

Before sampling using a syringe and hypodermic needle, the syringe body was filled and emptied five times with gas from the closed volume. Thereafter the sample was taken (60 ml), and injected into a 12 ml glass vial evacuated with nitrogen gas. At the start of sampling a cannula was used for drainage, but when 8 ml remained, the cannula was removed and the final gas was pressed into the syringe to give a positive pressure during storage. On each sampling occasion, reference samples were also taken from tubes containing three known concentrations of CH_4 and N_2O gas. The gas samples were analysed for CH_4 and N_2O with a gas chromatograph (HP 6890 Series, GC System, Hewlett Packard, USA) equipped with flame ionization (FID) and electron capture detectors (ECD) and a nickel catalyst for converting CO_2 to CH_4. The precolumn and analytical columns were 1.8 m and 3 m long steel columns packed with Hayesep Q (80/100 mesh). The temperatures of the GC oven, FID and ECD were 70°C, 300°C and 350°C, respectively. Nitrogen was used as the carrier gas and a mixture of argon and methane (5%) as a make-up gas (1.4 ml min^{-1}) to increase the sensitivity of the ECD. An autosampler (222 XL Liquid handler, Gilson Medical Electronics, France) fed the samples to the loop of the GC. The coefficients of variation for the analysis of atmospheric concentrations of N_2O, CH_4 and CO_2 were 0.3%, 0.6% and 0.8%, respectively. Gas volumes were corrected to standard temperature (0°C) and pressure (101.3 kPa).

2.2.4. Slurry

The cattle slurry was taken from a dairy farm in Uppland, Table 3. Before filling, the slurry was well-mixed and sampled for analysis of dry matter (DM), total nitrogen (Tot-N), ammoniacal nitrogen (NH_4-N), nitrate nitrogen (NO_3-N), pH, and volatile solids (VS) [20, 21].

Table 3. Properties of slurry used in tests during winter and summer in year 1; dry matter content (DM), volatile solids (VS), pH, and nutrient content.

Sampling date	DM, %	VS, kg ton^{-1}	pH	Tot-N, kg ton^{-1}	NH_4-N, kg ton^{-1}	P, kg ton^{-1}	K, kg ton^{-1}	Ca, kg ton^{-1}	Mg, kg ton^{-1}
Winter 23/02/06	8.5	66.1	8.0	4.6	2.8	0.7	5.0	1.3	0.6
Summer 19/06/06	7.9	62.7	7.5	4.0	2.3	0.7	4.5	1.2	0.4

2.3. GHG emissions from stored slurry during a full year

2.3.1. Construction of pilot-scale plant

For the full-year experiment during year 2, a pilot-scale plant with nine containers was constructed and placed outdoors at JTI in Uppsala (59°82′N, 17°65′E), figure 2. The containers were half-buried in the ground and had base area 2.0 m^2. Plastic sacks designed to fit the container volume were inserted into the containers in order to provide a non-reactive cover inside the steel containers and to protect the steel from the slurry. Three roofs were constructed so that three containers could be closed simultaneously during gas sampling, figure 2. These roofs were lifted using an electric winch attached to an I-joist and moved to the next three containers by rolling them along the I-joist. Each roof had four adjustable attachments on which it could be suspended on the edge of a container before sealing with a rubber tyre (pressurised to 210 kPa). The results presented below for year 2 are from three of these containers.

2.3.2. Slurry

The slurry used in containers was collected from the farm in Uppland, table 1. The average daily ration per dairy cow consisted of 9 kg DM silage, 5 kg cereals and 3.7 kg concentrate (Unik 72 P BK). At the start of the experiment, stored slurry was brought from the main slurry tank on the farm to the pilot-scale plant, and thereafter fresh slurry was taken monthly from the pumping pit on the farm. Before filling, the slurry was well-mixed and sampled for analysis of DM, Tot-N, NH_4-N, NO_3-N, pH, VS and total carbon (Tot-C) [20, 21]. Maximum methane production (B_0) was analysed for three samples of slurry using a laboratory-scale batch digestion test in 1-litre bottles together with inoculum at 37°C. The inoculum came from a production-scale biogas plant in Västerås that is used for digesting energy crops and municipal solid waste. The load used in the tests corresponded to 5 g VS/l reactor volume (active digester volume 600 ml, including 200 ml inoculum). The bottles were placed on a shaking table in a room at a constant temperature of 37°C to ensure gentle mixing and heating of the slurry. A triplicate digestion test was made for each slurry, using samples taken on three occasions during the year. The batch digestion test was run for 100 days. The DM and VS contents of both slurry and inoculum were analysed. The gas production was calculated by pressure measurements in each bottle using a digital pressure meter (GMH 3110) with a pressure sensor (GMSD 2BR; -1000 to 2000 mbar). The pressure was then converted to produce standard gas volume (0°C, 101.3 kPa). Gas samples were extracted and analyzed using gas chromatography (PerkinElmer ARNEL, Clarus 500, and column 7' HayeSep N 60/80, 1/8" SF; FID detector 250°C; carrier gas helium, flow 31 mL/min; injector temperature 60 °C; injection made by a Headspace sampler Turbo Matrix 110).

2.3.3. Dynamics of filling and emptying storage
The dynamics of slurry storage are shown in figure 3. At the start, on 6 October 2006, the three containers were filled with 0.3 m of slurry. Thereafter, 0.1 m fresh slurry was added once a month, except during summer when the amount was less to account for the cows grazing outdoors during part of the day. In spring, the containers were almost emptied, with a small residue 0.4 m left in the bottom of each to reflect the situation in full-scale storage.

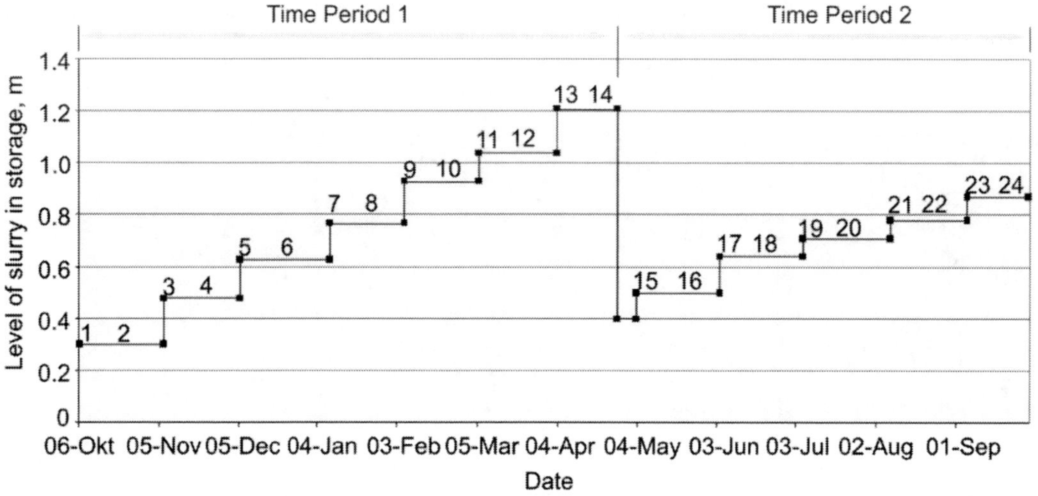

Figure 3. Level of slurry in storage container during the experimental year. Gas sampling occasions are numbered (1 to 24) and the storage year is divided into two periods for summarising winter/summer emissions and for calculating the methane conversion factor (MCF).

2.3.4. Conditions during storage
Temperature in the slurry was measured with thermocouples at 0.1 m from the bottom and 0.1 m beneath the surface in each of the containers, and registered every hour (Intab PC-logger AAC-2f). Weather data were collected from the nearby Ultuna meteorological station, about 900 m from the plant. The thickness of the crust (if any) was measured several times during storage and sampled for DM analysis at the end of the storage period. The crust was also documented with photos.

2.3.5. Gas sampling methodology
Gas was sampled on two occasions per month, with the first being taken one day after filling and the other about a fortnight later, figure 3. Sampling was done between 10 am and 2 pm. Three samples were taken per occasion; the first directly after closing the roof, the next 15 minutes later, and the last 30 minutes after roof closure. In summer, the headspace was mostly 0.3 m, while during colder periods the roof was placed 0.2 m, 0.15 or 0.1 m from the slurry surface in order to achieve a linear increase in gas concentrations of CH_4 and N_2O. A small mixing fan installed in the roof underside was run during closure in order to obtain a uniform distribution of gases. During sampling the fan was stopped. Analysis were made according to 2.2.3.

2.3.6. Calculation of MCF and EF_{nit}.
For calculating the MCF, two storage periods were established, figure 3. Period 1 lasted from 6 October 2006 to 3 May 2007 (210 days) and period 2 from 4 May to 7 October 2007 (157 days). For each period, the MCF was calculated from the sum of methane emissions per average amount of VS added to the containers, divided by B_0 per kg VS [4, 5] as:

$MCF = \Sigma CH_4\text{-}C\ (B_o\ \ VS_{Mean\_in\_storage})^{-1} 100$ (1)

where MCF is the methane conversion factor, %; CH_4-C is the sum of methane-C emissions for the period, kg; B_0 is the maximum methane producing capacity for manure, kg CH_4-C (kg VS)$^{-1}$; and $VS_{Mean\_in\_storage}$ is the average amount of VS added to the container during the storage period, kg. The emission factor for nitrous oxide (N_2O) was then calculated with the equation:

$EF_{nit} = \Sigma N_2O\text{-}N\ (total\text{-}N)^{-1} 10$ (2)

where EF_{nit} is the emission factor for nitrous oxide, %; ΣN_2O-N is the sum of nitrogen (N) lost as nitrous oxide during the period, kg; and total-N is the total amount of N in slurry added to the container during the storage period, kg.

3. Results

3.1. Conditions in full-scale slurry storage in different regions of Sweden

In full-scale storage of cattle slurry, mean slurry temperature measured at two levels of the liquid during one year was 9.7°C in south-west Sweden (Halland county), 8.1°C outside Uppsala (Uppland county) and 5.6°C in northern Sweden (Jämtland county), table 4. The air temperature fluctuated much more than the slurry temperature. In Jämtland during winter, the slurry was insulated with snow cover from December to April.

Table 4. Mean temperature (°C) at 0.5 and 1.5 m below the surface in stored slurry and in air in different regions of Sweden during one year. The range of recorded temperatures in slurry is given in brackets.

	Mean temperature (range), °C				
Region	Slurry, 0.5 m below surface	Slurry, 1.5 m below surface	Slurry, mean of two levels	Air**	Air, normal 1961-1990
Halland 2006	9.9 (1.4-19.3)	9.6 (2.9-15.6)	9.7	7.6 (-15.8-22.6)	7.3
Uppland 2006*	7.9 (-0.1-18.5)	8.4 (2.2-15.3)	8.1	7.4 (-15.1-23.5)	5.6
Jämtland 2007-08	5.3 (0.2-15.3)	6.0 (2.2-11.7)	5.6	4.0 (-14.3-22.3)	2.5

* Calculated on interpolated values for the period 05/09 - 09/11/2006
** Daily mean values

Figure 4 shows the temperature fluctuations in air and slurry as well as the level of slurry in the storage tank on the most southerly farm of the three studied. The tank was partly emptied twice or three times a year, figure 4 and table 1.

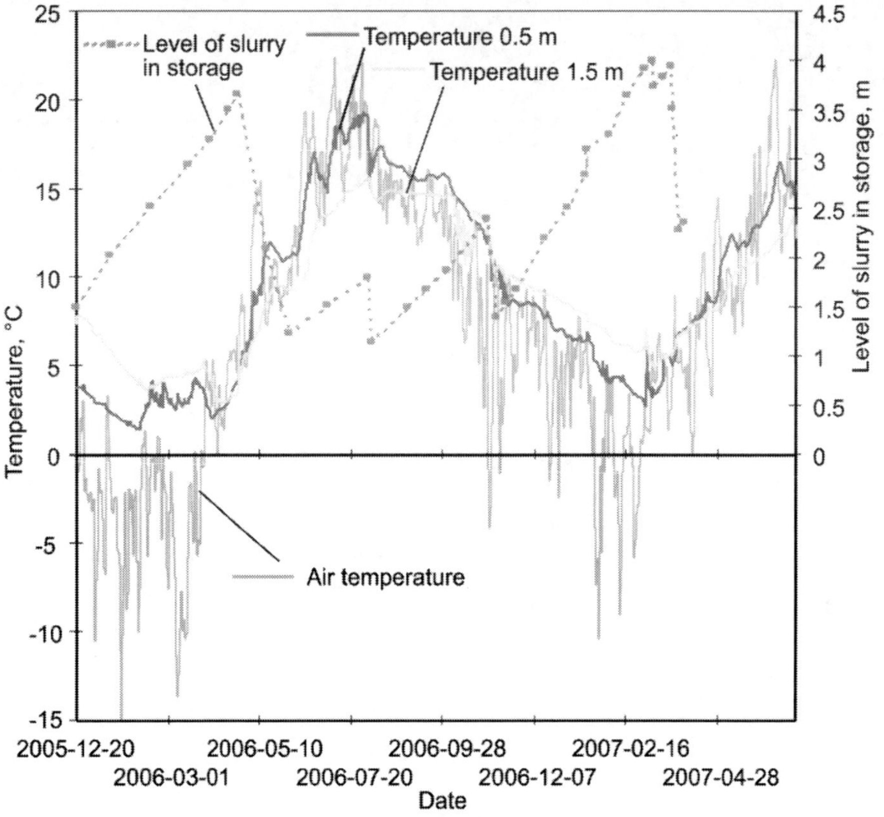

Figure 4. Halland. Air temperature (°C), temperature at 0.5 and 1.5 m below the slurry surface and level of slurry in tank, m.

3.2. Development of pilot-scale methodology for closed chamber gas measurements from stored slurry

3.2.1. Conditions in test containers

The temperature in the outdoor insulated slurry containers placed on the ground varied within a smaller range than the air temperature. However, during very cold periods there was a risk of total freezing of the slurry, and in order to prevent this, the insulated roof was placed on container 2 and the low-intensity heating was turned on for container 3. This prevented total freezing, but in container 2 the roof also reduced exposure to the weather and in container 3 the heater increased the slurry temperature rapidly. In addition, by mid-June the slurry temperature in the containers had already exceeded the maximum value recorded in the full-scale storage tank on the farm in Uppland (18.5°C).

3.2.2. Gas sampling

The flexible roof worked well and provided the possibility to place the roof at the desired height above the slurry surface and to close the gas volume airtight. In the winter experiments, the emissions from the outdoor containers were small (1 to 4.3 10^{-3} CH_4-C m^{-3}, d^{-1}). A linear increase in CH_4 and N_2O concentrations was achieved during the first half-hour after closure with the headspace of 0.1 m, but with a larger headspace the increase was hard to detect in this interval. The concentrations differed slightly between the two sampling valves and therefore a fan was placed on the underside of the roof before sampling in the summer. In the summer experiments, the emissions from the two outdoor containers were about the same at 0.6 m level, figure 5. However, with less slurry in the container the emissions differed more between treatments, especially when the smaller headspace was used. More stable values were obtained when the larger headspace of 0.3 m was used during summer, especially with the smaller amount of slurry in the container.

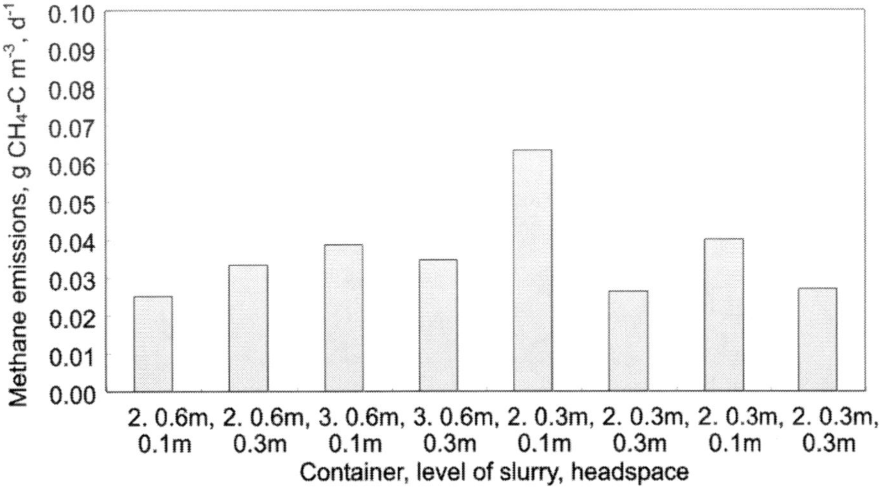

Figure 5. Summer period. Emissions of methane from two outdoor containers with different levels of slurry (0.3 and 0.6 m) and different headspace (0.1 and 0.3 m).

Figure 6 shows the increase in methane concentration over time during closed storage for the 0.3 m level of slurry with a headspace of 0.1 m (left) or 0.3 m (right). The correlation (R^2) was lower for 0.1 m headspace than for 0.3 m headspace.

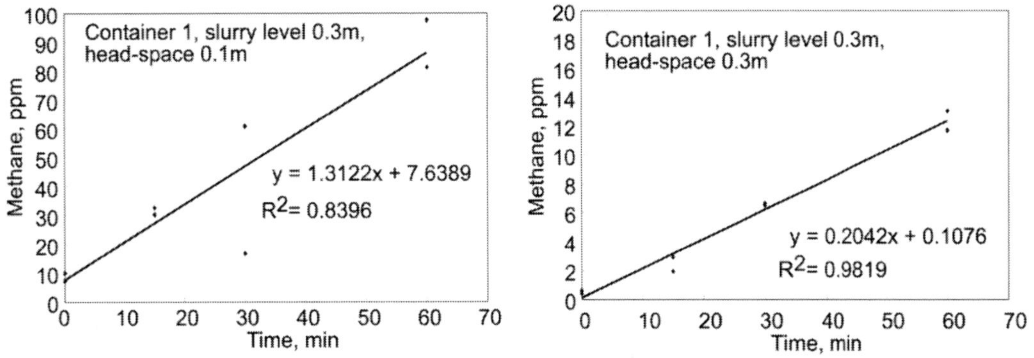

Figure 6. Summer period. Gaseous concentrations of methane in samples taken at different times after closure of the chamber over a slurry container for two different headspace values. Left: 0.1 m headspace, right: 0.3 m headspace.

For N_2O, the concentration increase tailed off after about 30 minutes with the 0.1 m headspace as shown in figure 7 (left), while with the larger headspace of 0.3 m, there was a linear increase in N_2O concentration for up to 1 hour.

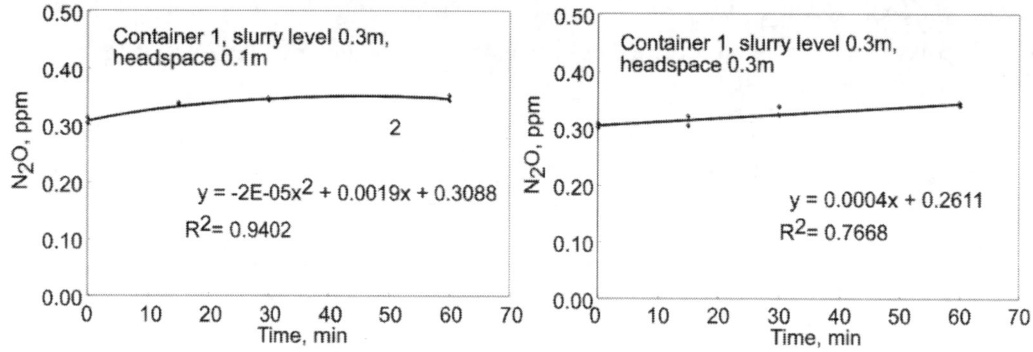

Figure 7. Summer period. Gaseous concentrations of nitrous oxide in samples taken at different times after closure of the chamber over a slurry container for two different headspace values. Left: 0.1 m headspace, right: 0.3 m headspace.

3.3. GHG emissions from stored slurry during a full year

3.3.1. Slurry properties

As shown in table 5, the concentrations of DM and most nutrients were higher in the slurry taken from the pumping pit (filling 2 to 12) compared with slurry from the main storage tank (filling 1). At the time of emptying the containers, the N and C concentrations were about the same as in the main storage tank on the farm.

Table 5. Properties of cattle slurry taken from the main storage tank (filling 1) or the pumping pit (filling 2-12) and at emptying in spring and autumn. Standard deviation in brackets.

					kg ton^{-1}					
	DM, %	VS, % of DM	VFA, g/L	pH	Tot-N	NH_4-N	P	K	Tot-C	C/N
Start:										
Filling 1	5.9	78.1	8.2	6.9	3.6	2.0	0.5	3.6	19.8	5.4
Filling 2-12 (sd)	8.1 (1.1)	80.6 (1.1)	7.7 (0.9)	7.0 (0.2)	4.7 (0.65)	2.2 (0.52)	0.6 (0.13)	4.7 (0.96)	27.8 (3.97)	5.9 (0.64)
Emptying spring	5,2 (0,3)	79.2 (0.4)	4.8 (0.5)	7.4 (0.1)	3.4 (0.14)	1.9 (0.03)	0.4 (0.02)	3.8 (0.08)	17.7 (1.08)	5.2 (0.22)
Emptying autumn	5,4 (0,5)	76.9 (1.5)	3.7* (0.7)	7.2* (0.4)	3.0 (0.31)	1.5 (0.17)	0.4 (0.01)	3.4 (0.20)	18.2 (1.90)	6.0 (0.10)

*Only two values

Slurry samples from three different filling occasions were analysed for the maximum methane producing capacity of the slurry (B_0), table 6. The mean value found of 294 normal-ml CH_4 g^{-1} VS is equivalent to 159.6 g CH_4-C (kg VS)$^{-1}$.

Table 6. B_0 values (normal-ml CH_4 g^{-1} VS) for different samples of cattle slurry digested for 100 days at 37°C. Standard deviation in brackets.

	100 days
Sample 1 (March 07)	264 (22)
Sample 2 (June 07)	285 (20)
Sample 3 (Aug -07)	334 (50)
Mean	294

The stored slurry usually had a thin crust, table 7.

Table 7. Mean thickness (m) of slurry crust measured five times during experimental year 2 for the three slurry storage containers. Standard derivation in brackets.

4 April 2007 (before emptying)	4 May 2007	15 June 2007	7 Aug 2007	7 Oct 2007 (before emptying)
0.08 (0.015)	0.01 (0)	0.05 (0.012)	No crust	0.06 (0.031)

Crust samples of a 0.2 by 0.2 m area was taken at two depths and analyzed for DM and VS before emptying on 4 April (spring) and 7 October (autumn), table 8.

Table 8. Mean dry matter content (%) of the crust before emptying of the containers in spring and autumn, 2007. Sampling at two levels; 0-0.05 m and 0.05-0.10 m below slurry surface.

	DM, %		VS, % of DM		Ash, % of DM	
Time	0-0.05 m	0.05-0.10 m	0-0.05 m	0.05-0.10 m	0-0.05 m	0.05-0.10 m
Spring	17.2	12.6	86.8	84.6	13.2	15.4
Autumn	11.7	11.7	81.8	83.4	18.2	16.6

3.3.2. Conditions during storage

The mean air temperature was 8.4°C (range -13.4 to 24.1°C) and mean precipitation was 625 mm during the experimental year 2. Slurry temperature is presented in figure 8 for the three different slurry containers during the year. The upper sensors had to be removed during winter (5 February to 20 March) in order to prevent them from freezing fast and thereby preventing the roof from closing. By the end, the sensors in container 3 had ceased to function. The mean temperature in containers 1 and 2 for the whole year was 8.6°C (range -0.4 to 20.8°C).

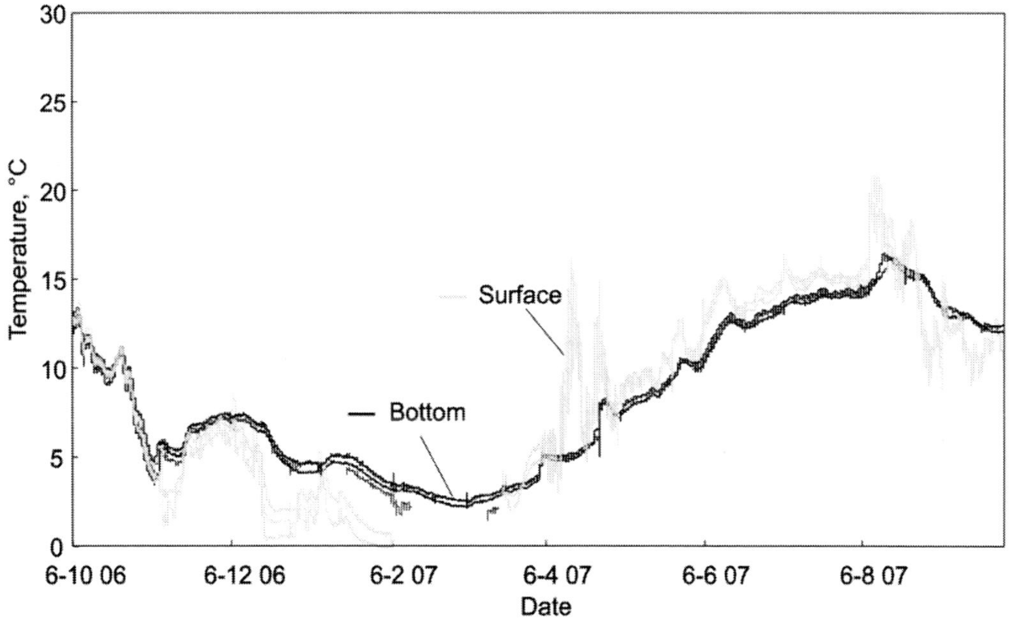

Figure 8. Slurry temperature in the three containers at two levels, 0.1 m from the bottom and 0.1 m below slurry surface, during the experimental year 2006-07.

3.3.3. Methane emissions

During the period October to January, GHG emissions were rather moderate (figure 9). In February, when the crust of the slurry was thawing, there was an occasional peak in emissions. From the start of March until the start of July, the emissions increased from 0.2 to 3.5 g CH4-C m^{-3} slurry and day^{-1}. A dip in emissions was measured in the middle of July and at the beginning of August, followed by a second increase. The total methane emissions for the first period (Oct to April; 210 days) amounted to 3.6 g CH_4-C $(kg\ VS)^{-1}$. For the second period, starting after emptying in April (May to Oct; 147 days), the rate of emissions approximately doubled, to 6.5 g CH_4-C $(kg\ VS)^{-1}$. On average during one year, this meant a loss of 4.8 g CH_4-C $(kg\ VS)^{-1}$ from fresh slurry. In total, 2.08 kg CH_4-C were lost from the three storage containers during the year.

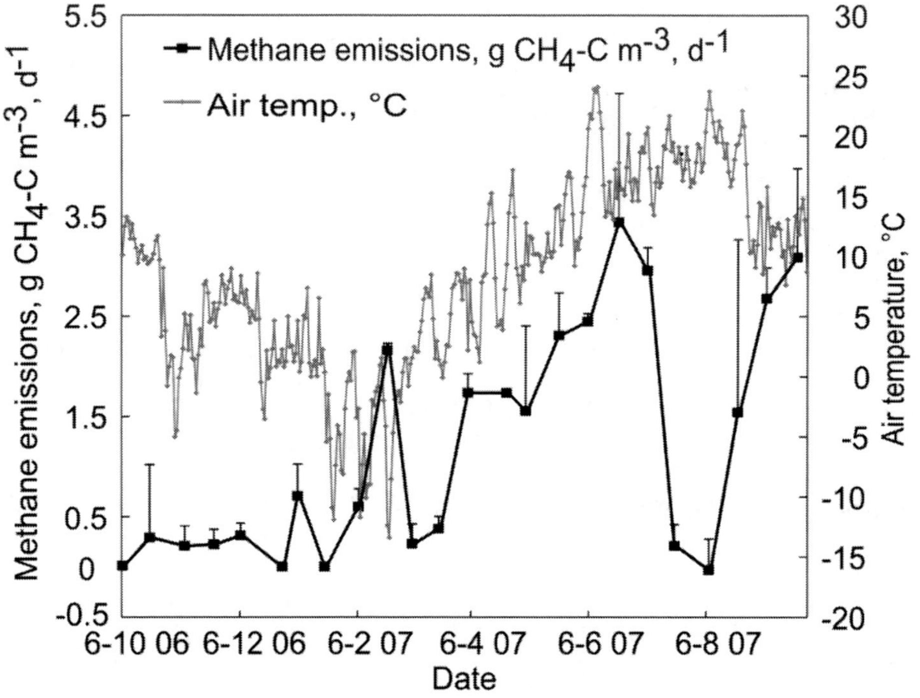

Figure 9. Air temperature and mean methane emissions from slurry storage containers during the experimental year (n=3). Standard deviation shown by error bars.

3.3.4. Calculation of MCF

The MCF as calculated according to equation (1) was 2.2% during the cold storage period (Oct-April) and 4.1% during the warm period (May-Oct). This amounted to an average MCF factor of 3.0% for the whole year.

3.3.5. Nitrous oxide emissions

The losses of N as N_2O were very low during the whole experimental year, figure 10. The concentrations occasionally decreased with time, possibly owing to nitrogen consumption. However, the values were so small that they were within the range of measuring accuracy. During the first (cold) period, the EF_{nit} factor was 0.0007%, while in the second period it was -0.0007%.

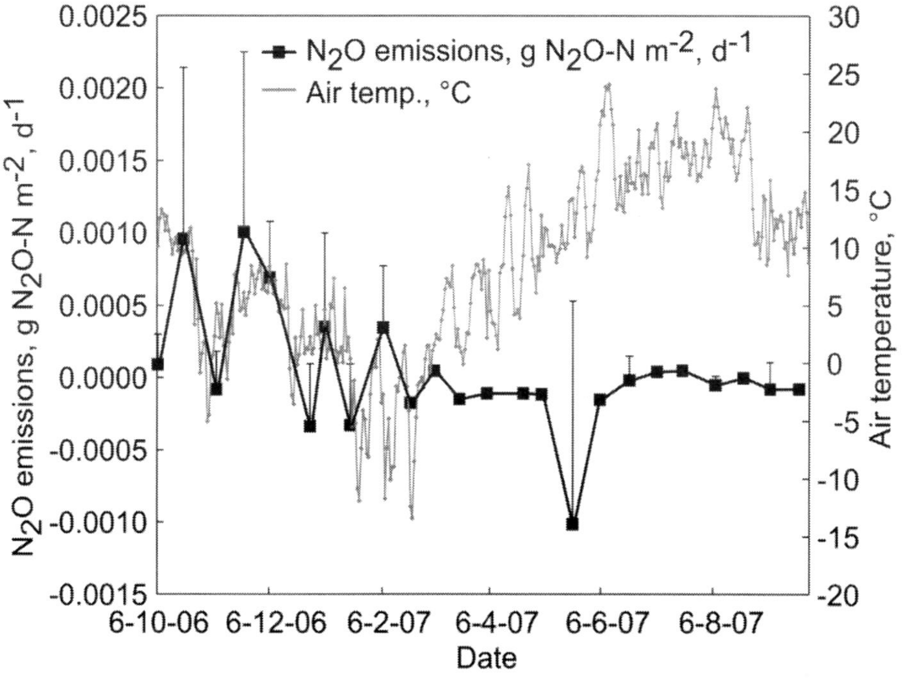

Figure 10. Air temperature and nitrous oxide emissions (g N_2O-N m^{-2} day^{-1}) during one year of storage. Standard deviation shown by error bars.

3.3.6. Balance of mass, total N, total C and VS in storage containers

During storage with no cover, the mass balance was positive (+29%) because of addition of rain. However, about 6% of the total-N, 9% total-C and 11% of VS were lost during storage, table 9. In total, 2.08 kg CH_4-C were lost from the three storage containers during the year, which was about 10% of the total C lost.

Table 9. Balance of total mass, total-N, total C and VS for the three containers.

	Weight, tons	Tot-N, kg	Tot-C, kg	VS, kg
In	8.27	37.71	217.83	510.49
Out	10.70	35.47	197.94	454.02
Increase (+)/Loss (-)	2.43	-2.24	-19.89	-56.47
Increase (+)/Loss (-)	29%	-6%	-9%	-11%

4. Discussion

4.1. Conditions in full-scale storage in different regions of Sweden

The studies in different regions of Sweden showed that farm slurry tanks were partly emptied two to three times a year. This means an average storage period for the slurry of about three months. With daily removal of slurry from animal houses with slatted floors, as required by law, the methane from slurry handling in Sweden originates almost solely from storage tanks.

Mean slurry temperature, table 4, was somewhat higher than the mean air temperature in all regions. Measurements in Ontario, Canada [14] also showed higher monthly mean temperature at 0.6 m depth below surface compared with air temperature. The thick snow cover in the north site probably had a great insulating effect and contributed to the smaller temperature range observed on that farm compared with those in the other regions. The mean air temperature for the experimental year was much higher than the 30-year norm in Uppland and Jämtland. In a 'normal' year, therefore, lower slurry temperatures than those measured in Uppland and Jämtland could be expected. As methane emissions are strongly related to temperature [22], especially at temperature levels above 15°C, this could mean lower methane emissions during a 'normal' year.

4.2. Pilot-scale methodology

The pre-study in year 1 provided valuable information concerning how to create conditions similar to those in full-scale storage for the pilot-scale slurry containers. The winter and summer samplings also provided target values for headspace and sampling times after roof closure.

4.2.1. Conditions in pilot-scale vs full-scale storage

In the full-scale storage in Uppland (70 km north-east of Uppsala) the range of slurry temperature was -0.1 to 18.5°C. In the pilot-scale storage in the same region in year 1, there was a risk of the slurry totally freezing in winter, while in summer the temperature already far exceeded the highest measured value in the full-scale storage by 18 June, when the pilot-scale experiment ended. Insulation with mineral wool was not enough to prevent the base from freezing in the containers placed on the ground, while low-intensity heating made the slurry temperature rise too quickly. The conclusion was that the containers had to be partly buried in the ground to get a more stable temperature.

4.2.2. Gas sampling methodology

The flexible roof made it possible to adjust the position of the roof to the actual slurry surface level and to use an appropriate headspace for the temperature conditions. Preliminary testing indicated that a fan was needed to give the same gas concentrations independent of where samples were taken below the roof. Specific differences in concentration can arise for instance due to cracks in the crust or temperature differences caused by inversion. At some occasions, the slurry was covered with snow and the snow cover was considered to be included in the headspace. In general, there was a linear increase in the concentrations of methane and nitrous oxide during the first hour after roof closure. The N_2O concentrations were very low, and during summer conditions with a small headspace of 0.1 m, the linearity prevailed only for the first 30 minutes. For methane a 0.3 m headspace was also preferable in summer, as it seemed to give more stable values.

From this first pre-study, target values for headspace and times for gas sampling in pilot-scale slurry storages were established, table 10. With a large number of slurry containers to sample, it is advantageous to limit the time per container. In addition, within the first half-hour a linear increase in concentration was achieved in most cases, although for winter conditions only with a headspace of 0.1 m.

Table 10. Target values identified for headspace height and time of gas sampling for pilot-scale slurry containers using a closed chamber technique in winter and summer conditions. Temperature values for slurry and air when the studies were conducted are also presented.

	Slurry temperature, °C	Air temperature (measuring point), °C	Distance between roof and slurry surface, m	Time of sampling, min after closing
Winter	0	-3.0 (outdoor)	0.1	0, 15, 30
Summer	20	30.0 (below roof)	0.3	0, 15, 30

4.3. GHG emissions from stored slurry during a full year

4.3.1. Slurry properties

The cattle slurry used came from a loose house system, currently the most common system for dairy cows in Sweden. The nitrogen contents were similar to earlier mean values reported from 17 farms in Sweden [23]. The maximum methane production during 100 days of incubation of the slurry was 159.6 g CH_4-C per kg VS, which is slightly higher than the default value of 129 g CH_4-C per kg VS (0.24 m^3 CH_4 per kg VS) given in [5]. In Sweden, milk production per cow was on average 8378 kg ECM (cow and year)$^{-1}$ in 2008, which is among the highest in the world [24]. This is partly the result of high consumption of fodder based on good quality roughage (silage) but the analytical method could have influenced the value of the B_0, as could the type of inoculum used.

4.3.2. Conditions in storage
The mean and range of temperature in the partly buried pilot-scale storage containers were roughly similar to those in full-scale storage on the farm in Uppland. At the time of emptying the containers, the slurry also had roughly the same properties as that in full-scale storage, indicating that changes caused by weather and processes in the slurry were similar to those on farm scale. The monthly addition of fresh slurry to containers was intended to reflect the daily addition to tanks at full-scale, which may stimulate methane formation as new, easily degradable VS are supplied. A thin crust was present most of the time, but this crust had a rather low DM content at sampling just before emptying the containers in spring and autumn. The amount of rain corresponded to 1250 kg water per container.

4.3.3. Emissions of methane and nitrous oxide
During the colder period, methane losses were low. The temporary peak on 6 February may have been a consequence of the crust thawing, releasing encapsulated methane gas. On this occasion, the crust was covered with 0.10-0.19 m snow and the slurry temperature 0.1 m below the surface was around 3°C. From March, the methane emissions increased with increasing temperature, with some exceptions. On the sampling occasions in mid-July and early August, there was a dip in emissions that is difficult to explain since the same routines were used at sampling throughout. A sensitivity analysis using interpolated values for measurements 20, 21 and 22 during the warm season resulted in emissions of 5.8 CH_4-C (kg VS of fresh slurry)$^{-1}$, corresponding to 20% higher emissions of methane for the whole year. The standard deviation for the three replicate measurements of methane emissions was on average ±41% (samplings 6, 8 and 14 excluded as the emissions values were very low), indicating the need for replicates. The emissions of nitrous oxide were low, at most 0.001 g N_2O-N m^{-2} day^{-1}. It has been reported previously [16-18] that during periods with drying conditions (negative water balance), there are N_2O emissions caused by nitrification and denitrification processes taking place in the porous crust. In the present study, the crust was rather thin, compact and wet, and no such N_2O emissions were observed.

4.3.4. Calculations of MCF
The calculated MCF (Equation 1) depended on both measured methane emissions and potential/maximum methane producing capacity of the manure. For the fresh slurry added to the storage containers, the MCF was estimated to be approx. 3.0%. Also, when the higher methane emissions value calculated by interpolation of measurements 20, 21 and 21 was used, the MCF value of 3.6% obtained was much lower than the IPCC default value of 10% for annual average temperatures below 10°C [5]. When the interpolated value was used together with the IPCC default B_0 value [5], the MCF value increased somewhat to 4.5%. Since the B_0 value was measured for the slurry, it is preferable to use the measured values. Factors such as low annual temperature, slurry properties and the experimental set-up with dynamic filling of containers as in full-scale storage all contributed to give a MCF value more appropriate for conditions prevailing in Sweden and for areas with similar conditions.

5. Conclusions
The mean annual temperature in stored slurry was 9.7°C in south-west Sweden (57°1.3′N), 5.6°C in the most northerly agricultural region of the country (62°57.6′N) and an intermediate 8.1°C in Uppland (60°0.9′N). Methodology and equipment for measuring GHG emissions with a closed chamber technique were developed and implemented in a pilot-scale plant, where similar conditions as in full-scale storage on a farm in Uppland in terms of slurry properties, temperature, climate and filling/emptying routines were simulated during one year. During the winter period (Oct-April), methane emissions were 3.6 g CH_4-C per kg VS of fresh slurry entering storage and during summer (May-Sept) 6.5 g CH_4-C per kg VS, giving a mean annual

value of 4.8 g CH_4-C per kg VS. This corresponds to an annual MCF factor of 3.0%, much lower than the IPCC default value of 10%. Emissions of nitrous oxide were close to zero.

Acknowledgements
Financial support by the Swedish Farmers' Foundation for Agricultural Research (SLF) and the Swedish Research Council FORMAS is gratefully acknowledged.

References

[1] Davidson E A 1991 Fluxes of nitrous oxide and nitric oxide from terrestrial ecosystems. In: Microbial Production and Consumption of Greenhouse Gases: Methane, Nitrogen Oxides, and Halomethanes. American Society for Microbiology, Washington D.C., 219-235.

[2] Swedish Board of Agriculture 2004 Förutsättningar för en minskning av växthusgasutsläppen från jordbruket. [Conditions for decreasing GHG emissions from agriculture]. *Report 2004:1*, Jönköping, Sweden.

[3] SCB 2008 Use of fertilizers and animal manure in agriculture in 2006/07. (In Swedish – with English table headings). *Statistical Message MI 30 SM 0803*, Stockholm, Statistics Sweden.

[4] Swedish EPA 2007 Sweden's National Inventory Report 2008. Swedish Environmental Protection Agency, Stockholm, Sweden.

[5] IPCC 2006 Guidelines for National Greenhouse Gas Inventories. Editors: Eggelston S, Leandro Buendia L, Miwa K, Ngara T and Tanabe K. Chapter 10: Emissions from Livestock and Manure Management. Intergovernmental Panel on Climate Change. IGES, Japan. ISBN 4-88788-032-4.

[6] Mangino J, Bartram D and Brazy A 2001 Development of a methane conversion factor to estimate emissions from animal waste lagoons. Presented at U.S. EPA's 17th Annual Emission Inventory Conference, Atlanta GA, April 16-18, 2002.

[7] Sommer S G, Petersen S O and Sogaard H T 2000 Greenhouse gas emissions from stored livestock slurry. *J. Eniron. Qual.* **29** 774-751.

[8] Külling D R, Dohme F, Menzi H, Sutter F, Lischer P and Kreuzer M 2002 Methane emissions of differently fed dairy cows and corresponding methane and nitrogen emissions from their manure during storage. *Env. Monit. Ass.* **79** 129-150.

[9] Massé D I, Masse L, Claveau S, Benchaar C and Thomas O 2008 Methane emissions from manure storage. *Trans. ASABE* **51**(5) 1775-1781.

[10] VanderZaag A C, Gordon R J, Jamieson R C, Burton D L and Stratton G W 2009 Gas emissions from straw covered liquid dairy manure during summer storage and autumn agitation. *Trans. ASABE* **52**(2) 599-608.

[11] Kasimir-Klemendtsson Å 2001 Metodik för skattning av jordbrukets emissioner av lustgas. [Methodology for estimating the nitrous oxide emissions from agriculture]. *Report 5170*, Swedish Environmental Protection Agency, Stockholm, Sweden.

[12] Dustan A 2002 Review of methane and nitrous oxide emission factors for manure management in cold climates. JTI – Swedish Instituet of Agricultural and Environmental Engineering. *JTI-report Lantbruk & Industri Nr 299.*

[13] Husted S 1994 Seasonal Variation in Methane Emission from Stored Slurry and Solid Manures. *J. Environ. Qual.* **23** 585-592.

[14] Park K-H, Thompson A G, Marinier M, Clark K and Wagner-Riddle C 2006 Greenhouse gas emissions from stored liquid swine manure in a cold climate. *Atmos. Environ.* **40** 618-627.

[15] Sneath R W, Beline F, Hilhorst M A and Peu P 2006 Monitoring GHG from manure stores on organic conventional dairy farms. *Agric. Ecosyst. Environ.* **112** 122-126.

[16] Sommer S G, Petersen, S O and Søgaard HT 2000 Greenhouse Gas Emission from Stored Livestock Slurry. *J. Environ. Qual.* **29** 744-751.

[17] Amon B, Kryvoruchko V, Amon T and Zechmeister-Boltenstern S 2006 Methane, nitrous oxide and ammonia emissions during storage and after application of dairy cattle slurry and influence of slurry treatment. *Agric. Ecosyst. Environ.* **112** 153-162.

[18] Clemens J, Trimborn M, Weiland P and Amon B 2006 Mitigation of greenhouse gas emissions by anaerobic digestion of cattle slurry. *Agric. Ecosyst. Environ.* **112** 171–177.

[19] SMHI 2005 Swedish Meteorological and Hydrological Institute – website with averaged temperature data for selected cities from 1860 to 1999, www.smhi.se.

[20] APHA 1985 Standard Methods 16th edition 1985, St. Meth. 417A+d. *American Public Health Association*, Washington D.C., USA.

[21] SIS 2000 Swedish standards: SS-EN 12176-1, SS 028113-1, SS 028101-1, SS – ISO 10694. *Swedish Standards Institute*, Stockholm, Sweden.

[22] Sommer S G, Petersen S O, Sørensen P, Poullsen H D and Møller H B 2007 Methane and carbon dioxide emissions and nitrogen turnover during liquid manure storage. *Nutr. Cycl. Agroecosyst.* **78** 27-26.

[23] Steineck S, Gustafson G, Andersson A, Tersmeden M and Bergström J 2000 Plant nutrients and trace elements in livestock wastes in Sweden. *Report 5111*, Swedish Environmental Protection Agency, Stockholm.

[24] Swedish Dairy Association 2009 www.svenskmjolk.se

IOP Conf. Series: Earth and Environmental Science **8** (2009) 012020 doi:10.1088/1755-1315/8/1/012020

Adaptation to and mitigation of climate change in agriculture in developing countries*

Speech, Aarhus 5 March 2009

Shivaji Pandey
Food and Agriculture Organization of the United Nations

***Opinion article (not peer-reviewed)**

1. Background

The number of hungry and malnourished in the world rose to almost 1 billion people at the end of 2008, from 842 M in 1990-92 when the World Food Summit and the Millennium Development Goals of halving hunger and poverty by 2015 were declared. Clearly, we are not headed in the right direction. FAO's Food Price Index rose from around 110 in September of 2007 to 215 in July of 2008, making both the rich and poor spend twice as much in less than a year to feed their families. The rich could afford, but the most vulnerable could not. It has been estimated that an increase of 33% in food prices lowers the standards of living in developing countries by 20% and in the developed world by 3%. Thus, the impact of the rise in food prices on the standards of living and hunger and poverty in developing countries has been huge!

Gurus predicted that higher food prices would increase profitability of agriculture so farmers would produce more and become rich. In 2008, farmers in the developed world did plant more and produced nearly 10% more than they did in 2007 and made money, but farmers in the developing countries in fact produced 1.6% less (excluding Brazil, China and India). Food prices at the global level have come down, but not in many developing countries. The rise in the input prices is continuing to block majority of developing country farmers from producing more to feed their families and their countries. Worryingly, there is a strong possibility that the experiences of 2008 could be repeated again.

Approximately 75% of the world's hungry and poor live in rural areas and derive their livelihoods from agriculture. It seems logical to think that solution to the problem lies in accelerating agricultural growth and development. Unfortunately, however, globally the rate of growth in agricultural productivity has been declining: it is expected to fall to 1.5% between now and 2030 and further to 0.9% between 2030 and 2050, compared with 2.3% per year since 1961. In developing countries, growth in wheat yields has gone down from about 5% in 1980 to 2% in 2005; growth in rice yield went down from 3.2% to 1.2% and in maize yields went down from 3.1% to 1% during the same period.

The per capita availability of agricultural land was 4.3 ha in 1960 and it will be about 1.5 ha in 2050. Significant declines are projected in the availability of another essential natural resource for agriculture - water. The world will need twice as much food in 2050 for its 9.2 billion people than it did in 2000, while preserving the natural resource-base and do it under limited water and land, increasing prices volatility, and climate change. Can we meet this challenge and why should it matter to a conference on climate change?

2. Agriculture and climate change

The factors that have caused current food insecurity are likely to get worse under climate change. Among other things, climate change would push crops, livestock and farmers out of their established niches, increase frequencies and severity of floods and droughts, shorten growth cycles and reduce grain filling rates, and increase incidence and severity of transboundary pests. All these effects would be more severe developing countries.

2.1. Impact of climate change on agriculture

The agriculture and forestry sectors are the major users – and managers - of global land and freshwater resources. About 1.4 billion ha of arable land (10 percent of total ice-free land) is used for crop production and an additional 2.5 billion ha are used for pasture. Roughly 4 billion ha are forested land, 5 percent of which is used for plantation forestry. Agriculture (including forestry) emits 13-15 billion tonnes CO_2 eq per year, about a third of the total from human activities. Overall, agriculture is responsible for 25 percent of carbon dioxide (largely from deforestation), 50 percent of methane (rice and enteric fermentation), and more than 75 percent of N_2O (largely from fertilizer application) emissions.

The Royal Society in 2001 estimated that terrestrial vegetation and soils are currently absorbing approximately 40% of global CO_2 emissions from human activities. UNFCC in 2008 estimated the global mitigation potential of agriculture from biomass to be around 5.5-6 Gt CO_2 eq per year. About 89% of this can be achieved by soil C sequestration. The economic potential in 2030 ranges from 1.5-1.6 Gt CO_2 eq per year and C price USD 20 per t CO_2 eq to 4.43Gt CO_2 eq per year and a C price of USD 100 per t CO_2 eq. 70% of this potential is said to exist in developing countries.

Climate change is already having a greater impact on agriculture than was anticipated, so it is vital that our mitigation and adaptation responses become swifter and more ambitious. In the past this challenge has been met by increasing agricultural productivity, and it has worked. However, it has also contributed to the situation we find ourselves in today.

There is evidence that climate change is altering the distribution, incidence and intensity of animal and plant pests and diseases. It creates new ecological niches and helps the establishment and spread of animal and plant pests and diseases and invasive alien aquatic species to new areas. It will also result in the emergence of new pests. Changes in rainfall, temperature, and relative humidity will result in further outbreaks and spread of plant and animal diseases and pests. Let me give you an example:

Spot Blotch disease of wheat is caused by *Cohliobolus sativus* and it present in Southern Brazil, Bolivia, Paraguay, and Eastern India. It causes heavy losses because there is not much resistance to this disease. The pathogen likes warm climate. As wheat-growing areas of Asia become warmer, the pathogen is likely to spread further and cause heavier losses. Observations over 6 years in India, Nepal and Bangladesh show that climate change is taking its toll – with rising night temperatures, rate and duration of grain filling are reduced and spot blotch severity is increased. Wheat-growing areas that are not yet affected are likely to be affected in the near future.

Adaptation to and mitigation of climate change in agriculture is complex and must involve a range of social and economic factors such as education and literacy. It must also involve creative financial and technological factors such as better understanding and application of indigenous knowledge and coping strategies. Let us look at some of these relationships in a little more detail, as the majority of the following can also serve as effective elements for financing mechanisms for climate change mitigation and adaptation:

2.2. Genetic resources and plant breeding

Plant and animal genetic resources are both affected by climate change, but they also play an essential role in adaptation. They are the reservoir from which new varieties that tolerate heat, drought, and pests will be developed to face climate change, but climate change also threatens their survival. Only about 150 plant species are cultivated and 30 of them provide most of today's food, feed, and fibre. About 1600 plant species are lost each year. Efforts are needed to accelerate their collection, preservation, and sustainable use. Fortunately, a Global Plan of Action was agreed on in 1996 to support these functions. An International Treaty also came into existence in 2006 for some of the same goals. A Global Crop Diversity Trust was recently created to support germplasm conservation. Other national and international efforts are also underway, but they are clearly not enough and much more needs to be done.

Similarly, a wide portfolio of animal genetic resources will be essential for adapting livestock production systems to climate change. They will provide new breeds that can cope with high temperatures, water shortages, and tropical diseases and parasites. However, the animal genetic resources are also under threat – 20 percent of breeds are classified as at risk of extinction, and climate change is likely to enhance this risk. Policies that help reduce greenhouse gas emissions from the livestock sector, reduce impact of increasing use of grains for biofuel production on feed, and contain and control livestock diseases are needed. The Global Plan of Action for Animal Genetic Resources, the first internationally agreed framework for the management of animal genetic resources, was adopted in September 2007. A Commission on Genetic Resources for Food and Agriculture also exists that deals with conservation and sustainable use of plant, animal and microbial genetic resources. Still, once again, this is not enough and much more needs to be done.

Most countries maintain a large number of traditional and improved crop varieties that have been selected for different uses and different production systems that have ability to resist to climate variability. The maintenance, multiplication, and distribution of seeds of this material are vital for improving the resilience to climatic risks of farmers. This requires research, training and efficient extension services, and that needs money. Not investing in them is a very costly proposition!

2.3. Reducing impact of climate change through agriculture

In addition to the genetic variability of plant and animal species, production systems in which they live and grow also help them escape or withstand the impact of climate change. A large range of good management practices already exists and needs to be fine-tuned at farm level, but there are also other technologies that need attention and support. I would highlight a few:

2.3.1. Conservation agriculture (CA)

Feeding the world in 2050 and beyond will need crop production intensification, but that has been talked about for over 40 years now. In the name of intensification in many places around the world, farmers over-ploughed, over-fertilized, over-irrigated, over-applied pesticides, and experts explained the harms caused by the harmful practices as an inevitable side-effect of agricultural production. Rates of productivity went up, and plenty of food was produced. Food became increasingly cheaper and agriculture was taken for granted. But in doing so, we also affected all aspects of the soil, water, land, biodiversity, and the services provided by an intact ecosystem. That began to bring yield growth rates down.

The new paradigm of "sustainable production intensification" recognizes the need for a productive agriculture, which at the same time conserves the natural resource base and environment, and positively contributes to environmental services. Sustainable crop production requires good and fertile soils and CA delivers on this goal of improved soil health and better soil management. So, what is CA? It is the application of three basic ideas to crop production: less to no ploughing, keeping the soil covered, and using crop rotations. CA reduces energy use in farming, helps sequester carbon in the soil at an average rate of 0.51t/ha/year; thus the world is sequestering about 54 million tons of carbon per year on the 106 million hectares under CA now. It produces more under drought and excess water and has potential to save 1,200 km^3/year of water by 2030.

2.3.2. Grassland management

Smith and his colleagues (2008) divided the world into four 'climatic types' based on the FAO Agro-Ecological Zones: cool-dry, cool-moist, warm-dry and warm moist. They estimated mitigation potentials for grassland management as 110 kg CO_2/ha/yr each for cool-dry and warm-dry ecologies and as 810 kg CO_2/ha/yr each for cool-moist and warm-moist ecologies.

In 28 countries, rangelands represent more than 60% land area. In Africa, rangelands cover around 40% of land surface. There are 20 million pastoralists in the world covering 5000 Mha rangelands which store 30% of the world C stocks according to Tennigkeit and Wilkes (2008). They also estimate that improved rangeland management can help sequester 1.3 to 2.0 Gt CO_2 eq worldwide by 2030. So, any serious efforts to explore role of agriculture in climate change mitigation must also consider the role of rangelands.

2.3.3. Legumes to reduce footprint from agriculture
Production of nitrogen fertilizers results in large GHG emissions. While there is no alternative to judicious use of chemical fertilizers to feed the growing world population, the efficiency by which crops remove and use mineral soil N has been the focus of much research. Work has focused on optimising timing and rate of fertilizer applications as well on formulations for higher fertilizer use efficiency. Breeding crops that use fertilisers more efficiently would help lower the agricultural footprint. Growing improved varieties of legumes, which "fix" their own nitrogen from the atmosphere, has the potential to reduce the carbon foot print of agriculture. New varieties of soybeans, pigeon pea, cowpea and groundnut are now available and they adapt well to climate variability, drought, intercropping, and double cropping. They can contribute to climate change mitigation.

Unfortunately, the global area under pulses dropped from approximately 5 M ha in 1968 to 3.9 M ha in 2007. Globally, consumption of pulses in terms of Kcal/capita/day also dropped from 73 in 1968 to 57 in 2007. But meat production has increased from 95 Mt in 1968 to 285 Mt in 2003.

Given the contribution that pulses can make to reduce footprint from agriculture and improve nutrition, they deserve greater attention in sustainable production systems.

3. Conclusions

The world has no alternative to scaling-up win- win agricultural practices to achieve the Millennium Development Goals 1 and 7. We are, however, currently headed in the wrong direction for both of them. To change this we need to:

- promote wider uptake of proven climate-change friendly practices and technologies by farmers which will require substantial policy support from governments, financial support from donors, and technical support from public and private sector;
- develop and validate new knowledge, experiences and practices for an agriculture that is based on reduced and renewable energy inputs;
- develop and agree upon sound methodologies to monitor GHG reductions and co-benefits and public goods including climate information and forecasting tools for farmers and policy makers;
- advocate for larger investments in agriculture and rural development;
- launch strategic partnerships and a global commitment so small farmers can turn agricultural growth into an engine for poverty reduction for improved rural livelihood, sustainable management of natural resources and greater climate change resilience.

Finally, ladies and gentlemen, climate change is not new; climate has been changing for a while now. What is new is the rate at which it is changing now. It is true that agriculture is both a contributor to and victim of climate change, but it is also true that we cannot survive without agriculture as we cannot survive without food. So, agriculture needs to be much higher on the agenda of policy makers and experts who deal with climate change. The Food and Agriculture Organization of the United Nations (FAO) has put food security under climate change very high on its agenda. We organized a Global Food Summit on the topic last June. We have a range of activities in support of ideas I have just mentioned. Nevertheless, the job is too big for any organization, whether UNFCC or FAO, to go alone. We need partnerships. You can count on FAO's full participation and support in any global action to mitigate impact of climate change on agriculture and mitigate impact of agriculture on climate change.

IOP Conf. Series: Earth and Environmental Science **8** (2009) 012021 doi:10.1088/1755-1315/8/1/012021

Spatial analysis and modeling to assess and map current vulnerability to extreme weather events in the Grijalva – Usumacinta watershed, México

D López L

Centro de Investigación en Geografía y Geomatica, Ing. Jorge L. Tamayo A.C., Contoy 137, col. Lomas de Padierna, del Tlalpan, México D.F.

Email: dlopez@centrogeo.org.mx

Abstract. One of the major concerns over a potential change in climate is that it will cause an increase in extreme weather events. In Mexico, the exposure factors as well as the vulnerability to the extreme weather events have increased during the last three or four decades. In this study spatial analysis and modeling were used to assess and map settlement and crop systems vulnerability to extreme weather events in the Grijalva – Usumacinta watershed. Sensitivity and coping adaptive capacity maps were constructed using decision models; these maps were then combined to produce vulnerability maps. The most vulnerable area in terms of both settlement and crop systems is the highlands, where the sensitivity is high and the adaptive capacity is low. In lowlands, despite the very high sensitivity, the higher adaptive capacity produces only moderate vulnerability. I conclude that spatial analysis and modeling are powerful tools to assess and map vulnerability. These preliminary results can guide the formulation of adaptation policies to an increasing risk of extreme weather events.

1. INTRODUCTION

There is general agreement that changes in the frequency or intensity of extreme weather and climate events would have profound impacts on both human society and the natural environment. At present, such events affect a wide variety of natural and human systems, and future changes in their frequency and or magnitude could have dramatic ecological, economic, and sociological consequences [1, 2, 3, 4]

In Mexico, the exposure factors as well as the vulnerability to extreme weather events have increased over the last the three or four decades. The numbers of events could still be within the limit of occurrence determined by the natural variability of the climate. Nevertheless, in the medium or long term, it is expected that these phenomena will undergo a gradual increase in intensity or frequency tied to the effects of global warming [5].

In southern Mexico, the climate change is beginning to be visible in the form of an increase in the intensity and number of extreme weather events. Thus, until some decades ago, strong precipitations events were on the order of 200 mm per day, while towards the end of the past century the intensity of severe storms reached around 300 mm per day; between October 28 and 30, 2007, heavy rains of between 300 and 400 mm occurred, showing that extreme events are now stronger and more frequent [5].

The Grijalva–Usumacinta watershed, located in southern part of Mexico[1] (see Figure 1), is one of the most important watersheds in the country; it contains between 27 and 40% of the stored water

[1] The part of the Usumacinta watershed that is located in Guatemala, is not considered in this study.

volume in the country, and represent 43% of the installed effective electricity capacity. One third of the watershed is used for agriculture. The watershed encompasses of 87,687 km^2 and includes Tabasco State, a large part of Chiapas State, and small parts of Campeche, Veracruz, and Oaxaca States; it consists of 119 "municipios" (counties). The area has an average annual precipitation ranging from 1200 to 3000 mm; and supports a population of about 5.300.000 inhabitants [6]. In most of the area (Chiapas State) the marginalization grade ranges from medium to very high.

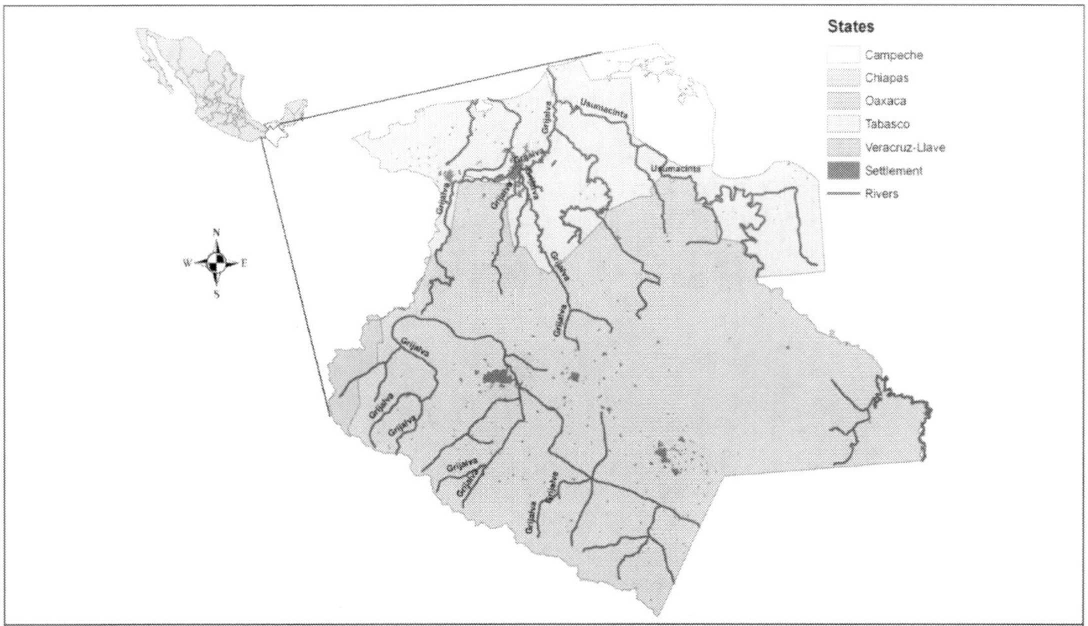

Figure 1 Study area: the Grijalva-Usumacinta watershed, Mexico

Tabasco State, located in the lower part of the watershed, is very susceptible to frequent floods; two of the most important occurred in 1999 and 2000. The floods are related to the high environmental degradation caused by intense deforestation activities and inadequate farming crop systems management that has taken place in the upper part of the watershed. In addition, the coastal zone of Tabasco is considered to be highly vulnerable to sea level rise in the delta zone. The most serious factor is that a large percentage of the population at high risk of flood lives in poverty conditions, mainly in Chiapas State (see figure 2).

In the Grijalva-Usumacinta watershed the two sectors that are particularly exposed to the adverse effects of extreme weather events, such as hurricanes, cyclones, and river flooding, are the crop systems and the settlement infrastructure. Table 1 lists some of the extreme hydro-meteorological events and their impacts registered over the last few decades for this region [5].

According to Landa [5], the floods of 2007 in Tabasco State caused by intense rains were a manifestation of the lack of social and institutional preparation of the country for extreme weather events. Under conditions of climate change, actions will need to go beyond aid to the victims, involving a long-term preventative vision that solves basic problems because events like the one in October 2007 may increase in intensity and frequency.

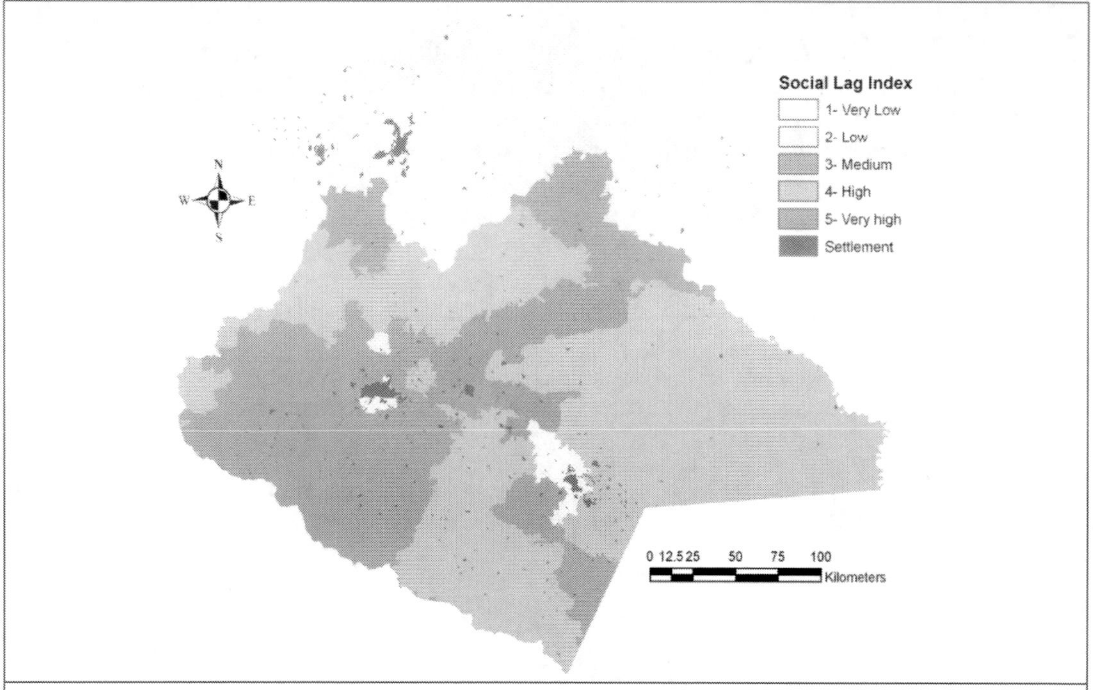

Figure 2. Social lag index in the Grijalva-Usumacinta watershed, Mexico (Source: CONEVAL, 2007)

Table 1.Some extreme weather events and their impacts in the Grijalva –Usumacinta watershed. Source: Landa, Magaña, Neri, 2008, after CENAPRED 2001-2005.

Event	Impact	Date
Heavy rains	417 dead , 353 settlements affected, and 30.000 persons homeless	1998
Heavy rains, landslide, flooding	95 dead, 126.854 persons affected, 10,000 houses damaged	2001
Heavy rains and flooding	800 persons affected, 171 houses damaged, 8.000 hectares of crops damaged	2002
"Larry " Storm, Heavy rains and flooding	52, 885 persons affected, 10,577 houses damaged	September-November 2003
"Stan" Hurricane, Heavy rains, flooding and Landslide	86 dead, 162.570 persons affected, 32.514 houses damaged, 305 schools damaged, 208.064 hectares of crops and grasslands damaged, 5.669 km roads affected. Total costs: 15.031 millions Mexican pesos	October 1- 5 - 2005
Heavy rains	617 persons affected, 1 road – bridge damaged. Total Costs: 3.3 million Mexicans pesos	2005
Heavy rains, flooding and Landslide	1.100.000 persons affected, 670 settlements affected, damage to the urban and road infrastructure. Total costs between 7.500 and 50.000 millions Mexican pesos.	October, 2007

In this context, assessment and mapping of the current vulnerability to climate change (in the context of the adverse effects of extreme weather events) in the Grijalva-Usumacinta watershed is an important part of formulating adaptation strategies both regionally and nationally. A vulnerability assessment can pinpoint areas and sectors where the vulnerability is high, and thus where adaptation strategies should be developed. A structured approach to the vulnerability assessment should provide comparability among regions (States, counties) so that common and integrated adaptation policies can be identified.

2. Assessing and Mapping Vulnerability

Vulnerability describes a central concept in climate change research as well as in the research communities dealing with natural hazards and disaster management, ecology, public health, poverty and development, secure livelihoods and famine, sustainability science, and land change [7, 8, 9 10,11, 12].

Vulnerability can be described as a lack of security against environmental threats. It results from a combination of processes that shape the degree of exposure to a hazard, sensitivity to its stress and impacts, and resilience in the face of those effects. All people, ecosystems, and regions confronting environmental or socio-economic stresses are potentially vulnerable to the impacts, of environmental hazards, but the level of vulnerability varies widely [13]. According to [14], the conceptual framework of vulnerability recognizes and builds upon the three major dimensions of vulnerability exposure, sensitivity, and adaptation/resilience; characterizing them in "causal maps" of the roots of vulnerability, where causal linkages are carefully depicted, is an important analytic task.

The assessment of vulnerability is important as it enables the identification of areas or resources at risk, as well as the threats posed by the reduction or loss of such resources or human lives, that could threaten future sustainable development. A number of studies have examined, either qualitatively or quantitatively, the individual components of vulnerability. This includes the sensitivity of agricultural systems [15, 16, 17] and human vulnerability [18, 19, 20] to climate change. In Mexico, several studies investigating vulnerability have been conducted, mainly related to the agricultural sector in general [21, 22, 23] and to specific crops such as rainfed maize [24, 25], coffee production [26], and water resources [27, 28]. However, in Mexico there are no reports regarding the spatial analysis of vulnerability using GIS and remote sensing environmental data.

Mapping the distribution of vulnerability—either in terms of the attributes of sensitivity, exposure, or adaptive capacity, or in terms of outcomes and impacts—has become a central tool for communicating the results of vulnerability research to other academics, researchers, policy makers, and the community at large [15, 23, 28, 29]. However, according to [29] the challenge remains of accounting for the dynamic nature of vulnerability and spatially representing some indicators (e.g., social capital, institutional relations) that may well be the determinant of vulnerability in particular places. Current challenges in future vulnerability analysis include: addressing multiple, interacting stressors, capturing socioeconomic and biophysical uncertainty, accounting for cross-scalar influences and outcomes, and emphasizing equity and social justice. In addition, when mapping vulnerability, some questions remain incompletely answered, such as: What is the role of spatial analysis as a whole in relation to analyzing and mapping vulnerability? How can we link global, regional and local scales? How should we treat the exposure unit as a coupled social ecological system and identify its interactions?

The goal of this work was to address the analysis and mapping of vulnerability in terms of the dynamic, cross-scalar influences interacting with and threatening the exposure unit as a coupled social ecological system. Some aspects included in the social system are poverty and inequity. The analysis presented in this article focused on assessing and mapping the current vulnerability of crop systems and settlement infrastructure to extreme weather events. One special focus of this work is the explicit, spatial illustration of the distribution of sensitivities and differential adaptive capacities, and how these two elements are spatially related.

3. Methodology for Vulnerability Assessment and Mapping

Methodological approaches to assess vulnerability (both sensitivity and adaptive capacity) that include cross-sectoral, multi-scale, and multi-stress relationships are an emergent area of research. Following Adger [30] the vulnerability of a particular community or livelihood system is a function of three main factors:

- ***Exposure***–the nature and extent of changes to which a place's climate is subjected including changes in climate variability, hereunder the magnitude and frequency of extreme events.
- ***Sensitivity***–how systems can be either positively or negatively affected by the change in climate.
- ***Adaptive capacity***–how much capability a society has to adapt to the changes so as to maintain, minimize loss of, or maximize a gain in welfare.

In this study, spatial analysis and modeling (inference modeling) were used to assess and map the vulnerability to climate change (in this case, based on the assumption of an increase in intensity and frequency of extreme weather events) in the Grijalva–Usumacinta watershed. Two sectors were considered: crop systems and settlement infrastructure. The evaluation of ***adaptive capacity*** was based on economic and demographic characteristics. The s***ensitivity*** of crop systems was evaluated based on the susceptibility of cropland to flood and hydric erosion. The settlement infrastructure **sensitivity** was evaluated based on the susceptibility of the population to floods and landslides. Figure 3 shows a diagram of the methodological approximation used in this study.

3.1. *Database*

The data used in this study include:

- A digital elevation model. Elevation data were obtained from the Shuttle Radar Topographic Mission/C-Band Synthetic Aperture Radar SRTM/C-SAR, with a spatial resolution of 90 meters. From this DEM, the topographic position [31], topographic wetness index [32], and slope were calculated [33].
- River segment, scale 1:250,000 [34]. From this layer several buffers (<2.5, 2.5-5.0 and >5.0 km) were calculated using the ArcMap GIS software.
- Land use and land cover map, scale 1:250,000, [35]. This map was reclassified in two categories, forest vegetation and agriculture (the latter includes grasslands).
- Soil map, scale 1:250,000 [36]
- Population data [37]
- Social lag index [38]

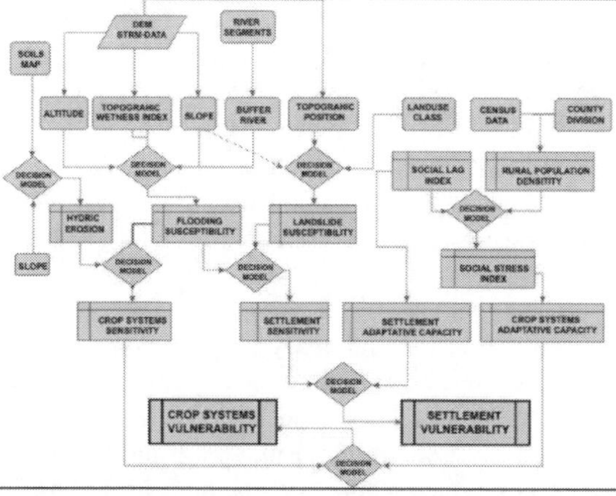

Figure 3. Methodological approximation for mapping vulnerability

3.2. *Adaptive Capacity Indicators*

In this study, the socio-economic conditions used to indicate adaptive capacity include demographic and economic characteristics. Economic capacity was estimated using a social lag index [38] and demography was estimated by population density. For settlements, the adaptive capacity was estimated by the social lag index and for crop systems the social stress index was used. The social lag index provides estimates for the multidimensional measurement of poverty. This index was calculated by CONEVAL [38] using the principal components technique, which numerically synthesizes different dimensions of poverty, including education, access to health services, quality of housing, and basic services. The indicators used to construct the social lag index are listed in Table 2. The rural population density was calculated taking into account the populations of localities with less than 2.500 inhabitants, referred to as the county area. Four grades of density were considered: low (<10 inhabitants), medium (10–50 inhabitants), high (50-100 inhabitants), and very high >100 inhabitants). Finally, the social lag index and population density were joined in a simple index (social stress index) according to the criteria listed in Table 3.

3.3. *Sensitivity Indicators*

Sensitivity was calculated for settlements and crop systems. For each sector, two indicators were selected that represent aspects of the sector that could be quantitatively modeled. For crop system sensitivity the indicators used were cropland at flood susceptibility and cropland at hydric erosion susceptibility; it was assumed that total or partial agriculture production could be affected by both of these processes. For settlement sensitivity, the indicators used were population at flood susceptibility and population at landslide susceptibility.

Three variables were considered for the construction of the landslide susceptibility map: slope, topographic position, and land use. These variables were combined in a decision model to produce a landslide susceptibility map with five categories of landslide susceptibility: 1: very low, 2: low, 3: medium, 4: high and 5: very high. Table 4 shows the rules used to combine the variables. The flood susceptibility was obtained by combining the following variables in a decision model: altitude, topographic wetness index, slope, and the distance from a river. The result was a flood susceptibility map with five categories of flooding susceptibility: very high, high, moderate, low, and no flooding. Tables 5 shows the rules used to combine the variables. The total potentially affected population for each unit of the flood and landslide susceptibility map was assigned using a GIS zonal operation.

Table 2. Indicators used to construct the social lag index [CONEVAL, 2007]

Education	Percentage of literate population 15 years or older. Percentage of the population between 6 and 14 years that does not attend school Percentage of households with members between 15 to 29 years of age, with one household member having less than 9 years of education. Percentage of the population 15 years or older with incomplete basic education
Access to health services	Percentage of the population without rights to health services.
Quality and space in the household	Percentage of occupied housing with bare floor Average number of persons per room
Basic services in the household	Percentage of occupied housing without toilet (w. c.) Percentage of occupied housing without clean water from a public aqueduct Percentage of occupied housing without sewer Percentage of occupied housing without electrical energy Percentage of occupied housing without clothe washing Percentage of occupied housing without a refrigerator

Table 3. Criteria to define Social Stress Index [23]

Criteria	Social stress index
1. If the social lag index is medium to very low and the population density is low	1. Low
2. If the social lag index is low - very low and the population density is medium 3. If the social lag index is high and the population density is low	2. Moderate
4. If the social lag index is low to very low and the population density is high 5. If the social lag index is medium to very high and the population density is low to medium	3. High
6. If the social lag index is medium to very high and the population density is high 7. If the social lag index is very high and the population density is medium	4. Very high

Table 4. Criteria used to define landslide susceptibility [31, 39, 40]

Criteria – Rules	Landslide susceptibility
1. Slope < 6 grades 2. Slope 6 – 12, Topographic position =2* and land use = forest.	1. No susceptibly
3. Slope 6 – 12, Topographic position =2 and land use = agriculture. 4. Slope 6 – 12, Topographic position =3	2. Very Low
5. Slope 6 – 12, Topographic position = land use = forest. 6. Slope 12 – 20, Topographic position = 2 and land use = forest. 7. Slope 12 – 20, Topographic position = 3 8. Slope > 20, Topographic position = 3 and land use = forest	3. Low
9. Slope 12 – 20, Topographic position = 1 and land use = agriculture. 10.Slope 12 – 20, Topographic position = 1 and land use = forest 11. Slope 12 – 20, Topographic position = 2 and land use = agriculture 12. Slope > 20, Topographic position = 2 and land use = forest 13. Slope > 20, Topographic position = 3 and land use = agriculture	4. Medium
14. Slope 12 – 20, Topographic position = 1and land use = agriculture. 15.Slope > 20, Topographic position = 1and land use = forest 16.Slope > 20, Topographic position = 2 and land use = agriculture	5. High
17. Slope > 20, Topographic position = 1and land use = agriculture	6. very high

*Topographic position: (1) valley and foot slopes, (2) Rectilinear slope, (3) Convex slope and crests

Table 5. Criteria used to define flood susceptibility

Criteria – Rules	Flood susceptibility
1. Slope <3%, altitude <5 m, river buffer <5 km 2. Slope <3%, altitude <5 m, river buffer >5 km and wetness* <3 3. Slope <3%, altitude 5-10 m and river buffer <2.5 km 4. Slope <3%, altitude 5-10 m, river buffer <2.5-5.0 km and wetness <3	1.Very high
5. Slope <3%, altitude <5 m, river buffer >5 km and wetness >2 6. Slope <3%, altitude 5-10 m, .river buffer 2.5-5.0 km and wetness >2 7. Slope <3%, altitude 5-10 m, river buffer >5.0 km and wetness <4 8. Slope <3%, altitude 10-25 m, river buffer <2.5.0 km 9. Slope <3%, altitude 10-25 m, river buffer 2.5-5.0 km and wetness <3 10. Slope <3%, altitude 25-50 m, river buffer <2.5 km and wetness <3 11. Slope <3%, altitude 25-50 m, river buffer 2.5-5.0 km and wetness =1 12. Slope <3%, altitude 50-100 m, river buffer <2.5 km and wetness <3* 13. Slope <3%, altitude 100-500 m, river buffer <2.5 km and wetness>1 14. Slope <3%, altitude 100-500 m, river buffer 2.5-5.0 km and wetness=1 15. Slope<3%, altitude >500 m and wetness <3 16. Slope =3-9%, altitude <500 m and wetness <3	2. High
17. Slope <3%, altitude 5-10 m, river buffer >5.0 km and wetness >3 18. Slope <3%, altitude 10-25 m, river buffer 2.5-5.0 km and wetness >2 19. Slope <3%, altitude 10-25 m, river buffer >5.0 km and wetness <4 20. Slope <3%, altitude 25-50 m, river buffer <2.5 km and wetness >2 21. Slope <3%, altitude 25-50 m, river buffer 2.5-5.0 km and wetness=2 22. Slope <3%, altitude 25-50 m, river buffer >5.0 km and wetness <4 23. Slope <3%, altitude 50-100 m, river buffer 2.5-5.0 km and wetness =2 24. Slope <3%, altitude 50-100 m, river buffer >5.0 km and wetness <3 25. Slope <3%, altitude 100-500 m, river buffer 2.5-5.0 km and wetness>1 26. Slope <3%, altitude >500 m and wetness=3 27. Slope 3-9%, altitude <500 m and wetness >2 28. Slope 3-9%, altitude >500 m and wetness<3	3. Moderate
29. Slope <3%, altitude 10-25 m, river buffer >5.0 km and wetness >3 30. Slope <3%, altitude 25-50 m, river buffer 2.5-5.0 km and wetness >2 31. Slope <3%, altitude 25-50 m, river buffer >5.0 km and wetness >2 32. Slope <3%, altitude 50-100 m, river buffer <2.5 km and wetness>2 33. Slope <3%, altitude 50-100 m, river buffer 2.5-5.0 km and wetness >2 34. Slope <3%, altitude 50-100 m, river buffer >5.0 km and wetness >2.0 35. Slope <3%, altitude 100-500 m, river buffer >5.0 km 36. Slope <3%, altitude >500 m and wetness>3 37. Slope 3-9%, altitude >500 m and wetness>2	4. Low

*Topographic wetness index: (1) >10.5, (2) 9.5 – 10.5, (3) 8 – 9.5, (4) >8.0 Non-dimensional

Hydric erosion susceptibility was estimated by combining the slope and soil map in a decision model according to the rules listed in Table 6. To assess the sensitivities of settlement and crop systems, the indicators for landslide susceptibility, flooding susceptibility, and hydric erosion susceptibility were combined in a decision model, according to the rules listed in Tables 7 and 8.

Table 6. Criteria used to define hydric erosion susceptibility [41, 42]

Criteria - Rules	**Hydric erosion susceptibility**
1. Slope < 3%, any soil, except lithosols	None
2. Slope of 3 -9 %, any soil, except litosols	Low
3. Slope of 10 – 15 %, litosols, or a combination of a slope of 10 – 15%, and acrisols or vertisols, other soil and slope of 3 -9 % and lithic contact < 50 cm	Moderate
4. Slope of 16 – 30% or a combination of slope 16 – 30% and acrisols or vertisols, or litosols and a slope 3 -9 %, other soils and slopes of 10 -15 %, and lithic contact < 50 cm	Moderate - high
5. Slope of 30 – 50% or a combination of a slope of 30 – 50% and acrisols or vertisols, or lithosols and a slope of 10 -15%,other soils and slopes of 16 -30% and lithic contact < 50 cm	High
6. Slope of 50 – 70%, or a combination of a slope of 50 – 70% and acrisols or vertisols, or lithosols and slope of 16 -30%, other soils and slopes of 30 -50%, and lithic contact < 50 cm	Very high
7. Slope > 70%, any soil	Extremely high

Table 7. Criteria used to define the settlement sensitivity

Criteria - Rules	**Settlement sensitivity**
1. Flood susceptibility is very high and landslide susceptibility is very high	Very high
2. Flood susceptibility is high and landslide susceptibility is high	High
3. Flood susceptibility is moderate and landslide susceptibility is moderate	Moderate
4. Flood susceptibility is low and landslide susceptibility is low to very low	Low

Table 8. Criteria used to define crop system sensitivity

Criteria - Rules	**Settlement sensitivity**
1. Flood susceptibility is very high and hydric erosion susceptibility is very high to extremely high	Very high
2. Flood susceptibility is high and hydric erosion susceptibility is high	High
3. Flood susceptibility is moderate and hydric erosion susceptibility is moderate to moderately high	Moderate
4. Flood susceptibility is low and hydric erosion susceptibility is low	Low

3.4. *Settlement and Crop Systems Vulnerability*

The Vulnerabilities of settlement and crop system was defined by combining the sensitivity and adaptive capacity indicators in a decision model according to the criteria indicated in Tables 9 and 10 Figure 4 shows an example of how these rules I combined in a decision model to assess, in this case, settlement vulnerability.

IOP Conf. Series: Earth and Environmental Science **8** (2009) 012021 doi:10.1088/1755-1315/8/1/012021

Table 9. Criteria used to define Settlement vulnerability

Criteria – Rules	**Grade of vulnerability**
1. Settlement sensitivity is high to very high and social lag index is very high 2. Settlement sensitivity is medium and social index is very high	1. Very high
3. Settlement sensitivity is high to very high and social lag index is medium 4. Settlement sensitivity is medium and social index is lag high 5. Settlement sensitivity is low and social index is very high	2. High
6.. Settlement sensitivity is high to very high and social lag index is low 7. Settlement sensitivity is medium and social lag index is medium to low 8. Settlement sensitivity is low and social lag index is high	3. Moderate
9. Settlement sensitivity is low and social lag index is medium to low	4. Low

Table 10. Criteria used to define the crop system vulnerability

Criteria	**Grade of vulnerability**
1. Crop sensitivity is high to very high and social stress index is very high 2. Crop sensitivity is medium and social stress index is very high	1. Very high
3. Crop sensitivity is high to very high and social stress index is medium 4.Crop sensitivity is medium and social stress index is high 5. Crop sensitivity is low and social stress index is very high	2. High
6.. Crop sensitivity is high to very high and social stress index is low 7. Crop sensitivity is medium and social stress index is medium to low 8. Crop sensitivity is low and social stress index is high	3. Moderate
9. Crop sensitivity is low and social stress index is medium to low	4. Low

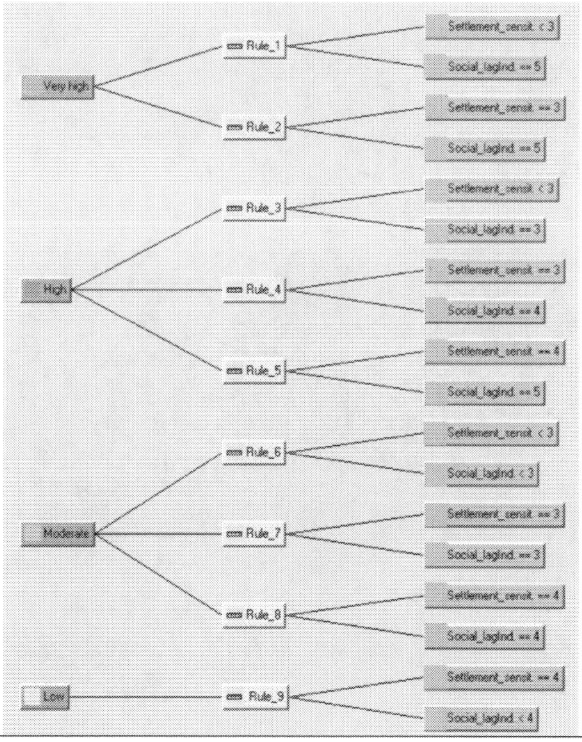

Figure 4. Decision model used to assess the settlement vulnerability

4. Results and discussion

4.1. *Coping and Adaptive Capacity*

Figures 5 and 6 show the rural population density and the social stress index of the Grijalva-Usumacinta watershed. In the low part of watershed the low rural population density is related to the presence of rangeland use, and in the upper part to the presence of conservation areas (forest use). The highest population density area, in general, corresponds to agricultural lands, which are located in the upper part of the watershed, including areas under shifting cultivation. Figure 6 shows the social stress index, a combination of the population density and the social lag index. This index is low in the lowlands of the study area (which includes most of Tabasco State), which can be interpreted as a high adaptive capacity and reflects good social conditions. On the other hand, in most of the upper watershed, this index ranges from high to very high; reflecting poor social conditions, and therefore, the adaptive capacity in this area can be interpreted as low.

Because the database used in this study to estimate the adaptive capacity was taken from a secondary, but credible source [37, 38], and very simple GIS operations were implemented, validation of these data was not considered here.

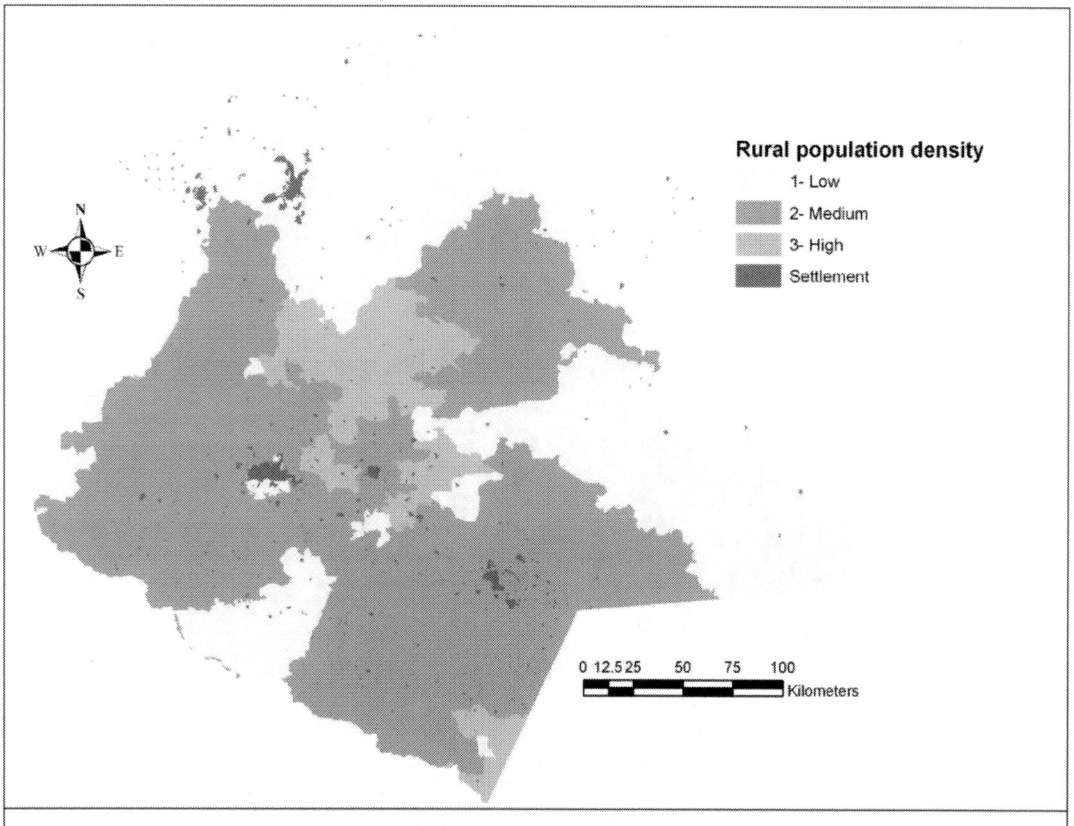

Figure 5. Rural population density in the Grijalva-Usumacinta watershed, Mexico

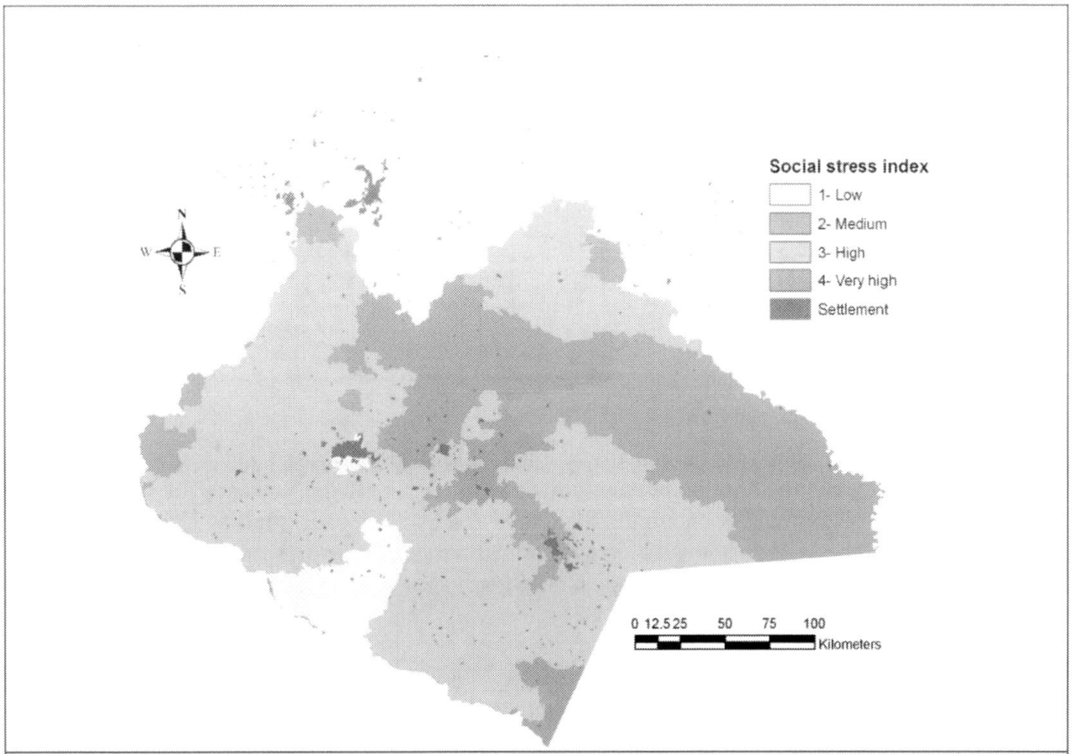

Figure 6. Social stress index in the Grijalva-Usumacinta watershed, Mexico

4.2. *Sensitivity*

4.2.1. Validation of Sensitivity Indicator. The indicators used to estimate the settlements and crop system sensitivity were hydric erosion susceptibility, flooding susceptibility, and landslide susceptibility. The results of the decision models implemented for each of these indicators were validated, as described in the following paragraph. The hydric erosion susceptibility map was compared with the soil degradation map [43] and it was found that the areas ranked from moderate to very high hydric erosion susceptibility correspond to areas of low productive capacity (as a result of the hydric erosion process), as reported in the soil degradation map. To validate the flooding susceptibility map, the flooded areas resulting from extreme precipitation (400 mm/day) in October 2007 were considered. The flooded area was mapped using MODIS and SPOT satellite images. When this map was compared with the flooding susceptibility map, the following relations were found (Figure 7): 75% of the inundated area was mapped as high to very high susceptibility and 88% of the inundated area was mapped as moderate to very high susceptibility. Finally, using a QuickBird satellite image, an inventory of landslides was developed and fifteen landslides were mapped. A comparison of this map with the landslide susceptibility map showed that 60% of the landslides mapped as high to very high susceptibility and 87% of the landslides as moderate to very high susceptibility (Figure.8). As described herein, the forecast capacity of the models to estimate sensitivity fluctuated about 90%. In future works regarding flooding and landslide susceptibility models, the forecast capacity can be increased by using DEMs with better resolutions, which are now available (5 and 30 meters).

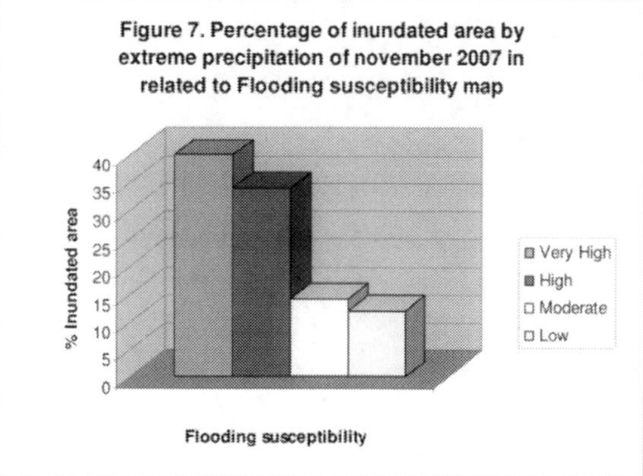

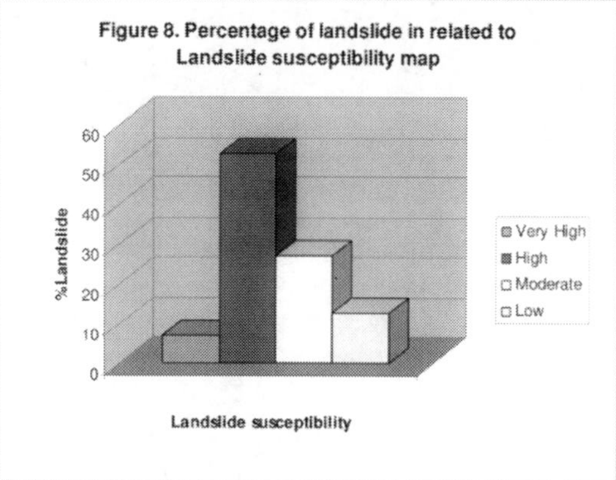

4.2.2. Settlements and Crop System Sensitivity. Figures 9 and 10 show the distribution of settlements sensitivity and the total population potentially affected. In the lowlands the high sensitivity is related to flooding risk and in the highlands to landslide susceptibility. The figures show that, in terms of area and the affected population, flooding poses the greater risk. Most of the potentially affected area corresponds to Tabasco State, which is located in the lower part of the watershed.

An important aspect of the sensitivity indices, particularly those based on flood susceptibility, is that the degrees of susceptibility may be interpreted as an indirect measure of a) flood height -- estimating the water table height for different zones gives an idea of the amount of water that would be needed to evacuate from those zones; and b) flood duration -- making a qualitative estimate of flood duration is possible. Information about these two aspects will be of great utility for decision makers: a) to make decisions about how to provide help to the people after the inundation takes place, and b) to implement measurements for adaptation.

Another important aspect of the flooding susceptibility map is the great contribution of the DEM (digital elevation model) data in the model. As can be seen in Figure 3, except for the river buffer, the information for the model is obtained from DEM data. Nevertheless, at the present time, this type of data is not being fully used in Mexico. The quality of this information can be substantially improved by the use of more detailed elevation data (5 and 30 meters of spatial resolution) that are now partially available.

In relation to the settlements potentially affected by inundation or landslides, the total population affected in each spatial unit was considered (e.g., very high or high), although it is also possible to obtain the location of each settlement or household unit. This is not shown in the map because it is not practical to display 35,000 localities. Figure 11 shows an example of the spatial distribution of localities in the area mapped as having a very high susceptibility to inundation. Figure 12 shows, for each spatial unit of susceptibility, the total number of localities. This spatial information is very useful for decision makers, as it provides an idea of the ease or difficulty associated with helping people; the more widely distributed the population, the more work there is to do.

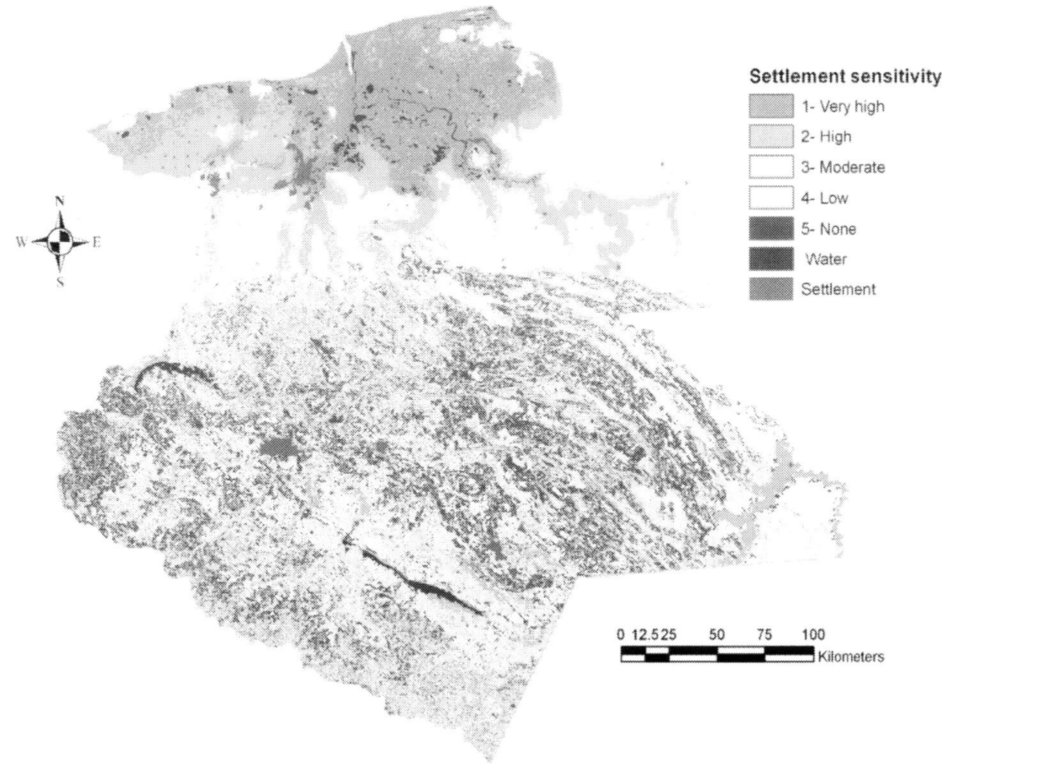

Figure 9. Settlement sensitivity in the Grijalva-Usumacinta watershed, Mexico

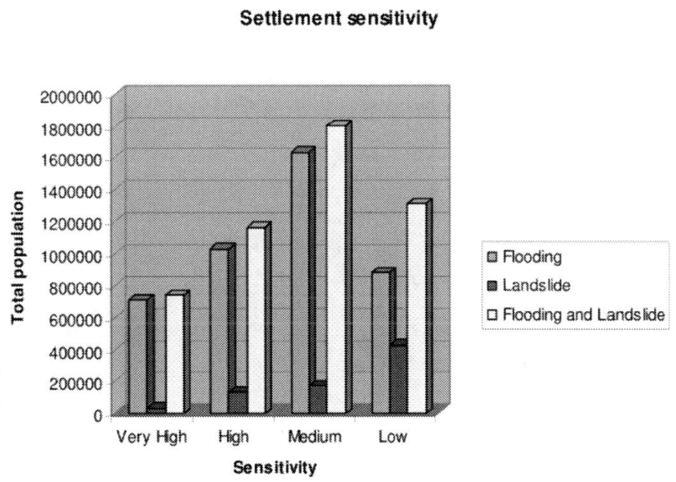

Figure 10. Settlement sensitivity and total population potentially affected for each grade of sensitivity, the Grijalva-Usumacinta watershed, Mexico

IOP Conf. Series: Earth and Environmental Science **8** (2009) 012021 doi:10.1088/1755-1315/8/1/012021

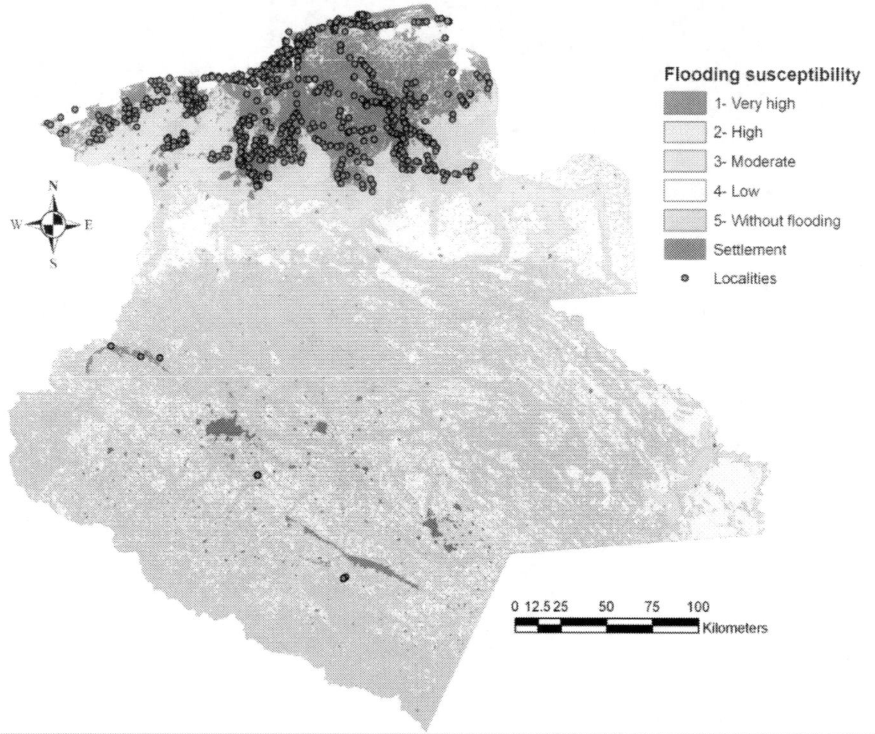

Figure 11. Spatial distribution of localities in areas with a very high susceptibly to flooding in the Grijalva-Usumacinta watershed, Mexico

Figure 12. Number of localities for each spatial unit of flooding susceptibility, the Grijalva-Usumacinta watershed, Mexico

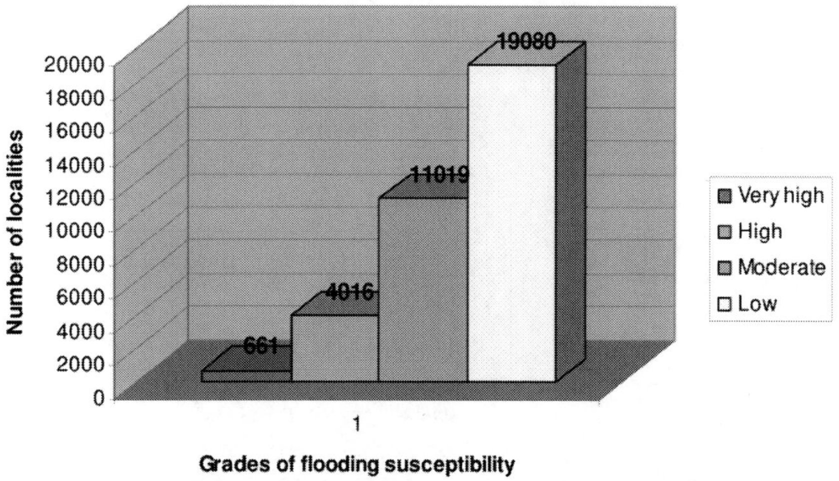

Figures 13 and 14 show the distribution of crop system sensitivities and the total area affected. In the lowlands the sensitivity is related to flooding, and in the highlands to hydric erosion. The figures show that both hazards are important for the cropland; about 500.000 hectares are potentially affected at high to very high grades by flooding or hydric erosion. These two processes are related to each other, because growing crops on unsuitable lands in the upper part of the watershed can contribute to an increased sensitivity of the lowlands. Information on flood severity would be immensely useful for decision-makers for evaluating the magnitude and cost of damages and implementing adaptation measures accordingly.

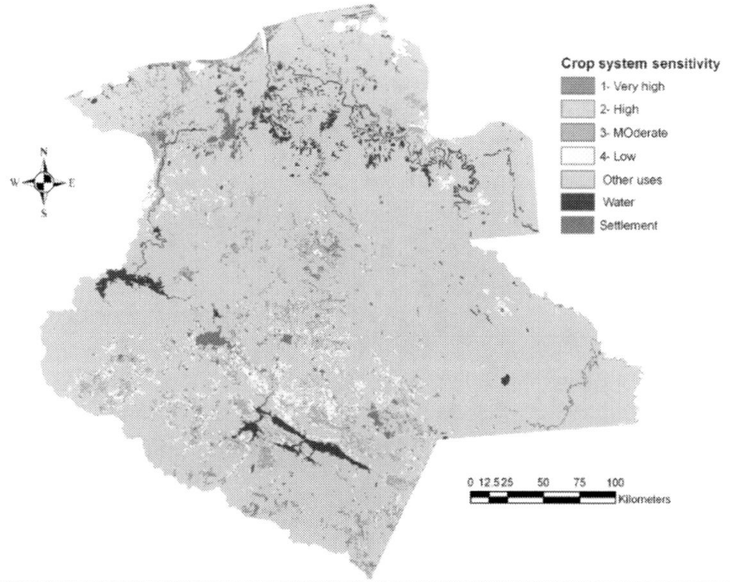

Figure 13. Crop system sensitivity in the Grijalva-Usumacinta watershed, Mexico

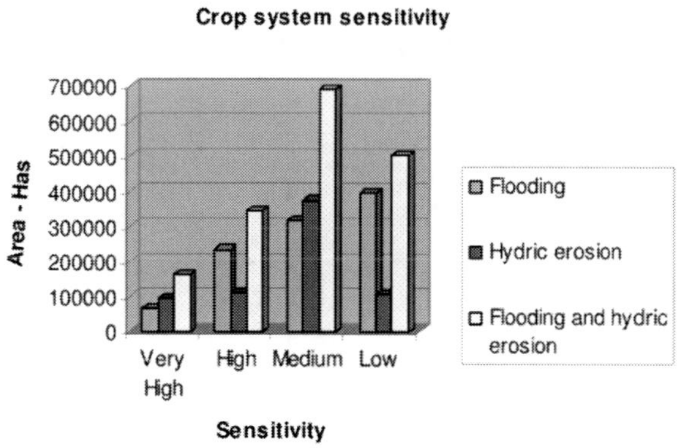

Figure 14. Crop system sensitivity and total area potentially affected for each grade of sensitivity, the Grijalva-Usumacinta watershed, Mexico

IOP Conf. Series: Earth and Environmental Science **8** (2009) 012021 doi:10.1088/1755-1315/8/1/012021

4.3. *Settlements and Crop System Vulnerability*

Figures 15 and 16 show the distribution of settlements and crop system vulnerabilities. For both sectors, the higher grades of vulnerability occur in the upper part of watershed as a consequence of a combination of high grades of *sensitivity* and a very low *adaptive capacity* (high grades of the social stress index or the social lag index).

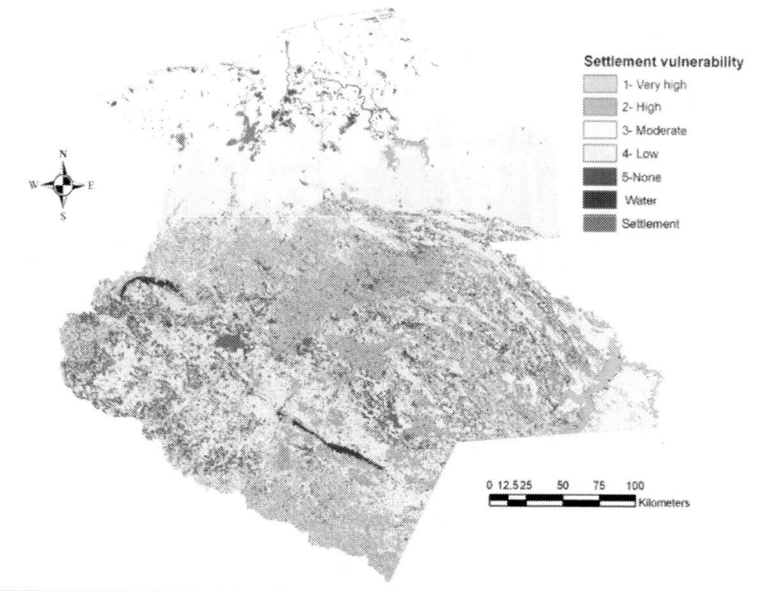

Figure 15. Settlement vulnerability, the Grijalva-Usumacinta watershed, Mexico

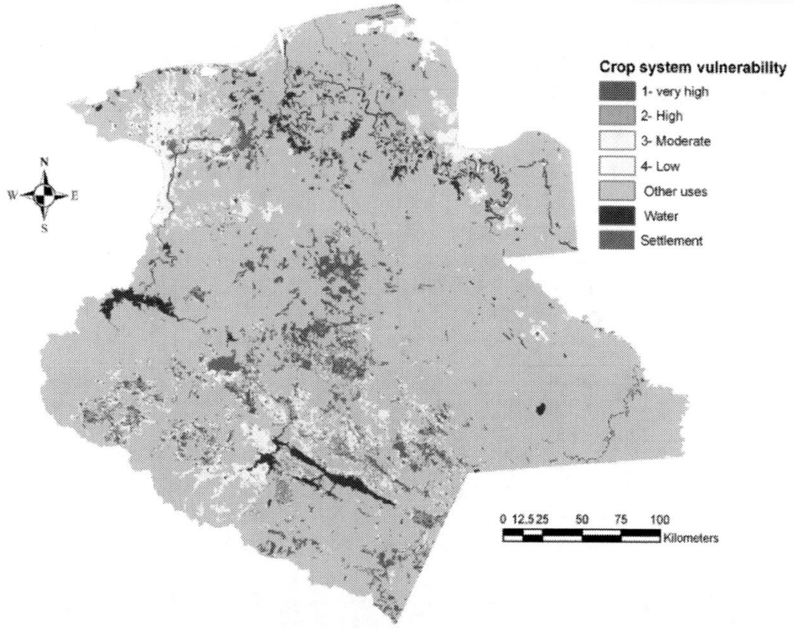

Figure 16. Crop system vulnerability, the Grijalva-Usumacinta watershed, Mexico

In the lowlands, although the *sensitivity* ranges from high to very high, the *adaptive capacity* is high (low grades for the social stress index or the social lag index); therefore, the final result is a moderate vulnerability. The vulnerability maps illustrate which areas are vulnerable at different grades, and the importance of theses maps relies on their ability to show the interactions between sensitive and adaptive capacities. In terms of social impacts and adaptation measurements theses maps help decision makers to establish priorities that help the affected people and implement adaptive measurements. However, because these indices occur in the final step of the synthesis process (the information is aggregated), they must be taken as a first approximation. For further analytical purposes, the constituents of vulnerability, the sensitivity (which has been discussed before) and the adaptive capacity index must be assessed separately. Finally, a more detailed analysis of each indicator must be carried out to achieve a better and more specific decision-making process (in terms of help and adaptive measurements).

The results of this study agree with Landa [5] who states that the Grijalva-Usumacinta catchment area exhibits conditions of high vulnerability to intense rainfall and that this vulnerability of rural areas is highly related to the stress that natural resources are under at present. This area is perhaps one of the regions of Mexico where the land use has changed drastically in last three decades, during which large areas of tropical forests have been transformed to agricultural use. The same author points out that Mexicans who lives under poverty conditions are particularly susceptible to the effects of extreme weather events; this condition increases the vulnerability in greater proportion than the population grows. On the other hand, if it is considered that in this area the most of the rural population depends on agricultural systems for their livelihoods, the vulnerability of crop systems has a special meaning for theses communities. It means that for these groups, food security is threatened. If it is considered that in this region the population has a high level of food poverty [38], and therefore the potential impacts of extreme weather events can cause additional inequities, special attention needs to be paid to indigenous people with subsistence livelihoods and groups with limited means of adaptation.

The identification and characterization of the way in which human and natural systems are sensitive is key input for targeting, formulating and evaluating adaptation policies [44]. The approach and partial results presented in this paper can hopefully contribute to the knowledge of the current characteristics of settlement and crop systems sensitivity and adaptive capacity in the Grijalva - Usumacinta watershed, and therefore guide the formulation of adaptation policies of these sectors to climate variability, focused on extreme weather events. Equally, it is expected that these results will help to improve the management of current and future climate risk. The development of adaptive capacity in relation to the potential impacts of climatic change depends on the decisions that we make today in the technological, social, economic and environmental fields; in the use of tools of climatic prognosis that allow us to deal with uncertainty, and in the development of preventative actions against extreme events [5]

5. Conclusions

The preliminary results presented here demonstrate a method to assess and map vulnerability to extreme weather events, considering both physical and societal aspects. I conclude that spatial analysis and modeling are powerful tools for assessing and mapping vulnerability. Compared to other current approaches to climate vulnerability assessment, the inference modeling approach has particular features that augment its transparency, reproducibility and comparability. The use of inference models allow us to analyze how different elements of vulnerability can contribute to the final vulnerability as well as the interactions that occur between the elements. Although at the present time the reduction of vulnerability is focused on constructing adaptive capacity for the population, in Grijalva-Usumacinta watershed, actions related to the adequate use of natural resources in concordance with sustainable development are needed to reduce the high sensitivity of natural resources, particularly the crop systems.

References

[1] Bates B C, Kundzewicz Z W, Wu S and Palutikof J P, Eds., 2008 *Climate Change and Water. Technical Paper of the Intergovernmental Panel on Climate Change (* IPCC Secretariat, Geneva) p 210
[2] Kundzewicz, Z W, Mata L J, Arnell N W, Döll P, Kabat P, Jiménez B, Miller K A, Oki T, Sen Z and Shiklomanov I A, 2007 *Freshwater resources and their management. Climate Change 2007: Impacts, Adaptation and Vulnerability. Contribution of Working Group II to the Fourth Assessment Report of the Intergovernmental Panel on Climate Change.* Parry ML, Canziani O F, Palutikof J P , van der Linden P J and. Hanson C E, Eds., (Cambridge University Press, Cambridge, UK) p 173-210.
[3] Diffenbaugh N S, Jeremy S Pal, Trapp R J and Giorgi F 2005 Fine-scale processes regulate the response of extreme events to global climate change. *PNAS,* **102 (44)** 15774–78
[4] Easterling D R, Meehl G A, Parmesan C, Changnon S A,. Karl T R, Mearns L O 2000. Climate Extremes: Observations, modeling, and impacts. *Sci.,* **289** 2068-74
[5] Landa R , Magaña V and Neri C 2008. *Agua y clima: elementos para la adpatación al cambio climático. Secretaria de Medio Ambiente y Recursos Naturales,* Centro de Ciencias de la Atmosfera, Universidad Nacional Autonoma de Mexico (Servicios y Publicaciones Grande, S A de C V , Mexico) p 133
[6] Instituto Nacional de Estadistica y Geografía. 2005. *II Conteo de población y vivienda.*
[7] Füssel H M 2007. Vulnerability: A generally applicable conceptual framework for climate change research. *Global Environ. Chang.,* **17** 155–167
[8] Füssel H M and Klein R J T 2006. Climate change vulnerability assessments: An evolution of conceptual thinking. *Global Environ. Chang.,* **75** 301-29
[9] Adger W N, 2006. Vulnerability. *Global Environ. Chang.,* **16** 268-81
[10] Eakin H and Luers A L 2006. Assessing the vulnerability of social-environmental systems. *Annu. Rev. Environ. Resour.***31** 365–94
[11] Cutter S, Boruff B and Shirley W L 2003. Social vulnerability to environmetal hazards. Social Science Quaterly, vol. **84** (2) 242-61
[12] Kavi Kumar K, Klein R, Ionescu C, Hinkel J and Klein R 2007.*Vulnerability to Poverty and Vulnerability to Climate Change: Conceptual Framework, Measurement and Synergies in Policy. Proceedings UNU WIDER Conference Fragile States – Fragile Groups: Tackling Economic and SocialVulnerability (*Helsinki). http://www.sei.se/publications.html?task=view&catid=2&id=1186
[13] Stockholm Environment Institute *2002.The SEI poverty and vulnerability programme.* http://www.sei.se/risk/poverty.html
[14] Kasperson J X and Kasperson R E 2001. *International Workshop on Vulnerability and Global Environmental Change.* Stockholm Environment Institute (Stockholm, Sweden) p 35
[15] Berry P M, Rounsevell M D A, Harrison P A and Audsley E 2006 Assessing the vulnerability of agricultural land use and species to climate change and the role of policy in facilitating adaptation. *Environ. Sci. & Policy***, 9** 189-204
[16] Wehbe M, et al., 2006. Local Perspectives on Adaptation to Climate Change: Lessons From Mexico and Argentina. *AIACC Working Paper* **39**
[17] Metzer, M J, Rounsevell M D A, Leemans R and Schroter D 2005. The vulnerability of ecosystem services to land use change. Agric. Ecosyst. Environ **114** 69-85
[18] Eakin H and Luers A L 2006. Assessing the vulnerability of social-environmental systems. *Annu. Rev. Environ. Resour.* **31** 365 – 94
[19] Thow A and de Blois M 2008. *Climate change and human vulnerability: Mapping emerging*

trends and risk hotspots for humanitarian actors Summary for decision makers. Discussion paper Edition 3 (UK) p 29

[20] Morton J F 2007. The impact of climate change on smallholder and subsistence agriculture. *PNAS*, vol. **104** (50)

[21] Conde C, Magaña V and Ferrer R M 1999. *On the use of climate forecast on the planning of agricultural activities in the State of Tlaxcala, Mexico. Articulo presentado en la 11 Conferencia sobre climatología Aplicada (*Dallas, Texas)

[22] Vasquez-Leon M, West C T and Finan T J 2003. A comparative assessment of climate vulnerability: agriculture and ranching on both sides of the US–Mexico border. *Global Environ. Chang.*, **13** 159–73

[23] López L D 2008. *Spatial analysis and and modelling for mapping environmental and social land vulnerability in México, EUROSOIL* (Vienna, Austria)

[24] Liverman D *et al.*, 1994 *Possible Impacts of climate change on maize yields in Mexico.* In: Rosenzweig C and Iglesias A (Eds.) Implications of Climate Change for International Agriculture: Crop Modeling Study. U.S. Environmental Protection Agency (Washington D.C) p 1-4.

[25] Conde C, Liverman D, Flores M, Ferrer R, Araujo R, Betancourt E, Villarreal G and Gay C 1997. Vulnerability of rainfed maize crops in Mexico to climate change. *Climate Research.* **9** 17-23.

[26] Gay C *et al,* 2004 Potential impacts of climate change on agriculture: A case of study of coffee production in Veracruz, Mexico". *Agr. Econ.*

[27] Mendoza V M, Villanueva E and Adem J 1997. Vulnerability of basins and watersheds in Mexico to global climate change. *Clim Res.***2** 139-145

[28] Leichenko R, O'Brien K, Aandahl G, Tompkins H, and Javed A 2004. *Mapping vulnerability to multiple stressors: A technical memorandum* CICERO (Oslo, Norway)

[29] Eakin H V et al., 2007. *A stakeholder driven process to reduce vulnerability to climate change in Hermosillo, Sonora, Mexico. In: Mitigation and adaptation strategiesd for global chang*e, (Springer Netherlands)

[30] Adger W N 2003. Social capital, collective action and adaptation to climate change. *Econ. Geogr.*, **79** 387–404

[31] Zimmermann N *2000 Topographic position mapping routines.Toposcale.aml* http://www.wsl.ch/staff/niklaus.zimmermann/programs/aml4_1.html

[32] Barling R D 1992 *Saturation zones and ephemeral gullies on arable land in southeastern Australia*. Ph.D. Thesis, University of Melbourne.

[33] Zebenbergen L W and Thorne C R 1987 Quantitative analysis of land surface topography. *Earth surface processes and landforms* **12** pp 47-56.

[34] Instituto Nacional de Estadistica y Geografía 1992 *Hidrologia superficial Serie I* (Mexico, D. F)

[35] Instituto Nacional de Estadistica y Geografía 2004 *Vegetacion y uso de suelo, SerieIII.* (Mexico, D. F)

[36] Instituto Nacional de Estadistica y Geografía 2000 *Mapa de edafologia.* (Mexico, D. F)

[37] Instituto Nacional de Estadistica y Geografía 2005 *II conteo de población.* (Mexico, D. F)

[38] Consejo Nacional de Evaluación de la Política de Desarrollo Social - CONEVAL 2005 *Mapas de Pobreza por Ingreso y Rezago Social.* Primera edición (Mexico D.F.)

[39] Ohlmacher G C 2004. Landslide coves, plan curvature, and landslide probability in areas with cohesive soils. Denver annual meeting. Paper no. 123-3. http://gsa.confex.com/gsa/2004AM/finalprogram/abstract_80710.htm

[40] López L D y Núñez J M 2006 *Análisis de la susceptibilidad a deslizamientos mediante el procesamiento de datos del programa de mapeo de radar topográfico SRTM. Memorias congreso SELPER* (Cartagena, Colombia)

[41] Instituto Geográfico Agustin Codazzi. Subdirección de Geografía 2001 *Guia metodologica para clasificacion de tierras por capacidad de uso* (Bogota D.C.) p 35

[42] Soil Survey Division Staff 1993. *Soil survey manual*. (United States Department of Agriculture) Handbook No. 18 p 437

[43] Secretaria de Medio Ambiente y Recursos Naturales y Colegio de Postgraduados.2002. *Evaluación de la degradación del suelo causada por el hombre en la república Mexicana, Escala 1:250,000. Memoria nacional* (Mexico, D. F)

[44] Burton I, Malone E and Huq S 2005 *Adaptation policiy Frameworks for climate change: Developming strategies policies and measures* (Cambridge University)

IOP Conf. Series: Earth and Environmental Science **8** (2009) 012022 doi:10.1088/1755-1315/8/1/012022

Impacts of 21st century sea-level rise on a Danish major city – an assessment based on fine-resolution digital topography and a new flooding algorithm

Jesper Erenskjold Moeslund[1,2,3], Peder Klith Bøcher[2], Jens-Christian Svenning[1], Thomas Mølhave[3] & Lars Arge[3]

[1] Ecoinformatics and Biodiversity Group, Department of Biological Sciences, Aarhus University, Denmark.
[2] Department of Agroecology and Environment, Aarhus University, Denmark.
[3] MADALGO - Center for Massive Data Algorithmics, Aarhus University, Denmark.

Corresponding author: jesper.moeslund@biology.au.dk

Abstract. This study examines the potential impact of 21st century sea-level rise on Aarhus, the second largest city in Denmark, emphasizing the economic risk to the city's real estate. Furthermore, it assesses which possible adaptation measures that can be taken to prevent flooding in areas particularly at risk from flooding. We combine a new national Digital Elevation Model in very fine resolution (~2 meter), a new highly computationally efficient flooding algorithm that accurately models the influence of barriers, and geospatial data on real-estate values to assess the economic real-estate risk posed by future sea-level rise to Aarhus. Under the A2 and A1FI (IPCC) climate scenarios we show that relatively large residential areas in the northern part of the city as well as areas around the river running through the city are likely to become flooded in the event of extreme, but realistic weather events. In addition, most of the large Aarhus harbour would also risk flooding. As much of the area at risk represent high-value real estate, it seems clear that proactive measures other than simple abandonment should be taken in order to avoid heavy economic losses. Among the different possibilities for dealing with an increased sea level, the strategic placement of flood-gates at key potential water-inflow routes and the construction or elevation of existing dikes seems to be the most convenient, most socially acceptable, and maybe also the cheapest solution. Finally, we suggest that high-detail flooding models similar to those produced in this study will become an important tool for a climate-change-integrated planning of future city development as well as for the development of evacuation plans.

1. Introduction

Anthropogenic climate change is causing a global sea-level rise, which is expected to accelerate over the 21st century [1, 2]. The potential effects of future sea-level rise on coastal zones are one example of the impacts of 21st century climate change that are likely to become a major challenge for society [1, 3, 4]. The total value of the economic assets located within 500 m of the European coastline is currently estimated at € 500,000,000,000 to 1,000,000,000,000, and several million people live in these areas [5]. The most recent predictions foresee a mean sea-level rise of up to 1.35 meter by 2090-2099 [2]. A sea-level rise of such a magnitude is likely to have drastic consequences for coastal cities in terms of regular and maybe even permanent flooding of low-lying areas as well as an increased risk of catastrophic flooding and associated losses of human lives. In 2005, as the hurricane Katrina reached southeast Louisiana it drew a path of destruction across the area and caused the flooding of nearly 80% of the city of New Orleans and an overall mortality of the exposed human population of ~1%, with the mortality risk increasing with water depth [6]. As numerous major cities around the World are located by the coast, the expected future sea-level rises will increase the threat of catastrophic flooding due to extreme weather events like the hurricane Katrina. Exacerbating this risk, global warming is predicted to cause extreme weather events to increase in intensity over the 21st century [1].

In Denmark, the ten largest cities are all coastal [7], typically located in topographically homogenous landscapes only a few meters above sea level (a.s.l.). Weather events have historically not become as extreme as the hurricane Katrina, but storms and even category 1 hurricanes regularly sweep over the country [8, 9]. The waters around Denmark frequently rises in response to these strong winds especially when they come from the west or northwest and subsequently pushes huge amounts of water into the Inner Danish Sea and the Baltic Sea. For example, this scenario was borne out on a day in November 2006, causing the water level to rise to 1.68-1.72 m above the normal water level in the harbor of Aarhus, a large Danish coastal city [10, 11]. The event caused no remarkable damage [11], but if the starting point had been 1.35 meters above current sea level (a.c.s.l.) consequences could have been massive economic losses. Furthermore, extreme weather events such as the above are foreseen to increase in intensity and frequency over the 21st century [1]

Most storms are often predicted relatively early [10], allowing timely evacuation of humans in the case of an emergency. Thus, humans are not in immediate danger in most coastal cities, at least in the more developed parts of the World, but buildings may be. Coastal real estate constitutes a relatively high value compared to real estate further inland, and, at the same time, is particularly at risk from future sea-level rise.

Predictive ability in terms of flood risk is crucial for preventing future catastrophes when extreme weather events combine with an increasing normal water level. For this reason, the European Commission recently identified flood risk mapping as a major component in the EU action programme for flood protection [5]. Under a given average or extreme weather scenario the most obvious way to conduct flood risk mapping would be to designate all areas below the estimated scenario sea-level as likely to become flooded. However, natural topographic features such as hills and artificial infrastructure such as embankments and roads may constitute barriers that will prevent water from reaching all areas that would otherwise become flooded. Consequently, detailed information about potential barriers is central for accurately predicting flooding risk. Rather than simply basing predictions on elevation, improved predictive capacity will result from implementing flooding algorithms that directly models water flow as a function of terrain features. So far the feasibility of this approach has been limited by its huge requirements in terms of elevation data accuracy and computing time [12].

Here we take on this challenge and use data on real-estate pricing, a completely new national Digital Elevation Model (DEM) in very fine spatial resolution (~2 meter), and a new highly computationally efficient flooding algorithm that accurately models the influence of barriers to assess the economic real estate risk posed by future sea-level rise to Aarhus, the second largest city in Denmark. Using several sea-level scenarios for 2090-2099, we specifically assessed: (1) Which areas will become subject to flooding during extreme weather events? (2) What is the real estate composition of these areas, in terms of residential vs. non-residential areas, and (3) high vs. low relative economic value? In addition, we make a short assessment of possible adaptation measures that can be taken to reduce the future flood risk.

2. Data and methods

2.1. Study area

Aarhus (N 56°9′, E 10°12′) is found on the eastern coast of the peninsula of Jutland (figure 1). The city's southern coastline and a small section of its northern coastline consist of relatively tall, steep bluffs, preventing flooding more than a few meters inland. However, the majority of the northern city is rather flat and low-lying, probably rendering this district especially vulnerable to flooding (figure 2). A 2.1-m tall dike prevents flooding of the area in the case of high waters [11]. At the city center, the Aarhus River valley holds some low-lying residential areas as well. The region of interest was delimited by the Aarhus city zone (figure 1; hereafter simply referred to as Aarhus).

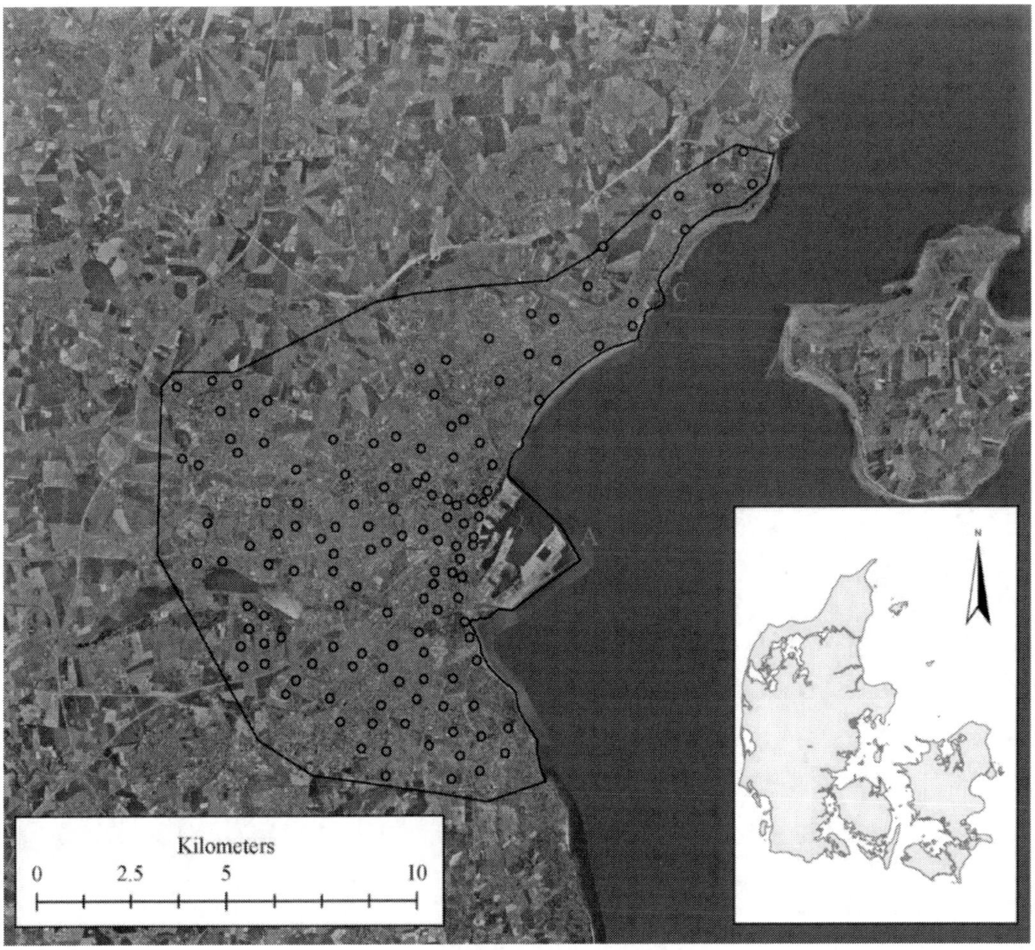

Figure 1. The city of Aarhus. The red dot on the insert map indicates the location of Aarhus. The analysis mask (black line) as well as the 135 buildings chosen for the study (violet dots) is indicated as well. The harbour of Aarhus is seen in the central part (A) of the image while the Brabrand Lake is located westernmost in the analysis mask (B) and the Egaa Marina 6-7 km north of the harbour (C).

2.2. Real estate pricing

To assess the geographic pattern of real-estate economic value in Aarhus we obtained data on real estate assessments from several Danish real-estate registers: Bygnings- og Boligregistret (BBR), Ejendomsstamregistret (ESR), Statens Vurderingsregister (SVUR), and Officielle Standardadresser og Koordinater (OSAK). The data were combined to a georeferenced point data set consisting of the official value estimate (DKK per residential m^2) for 2005 according to the Danish authorities for each building listed in the above registers. Only a subset of buildings was available, but geographically they were broadly distributed across Aarhus and, hence, probably representative of the general pricing pattern. Still, the data set contained numerous erroneous values due to the way the value per square meter was originally calculated, namely by dividing the value of a building by its residential area. In some cases this area is much less than the total resulting in unrealistically large value estimates for some buildings. These and other apparent errors were handled by selecting 135 buildings that were broadly distributed across Aarhus (figure 1). 10 buildings were picked within 7 of the city's postal districts (8200, 8210, 8220, 8230, 8240, 8250, 8310, and 8381) and in the large districts (8000, 8260, and 8270) 20-30 buildings were chosen. These were selected to present the district neighborhood by comparing a set of 3-8 neighboring buildings (depending on the number of available buildings in the neighbourhood). If these agreed with regard to value per square meter, one building was haphazardly chosen to represent the group. If they disagreed markedly, we used our thorough knowledge of real estate pricing in Aarhus to select the most representative value.

We used the log-transformed values for the 135 selected buildings to compute a continuous spatial surface of residential area square-meter values (DKK m^{-2}) using ordinary kriging. Kriging was implemented using the Geostatistical Analyst in ArcGis 9.3. Its default kriging settings were used, except that the following settings where changed to optimize the semi-variogram yielding the most statistically likely estimated surface: Second-order trend removal, neighbourhood of 25% global and 75% local data points, nugget = 0, lag size = 200, and number of lags = 12. We note that the estimated value surface is generally in line with our impression of real-estate prices in Aarhus, except that the central city zone is probably more expensive than our estimates.

Figure 2. A sea-side view of the northern part of the city. The forest to the left is a small urban beech forest. Its right (northern) limit marks the beginning of the northern residential areas at risk from flooding in the case of extreme weather events in 2090-2099. The enlarged segments of the photo illustrate a likely sea level during such events (approximately 2.5-3.5 m sea-level rise).

2.3. Digital Elevation Models (DEM)

Detailed topographic data was extracted from a new national DEM derived from airborne laser altimetry (LiDAR: Light Detection and Ranging) technology [13]. We used two available DEMs: (1) One DEM featured only the raw terrain, with all artificial infrastructure removed, while (2) another DEM included all bridges in the landscape as impervious barriers. We used the latter to assess the potential role of flood-gates at the bridges in flooding prevention. LiDAR produces high density and precision 3D points by emitting a laser pulse for which both first and last return of the backscatter is stored specifying the (x,y)-coordinates, the elapsed travel time for the signal and the intensity of the backscatter. In this study each pulse had a footprint of approximately 50 cm on the ground and the sampling interval was 1.6 m on average [14].

The LiDAR approach has been documented to be an effective and reliable method for terrain data collection in large areas [15-18], and it has the advantage that it can be used both day and night as well as under low-intensity cloud conditions. This facilitates the efficient capture of data, for which reason it is fair to assume that the resulting DEM is considered a "snapshot" in time; i.e. the acquisition is performed within a few weeks for the whole country and not over periods of years, as otherwise has been the standard procedure for recording national topographic data.

The DEMs used here are constructed as raster grids with 2-meter pixel resolution and come with a 80-cm horizontal and 10-cm vertical accuracy in vegetation-less areas with simple microtopography [14]. Each DEM only embraces Aarhus, take up approximately 350 MB, and consists of 90,000,000 raster cells.

2.4. TerraSTREAM

TerraSTREAM [19] is a series of tools that can perform various computational tasks on very large DEMs. The algorithms used have provable efficient performance, even on very large elevation models (hundreds of gigabytes) that do not fit in the main memory of a computer. In such cases, the model must reside on large, but slow disks and the transfer of data between a disk and the computer's main memory often becomes the primary bottleneck, rather than the CPU computation time. Thus, the algorithms that form the core of TerraSTREAM ensure that the disk is used efficiently and enables us to process terrains in hours or a few days that would make most GIS software crash or stall for weeks and months. TerraSTREAM contains a number of tools for hydrological analysis of terrains including flow accumulation computation and watershed hierarchy extraction. In this study we used TerraSTREAM to compute which cells become flooded when the sea level rises.

TerraSTREAM contains an implementation of a flooding algorithm where a cell is only marked as flooded if there is a path from that cell to the flooding source (the ocean in most cases) that never passes through a point that is higher than the new water level. This algorithm ensures that points in the terrain protected by dikes are not flooded unless the water level increases to a level higher than the dikes. TerraSTREAM enables us to compute flood risk maps on nationwide high-resolution terrain models on standard desktop computers, although here we focused on the relatively small Aarhus area.

2.5. Future climate scenarios and extreme weather events

Reflecting the trend that the future climate change scenarios are judged to most likely become increasingly severe as global temperature rise [1, 20], we focused on the two Intergovernmental Panel on Climate Changes (IPCC) scenarios predicting the most severe temperature increases in the 21st century, A2 and A1FI [1]. However, the corresponding sea-level rise predictions are probably too conservative [2, 21, 22], and we therefore used updated sea-level rise scenarios for the IPCC A2 and A1FI climate scenarios [2]. These scenarios predict a mean sea-level rise of 1.15 m and 1.35 m by 2090-2099, respectively.

The November 2006 high-water level of 1.68-1.72 m a.s.l. was the most extreme level observed during the last 118 years [10, 11]. To our knowledge no local extreme weather water-level rise scenarios exist. Therefore, we use the above high water level (1.68 m) to exemplify an extreme high water level during a strong storm. Adding this to the two future sea-level rise scenarios yields a future high-water level of 2.83 and 3.03 m a.c.s.l. for the A2 and A1FI, respectively. We contend that these estimates are rather conservative as extreme weather events are predicted to increase in intensity and frequency over the 21st century [1].

Flooding simulations were performed for both DEMs and the two future extreme high-water estimates using the TerraSTREAM flooding algorithm. This resulted in four flooding maps that were used for pin-pointing residential areas at risk from future flooding. In addition, flooding simulation was also performed for both DEMs using only the predicted mean sea levels (i.e., 1.15 m and 1.35 m).

An analysis mask was produced in ArcGIS 9.3 to delimit residential areas within the city zone of Aarhus, and the effects of flooding on real estate were primarily considered within this mask.

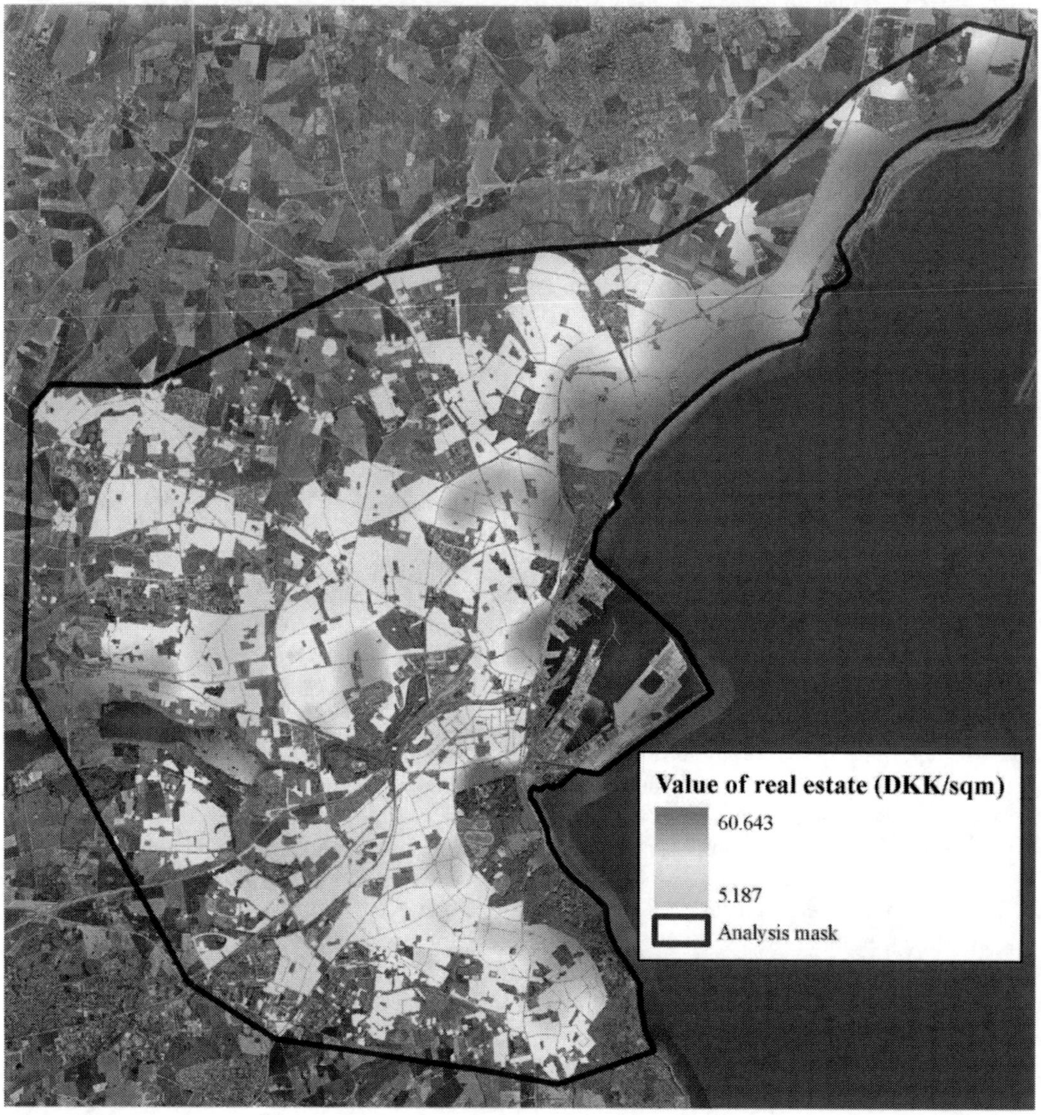

Figure 3. Real estate values represented in DKK per residential m^2. Only real-estate areas are shown.

3. Results

The most expensive residential areas in Aarhus tend to be found near the coastline, with the exception of the city center (figure 3). The flooding simulations based on the predicted mean sea levels showed that almost no contemporary land areas in Aarhus are at risk from permanent flooding in 2090-2099 (not shown). However, for all combinations of sea-level rise scenarios and DEMs a considerable part of the residential areas are estimated to become flooded in 2090-2099 under extreme high-water situations similar to the November 2006 storm (figure 4). The flooding simulations for the DEM that represented bridges as barriers (figure 4, A and C) revealed that flood-gates would prevent the flooding of the residential areas around the Aarhus River, while large residential areas in northern

Aarhus would still be likely to become flooded. The flooding simulations for the DEM without artificial infrastructures showed that relatively large residential areas in northern part of the city (figure 2) as well as areas around the Aarhus River are likely to become flooded (figure 4, B and D). In addition, most of the harbor and the areas fringing the Brabrand Lake would also risk flooding. Small but clear differences were seen between the extent of flooding under the A2 and the more extreme A1FI climate scenarios. About ⅔ of the residential areas likely to become flooded were located in areas of high-value real-estate, i.e., areas with real-estate values greater than the general mean value for Aarhus (figure 4).

4. Discussion

Our results exemplify the challenge posed upon the society by potential effects of future sea-level rise. We found that 21st century sea-level rise may well put large residential districts in Aarhus, the second-largest city in Denmark, at risk from flooding during likely extreme weather events. Much of the area at risk represents high-value real-estate neighbourhoods. Our results, however, also show that strategic placement of flooding barriers could dramatically reduce the flooding risk for some areas.

The flooding simulations performed on the DEM including bridges as impervious barriers provided an instructive illustration that strategically placed flooding barriers, in this case flood-gates, could form an important defence against future catastrophic flooding. We note that recently a proposal in fact has been made to construct a lock at the mouth of the Aarhus River, with exactly this purpose in mind [23]. We also note that the use of artificial infrastructure to prevent flooding is likely to have further complexities, notably requiring the installation of pumps to handle water inflow through underground connections in the sewer system as well as river water damming up behind the flood-gates.

As flood-gates are not currently installed at the mouth of the Aarhus River, we focus in the following on the flooding results from the flooding simulations for the DEM without artificial infrastructures (figure 4, B and D). Approximately ½ the areas identified in this study as being prone to flooding are residential zones and about ⅔ of these are relatively high value real-estate neighbourhoods. If these areas have to be abandoned following future sea-level rises relatively large values will be lost. However, it seems fair to anticipate that society, notably the municipality, or local citizen groups will attempt to prevent flooding of these areas by a number of measures (e.g. [24]). At least three preventive measures are possible: (1) constructing or enhancing existing dikes, (2) elevating [25] or (3) sealing houses [26]. The construction or retrofitting of dikes has an estimated cost of €0.9 million to €9 million per linear km [25]. In Aarhus, at least 5-7 km of dike protects the zones vulnerable to flooding. These are 2.1 meter tall [11] and will have to be elevated by at least one meter and preferably more under the scenarios used here. At some locations it is possible that entirely new dikes have to be constructed as well. The costs of such a solution will be shared among all >1000 households in the neighbourhood [27] or perhaps paid by the municipality. Elevation of the exposed houses is a relatively costly process: elevating a 90 m^2 house some 60 cm costs approximately €41500 [25]. The real houses in the northern part of Aarhus are at least double or triple that size and according to the DEM many of them will have to be elevated more than 60 cm. Preventing water to enter a 90 m^2 house by sealing only costs around €5000 [25], but the sealing inevitably will have to be renewed frequently. Adapting the two last solutions entails regular flooding of gardens and neighbouring roads, a scenario that is unlikely to be acceptable to citizens living in the area.

Apart from the residential areas, large parts of the areas flooded in the event of extreme future weather is on the harbour of Aarhus, with a smaller area at the marina in Egaa as well. Here, the water masses might induce great damage to buildings and on-ground cargo [28] as well as cause the leakage of harmful chemicals to the surroundings. Because of space limitations, dike construction is probably not possible on the harbour, in which case parts of this would have to be either elevated [11, 29] or sealed.

IOP Conf. Series: Earth and Environmental Science **8** (2009) 012022 doi:10.1088/1755-1315/8/1/012022

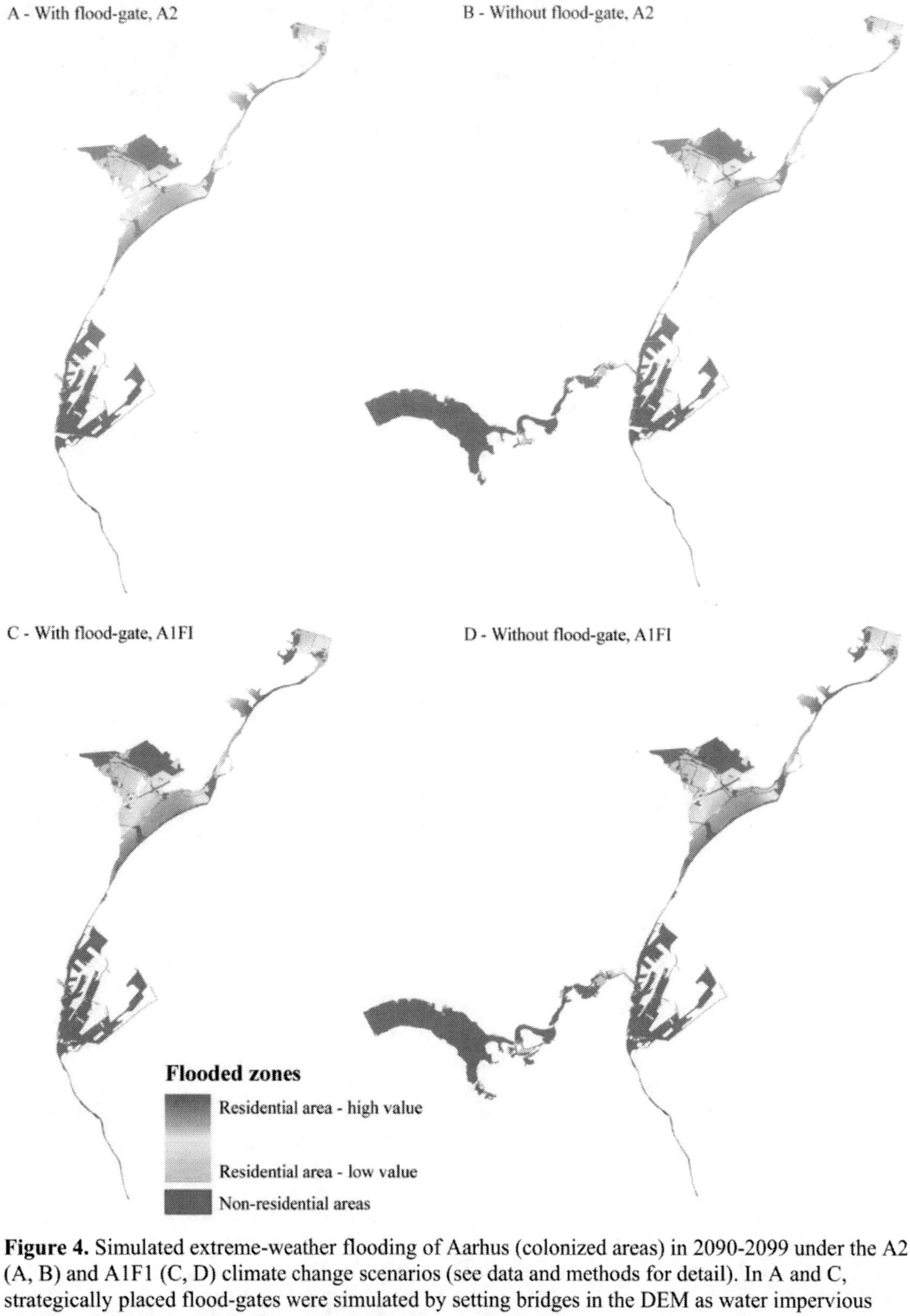

Figure 4. Simulated extreme-weather flooding of Aarhus (colonized areas) in 2090-2099 under the A2 (A, B) and A1F1 (C, D) climate change scenarios (see data and methods for detail). In A and C, strategically placed flood-gates were simulated by setting bridges in the DEM as water impervious barriers. The colour code indicates the real-estate value (see figure 3).

In addition to the predicted sea-level rises in the 21st century, current extreme weather phenomena are foreseen to become even more extreme as well as frequent [1]. From this study it is therefore clear that huge areas especially in the northern part of Aarhus and around the Aarhus River will become vulnerable to extreme weather flooding in the 21st century. As much of the area at risk represents high-value real estate, it seems clear that proactive measures other than simple abandonment should be taken. Among the different possibilities for dealing with an increased sea level, the construction or elevation of existing dikes seems to be the most convenient, most socially acceptable, and maybe also the cheapest solution. As fine resolution DEMs logically reveal small features in the landscape not necessarily found in coarser terrain models, flooding simulations using fine-resolution topography generates detailed predictive maps of exactly where flooding will take place if the water reaches a given level. Therefore, in line with previous studies [5, 30], we suggest that large editions of maps similar to those produced here will become an important tool for the planning of future city development in relation to expected climate changes, hereunder the strategic placement of flooding barriers to efficiently reduce flooding risks, as well as for the development of evacuation plans.

5. Acknowledgements

We thank Inge T. Kristensen for providing real estate pricing data and gratefully acknowledge economic support from the Danish Natural Science Research Council (grant #272-07-0242 to JCS). We also acknowledge support by an Ole Roemer Scholarship from the Danish National Science Research Council, a NABIIT grant from the Danish Strategic Research Council, and by the Danish National Research Foundation, and in part by MADALGO: Center for Massive Data Algorithmics, a Center of the Danish National Research Foundation.

References

[1] IPCC 2007 *The Physical Science Basis. Contribution of Working Group I to the Fourth Assessment Report of the Intergovernmental Panel on Climate Change* (Cambridge, United Kingdom and New York, NY, USA : Cambridge University Press)

[2] Grinsted A, Moore J and Jevrejeva S 2009 Reconstructing sea level from paleo and projected temperatures 200 to 2100 ad *Clim.Dyn.* **OnlineFirst**

[3] Titus J G, Park R A, Leatherman S P, Weggel J R, Greene M S, Mausel P W, Brown S, Gaunt C, Trehan M and Yohe G 1991 Greenhouse effect and sea level rise: The cost of holding back the sea *Coast.Manage.* **19** 171-204

[4] Lewis J and Kelman I 2009 Housing, Flooding, and Risk-Ecology: Thames Estuary South-Shoreland and North Kent *Journal of Architectural and Planning Research* **26** 14-29

[5] Commission Of The European Communities 2004 *Communication From The Commission To The Council, The European Parliament, The European Economic And Social Committee And The Committee Of The Regions - Flood risk management: Flood prevention, protection and mitigation* (Brussels, Belgium : Commission Of The European Communities)

[6] Jonkman S N, Maaskant B, Boyd E and Levitan M L 2009 Loss of Life Caused by the Flooding of New Orleans After Hurricane Katrina: Analysis of the Relationship Between Flood Characteristics and Mortality *Risk Analysis* **29** 676-98

[7] Wikipedia *Danmarks største byer* Last updated: 2009 01/01. Accessed: 2009 06/25 Available from: http://da.wikipedia.org/wiki/Danmarks_st%C3%B8rste_byer

[8] Cappelen, J 2008 *Danmarks klima 2007 med Tórshavn, Færøerne og Nuuk, Grønland - with English translations* (Denmark : Danmarks Meterologiske Institut)

[9] Cappelen, J 2009 *Danmarks klima 2008 med Tórshavn, Færøerne og Nuuk, Grønland - with English translations* (Denmark : Danmarks Meteorologiske Institut)

[10] Sørensen P, Sørensen C, Ingvardsen S M, Andersen I and Kloster B B *Højvandsstatistikker 2007* (Denmark : Kystdirektoratet)

[11] Nørby, E 2009 *Århus skal sikres mod stigende vandstand* (Denmark, Aarhus : Jyllands Posten)

[12] Apel H, Aronica G, Kreibich H and Thieken A 2009 Flood risk analyses - how detailed do we need to be?*Nat.Hazards* **49** 79-98

[13] Vierling K T, Vierling L A, Gould W A G, Martinuzzi S and Clawges R M 2008 Lidar: shedding new light on habitat characterization and modeling *Front. Ecol. Environ.* **6** 90-8

[14] COWI 2008 *Kvalitet DDH®* (Aarhus, Denmark : COWI)

[15] Baltsavias E P 1999 Airborne laser scanning: basic relations and formulas *ISPRS Journal of Photogrammetry and Remote Sensing* **54** 199-214

[16] Baltsavias E P 1999 Airborne laser scanning: existing systems and firms and other resources *ISPRS Journal of Photogrammetry and Remote Sensing* **54** 164-98

[17] Wagner W, Ullrich A, Ducic V, Melzer T and Studnicka N 2006 Gaussian decomposition and calibration of a novel small-footprint full-waveform digitising airborne laser scanner *ISPRS Journal of Photogrammetry and Remote Sensing* **60** 100-12

[18] Luo Y, Trishchenko A P and Khlopenkov K V 2008 Developing clear-sky, cloud and cloud shadow mask for producing clear-sky composites at 250-meter spatial resolution for the seven MODIS land bands over Canada and North America *Remote Sensing of Environment* **112** 4167-85

[19] Danner, A, Mølhave, T, Yi, K, Agarwal, P K, Arge, L and Mitasova, H 2007 GIS '07: Proceedings of the 15th annual ACM international symposium on Advances in geographic information systems *TerraStream: from elevation data to watershed hierarchies* (New York, NY, USA : ACM) pp 1-8

[20] Meinshausen M, Meinshausen N, Hare W, Raper S C B, Frieler K, Knutti R, Frame D J and Allen M R 2009 Greenhouse-gas emission targets for limiting global warming to 2°C *Nature* **458** 1158-62

[21] Rahmstorf S 2007 A Semi-Empirical Approach to Projecting Future Sea-Level Rise *Science* **315** 368-70

[22] Horton R, Herweijer C, Rosenzweig C, Liu J P, Gornitz V and Ruane A C 2008 Sea level rise projections for current generation CGCMs based on the semi-empirical method *Geophys.Res.Lett.* **35** L02715

[23] Nørby, E 2009 *Sluse skal holde Århus tør* (Denmark, Aarhus : Jyllands Posten)

[24] Burrel B C, Davar K and Hughes R 2007 A Review of Flood Management Considering the Impacts of Climate Change *Water Int.* **32** 342

[25] Ackerman, F, Stanton, E A, Hope, C, Alberth, S, Fisher, J, Biewald, B, Perera, E M and Lashof, D 2008 *The Cost of Climate Change: What We'll Pay if Global Warming Continues Unchecked* (New York, USA : Natural Resources Defense Council (NRDC))

[26] Federal Emegency Management Agency 1998 *Homeowner's Guide To Retrofitting: Six Ways To Protect Your House From Flooding* (Washington, USA : Federal Emegency Management Agency)

[27] Digelauget Vejlby Fed *Digelauget Vejlby Fed er nu en realitetet* Accessed: 2009 06/21 Available from: http://www.fedet-risskov.dk/index.php?page=Digelaug01

[28] Danmarks Meteorologiske Institut *Stormfloder - historiske stormfloder* Last updated: 2001 12/20. Accessed: 2009 06/21 Available from: http://www.dmi.dk/dmi/index/hav/stormflodstema-2/historiske_stormfloder.htm

[29] Jensen, B C, Brøndum, S, Christensen, J, Fenger, J, Hansen, S O, Jakobsen, J, Jørgensen, A M K, Linde, J J, Tønnesen, C *et al.* 2003 *Effekter Af Klimaændringer - Tilpasninger I Danmark* (Kgs. Lyngby, Denmark : Akademiet For De Tekniske Videnskaber (ATV))

[30] Vidal J and Wade S 2008 A framework for developing high-resolution multi-model climate projections: 21st century scenarios for the UK *Int.J.Climatol.* **28** 843-58

IOP Conf. Series: Earth and Environmental Science **8** (2009) 012023 doi:10.1088/1755-1315/8/1/012023

Energy mitigation, adaptation and biodiversity: synergies and antagonisms

P M Berry and J S Paterson

Environmental Change Institute, University of Oxford, Dyson Perrins Building, South Parks Road, Oxford, OX1 3QY, UK.

E mail: pam.berry@eci.ox.ac.uk

Abstract. In this paper we review the current impacts of different energy producers (and energy conservation) on biodiversity and investigate the potential for achieving positive biodiversity effects along with mitigation and adaptation objectives. Very few energy producers achieve all three aims – although it may be possible with careful choice of location and management. In some instances, energy conservation can provide mitigation, adaptation and biodiversity benefits. There is still a gap in knowledge regarding the effects of newer energy technologies on biodiversity. There is an additional concern that many supposedly 'green' renewable energy projects may actually harm biodiversity to such a degree that their overall human benefits are negated. The increasing understanding that ecosystem services are vital for human well-being though means that attempting positive mitigation, adaptation and biodiversity conservation in the energy sector should be an imperative goal for international policy. Whilst research into synergies between mitigation and adaptation is established, there has been very little that has examined the impacts on biodiversity as well. Further work is required to identify and provide evidence of the best ways of optimising mitigation, adaptation and biodiversity in the energy sector.

1. Introduction

The multitude of observed changes in aspects of biodiversity that are attributable to climate change combined with its projected impacts mean that climate change is of increasing importance on a number of agendas. Already many actions are being taken or are proposed to mitigate climate change and, as these are often viewed as being insufficient to significantly reduce undesirable impacts, especially in the short-term, adaptation measures are also being proposed [1]. These are usually devised by individual sectors to meet particular circumstances, without identifying their possible wider ramifications. These will be examined with particular reference to the energy sector, where the mitigation to reduce greenhouse gas emissions has implications for all energy types. We argue for the necessity of integrating mitigation and adaptation responses to climate change, and their impacts on biodiversity, in order to avoid the degradation of biodiversity and the services it provides, and to ensure the efficient use of resources.

1.1 Observed impacts of climate change on biodiversity

Since the Fourth Assessment Report (AR4) of the Intergovernmental Panel on Climate Change (IPCC) was published in 2007, ongoing observations continue to support their conclusions that recent warming is strongly affecting natural biological systems in a multitude of ways [2]. Changes in phenology, such as earlier budburst [3], longer active periods for ectothermic organisms [4] and delayed autumn fruiting of mushrooms by 6.4. days per decade [5], continue to be recorded, although not all observations are in the direction expected [6]. A few studies have shown also that climate warming may generate temporal mismatches among mutualistic partners [7]. Evidence of latitudinal [8, 9] and altitudinal [10, 11] range shifts continue. In many cases these responses are too slow for species to track the projected changes in climate [12]. Thus there is a need to both reduce the projected rate of climate change through mitigation and to assist species in their adaptation, through such measures as increasing their habitat availability or improving landscape connectivity.

1.2 Climate change mitigation

Mitigation is aimed at reducing the magnitude of climate change and thus the risks associated with it and particularly the vulnerability of human systems. Reductions can be achieved through two complementary pathways: a net reduction of greenhouse gas emissions, and the protection and promotion of carbon sinks, through land-use and habitat management or geo-engineering [13]. Efforts to reduce the former have affected all forms of energy production [14], with many, even supposedly green energy sources, having wider environmental impacts [15]. With energy demands expected to rise, the urgent need to reduce both emissions *and* environmental impacts is paramount [16]. Biodiversity is critical in the second form of mitigation and it has been estimated that terrestrial ecosystems absorb about 30% of all CO_2 emissions from fossil fuel burning and net deforestation [17], with forests and wetlands being particularly important. The loss of this climate regulation service would have important repercussions on the efforts to address global warming.

1.3. Climate change adaptation

The IPCC suggested that adaptation practices can be differentiated along several dimensions, for example by: spatial scale, sector, actor, climatic zone, development level of the systems in which they are implemented or some combination of these [1]. Generally adaptation measures are applied to particular (local) situations or sector(s) [18], although they may have wider implications. Much adaptation research so far also has been sectoral and local, place-based [19] and this paper examines some of the interactions between adaptation and mitigation measures in the energy sector and with biodiversity and thus the wider environment. It has also been suggested that adaptation and development should not be viewed separately as they are often inter-related, as many adaptation measures are seen as part of sustainable development and there could be synergies [20, 21].

1.4. Mitigation and adaptation

Mitigation and adaptation are often also seen as necessary, complementary strategies for dealing with climate change [20, 22], while in economic terms they may be viewed as substitutes and only become complementary when the adaptation costs depend on the amount of mitigation. The relationship between mitigation and adaptation is complex, as the costs and benefits are not equally experienced between different actors or spatial and temporal scales [20]. This can make strategic planning and decision-making complicated. It should also be noted that mitigation benefits are lagged in time [23], unlike some adaptation benefits (i.e. mitigation in one period produces long-lasting benefits for future generations while adaptation measures are often specific to one time-period), mitigation benefits are more global and adaptation benefits are more localised, and that mitigation focuses on greenhouse gas sources and sinks, while adaptation focuses on climate sensitive sectors and activities.

1.5. Mitigation, adaptation and biodiversity

Mitigation and adaptation, however, not only interact with each other, but they also have important environmental implications and especially for biodiversity, which may in turn affect its ability to mitigate or adapt to climate change. A review of these interactions for eight sectors, including energy (production and conservation), showed that there were a number of different possible outcomes. These ranged from opportunities for biodiversity to be enhanced by mitigation and adaptation measures, through no discernible impact, to potential threats to the existence or functioning of ecosystems [24].

This range of outcomes will now be explored, in order to examine for one sector, energy, what affects the outcome and whether there are possibilities to enhance the synergies and avoid/minimise antagonistic actions.

2. Impacts of mitigation and adaptation on biodiversity

2.1. Neutral outcomes for biodiversity

There are a number of energy schemes that offer carbon mitigation without affecting biodiversity either negatively or positively or providing any climate change adaptation. Mostly, these options are very small-scale or do not alter an existing management regime. For example, the addition of an annual biofuel crop (e.g., oilseed rape – *Brassica napus*) to an existing arable crop rotation is unlikely to make a difference to biodiversity due to similar management and inputs (fertiliser, pesticides) requirements [25]. Small-scale systems include solar electric panels or wind turbines added to existing infrastructure (e.g., housing) [26]; or, run-of-river hydropower schemes, which, if well designed, can have minimal or no effects on biodiversity by allowing safe fish passage [27].

Improving energy efficiency is also seen as a major component of carbon mitigation either in industry, transport or other sectors. Improving fuel efficiency in cars, for example, is unlikely to have any effect on biodiversity although large-scale improvements in efficiency may lead to less air pollution and a reduction in the detrimental effects on biodiversity.

2.2. Negative outcomes for biodiversity

2.2.1. Positive for mitigation, neutral or negative for adaptation. In the last few years a number of major schemes have been promoted by governments that have been designed to offer huge mitigation gains; however, many of these schemes, upon closer inspection, are potentially very damaging to biodiversity and offer no adaptation potential either. In the EU, two such examples are large dam hydropower and tidal barrages, both of which could have catastrophic consequences for local and regional wildlife including changes to aquatic habitats and water flow, and interruption of fish migration patterns amongst other effects [28, 29, 30].

In sunny, arid regions the development of large-scale solar thermal towers is gaining pace but they can have major negative impacts on local ecosystems because of their huge water demands for cooling [31]. An additional problem is the distribution and transfer of energy away from high solar insolation areas to areas of high energy demand. Likewise, wind-farms can have localised negative effects on bat and bird populations and if badly positioned can become a barrier to bird migration; however, this should be put in the context of overall species' population dynamics [32]. In fact, according to one study, the mortality of birds caused by wind farms is far lower (0.3 - 0.4 fatalities per giga-watt hour) than fossil-fuelled power stations (5.2 fatalities per giga-watt hour) [33]

Nuclear power is also promoted heavily in many countries and although operational mitigation benefits can be achieved, power plants demand huge quantities of water and new sites need to be located near major rivers or coastlines often resulting in aquatic species mortality from water outflows (temperature, chemical and turbidity changes) and entrainment [34, 35]. Adaptation to rising sea levels through hard coastal defences would lead to negative impacts on biodiversity as habitats, such as salt marsh, are prevented from adapting to climate change by inland migration. Mining uranium also has numerous environmental costs [36].

Carbon capture and storage systems for fossil fuel plants are proposed as a viable mitigation strategy and although there are very few examples working, the widespread adoption of these technologies will not reduce (and may even increase) the negative effects on biodiversity currently associated with fossil fuels [37]. These systems are very energy intensive and are therefore likely to have a knock-on effect on the environmental impact of the energy source; furthermore biodiversity could be impacted at the capture site, carbon storage site and transporting the carbon between sites.

The widespread adoption of *some* biofuel crops can be particularly harmful to biodiversity; these include invasive grasses or late-cropped arable plants such as sugar-beet (*Beta vulgaris*) in the EU but also imports of palm oil derived biofuels [38, 39, 40]. Furthermore, the increasing use of food-crop by-products (haulm, straw, corn stovers, etc) for biofuels will reduce field fertility and hence increase agro-chemical use leading to biodiversity loss [41].

2.2.2. Neutral or negative for mitigation, positive for adaptation. In drier regions, hydropower schemes with dammed reservoirs are used to supply water for drinking and agricultural irrigation and are being touted as possible combined mitigation and adaptation strategies in areas predicted to have water shortages (e.g., in the Iberian peninsula [42]). However, if future climate change water shortages are realised many of these dams will not meet their predicted energy capacity production, which, along with the carbon costs of construction (mainly from concrete), may result in no overall mitigation effect.

2.3. Positive outcomes for biodiversity

2.3.1. Positive for mitigation, neutral or negative for adaptation. Biofuels crops vary widely in type and species used and some species may have positive biodiversity effects in certain circumstances. For example, growing low-input perennial species (e.g., poplar and willow for Short Rotation Coppice or native grasses) on contaminated land may provide habitats for wildlife reduce erosion and improve soil biodiversity. Similarly, restoration of derelict semi-natural woodland coppice systems (i.e., woods that have reverted to high forest) would provide a welcome boost to woodland biodiversity adapted to the cycle of periodic coppicing as well as provide a replacement (of fossil fuels) energy source for domestic heating [43].

2.3.2. Positive for mitigation and adaptation to varying degrees. Some woody perennial biofuel crops may offer better drought tolerance than annual crops, as well as provide habitats for biodiversity. Additionally, agro-forestry systems can provide similar benefits as well as shade for the annual crops in hot regions [44, 45]. Similarly, semi-natural grassland biofuel crops may provide biodiversity adaptation (e.g., stepping stones or corridors for migrating grassland species) as well as biodiversity conservation and mitigation [46]. The restoration of derelict coppice systems discussed above may have benefits in hot regions where coppicing would provide a fire-break (and is therefore an adaptation strategy in ecosystems likely to be more fire-prone in the future). Energy conservation measures for buildings include the creation of green roofs which can increase insulation in buildings thereby reducing energy demand; furthermore, they can increase carbon sequestration in the soil media and provide a useful habitat for a range of species, particularly in urban areas [47]. These represent win-win-win situations [24, 30] whereby the objectives of mitigation, adaptation *and* biodiversity can all be met, albeit on a small-scale.

3. Discussion and conclusions

The above mentioned examples have shown that the various forms of energy production have different mitigation and adaptation potentials and a range of possible outcomes for biodiversity. These will depend upon a range of factors including management, scale, location, species and the biodiversity status of site used; this is clearly seen in the case of biofuels [48]. For example, if an annual biofuel crop is grown in an existing (food) crop rotation, and the management and inputs are similar to those for other crops, there is little additional effect on biodiversity. However, if the farmer maximises the biomass potential from the crop rotation by removing by-products (e.g., straw) from non-biofuel crop harvests, then the deleterious consequences for biodiversity may be increased through the depletion of organic matter [41]. But growing biofuels on agricultural land is one thing; even more damaging is the clearance of high biodiversity value land, such as rainforests, to make way for biofuels such as oil palm [39].

A review of the environmental impacts of biofuel crops, provided recommendations to minimise biofuel crop impacts and to maximise their environmental and economic benefits and suggested that a strong emphasis should be placed on appropriate planning and management which should ensure returns are optimised, whilst avoiding unwanted environmental damage [48]. For example, low-input perennial biofuel crops grown on contaminated or degraded land can provide new habitats for wildlife, and possibly help in soil stabilisation and remediation. Short rotation coppice of alder and willow have been shown to be effective in reducing metal concentrations on contaminated land [50], as well as removing nitrates and other nutrients from municipal waste water, which is a cheap alternative to convention sewage treatment and provides water and fertiliser for the energy crop, thus representing a potential contribution to mitigation, adaptation and biodiversity, another win-win-win situation [51]. There is also evidence that compared to annual crops, perennial grasses, such as *Miscanthus* and *Panicum virgatum* (switchgrass), can lead to increased biodiversity, especially in their earlier stages of growth [51]. Additionally, locating them close to different types of vegetation (thereby increasing habitat mosaic diversity) and varying the harvesting schedule can be beneficial.

Energy policy will have a large part to play in driving the combination of energy sources at both the national and international levels. In Europe, for example, the Directive on Renewable Energy (http://ec.europa.eu/energy/climate_actions/doc/2008_res_directive_en.pdf) places a binding target to source 20% of energy needs from renewables such as biomass, hydro, wind and solar power by 2020. Also there is an ambitious 10% biofuels target by 2020 and, in order to qualify for inclusion in this target, it states that the use of biofuels must result in an overall greenhouse gas emissions saving of 35%. This increases to 50% from 2017 for existing production and 60% for new installation from this date. It has, however, now recognised the need to comply with environmental and social sustainability objectives, both within Europe and in exporting countries and thus the Directive stipulates that old forest, highly species rich grassland or land with a high carbon stock, e.g. wetlands and pristine peatlands, cannot be used for biofuels. Enforcement of these regulations will require a transparent accounting and certification system, as is needed for forests under REDD (Reducing Emissions from Deforestation and Degradation), to ensure that sustainability is not compromised. However, the Directive still allows scope for inappropriate management on current cropland which could see wide scale adoption of potentially harmful biofuel practice. Further refinement of EU policy to incorporate best practice guidelines or financial incentives for farmers is required (e.g., through agri-environment schemes).

Other sectors are beginning to realise the opportunities for policy responses to meet both mitigation and adaptation objectives. For example, a study of land-use plans and policies designed to address climate change, focused primarily on cases from the United States and Australia, showed that while nearly one half of the actions identified contain potential conflicts to achieving adaptation and mitigation simultaneously, a fifth were able to achieve this [52]. Nevertheless, this did not include the third dimension of interest in this paper - biodiversity impacts; there are, however an increasing number of examples of individual actions which not only combine mitigation and adaptation but also are positive for biodiversity. Many ecosystem-based river and coastal flood management schemes, such as wetland or salt marsh re-creation can meet all three objectives [24].

In the energy sector, the focus is very much on mitigation, given that 70% of the increase in CO_2 emissions over the last 30 years came from fossil fuel combustion and cement manufacture [53]. Mitigation is vital as climate change affects all human societies, as well as all natural ecosystems, but given the longer response time of the climate system to mitigation, the extent of the impacts already experienced and those which are projected, adaptation is an important complementary response[54]. Within the energy sector there are direct actions which are required in order to cope with climate change impacts, such as extreme events affecting power generation and installations, but also indirect ones resulting from changed patterns of energy usage. In the case of the former, then icing events should decrease under global warming, but where they do occur then solutions that include heating, e.g. for wind turbines blades [55], will contribute to further energy usage and those that use chemicals would be negative for biodiversity. A decentralisation of power generation could also be considered an appropriate adaptation response and generally small-scale schemes have lower biodiversity impacts. Reduced energy usage (or increased efficiency of production) is an important adaptation behaviour which also contributes to mitigation. There is little specific evidence of its consequences for biodiversity, but perhaps represents the most positive outcome.

For some energy producers, the options for optimising mitigation, adaptation and biodiversity may be limited (e.g., fossil fuels); for others, there is, however, an opportunity for climate change and related energy policies to seek those actions which are energy efficient, and also maximise mitigation and minimise damage to, or enhance biodiversity. Overall, there is limited evidence for the impacts of many forms of power production on biodiversity, and this needs to be addressed if important ecosystems services are not to be lost or disrupted. There is, therefore, still a long way to go before the energy sector truly recognises not only the intrinsic value of biodiversity, but also its role in providing cost-effective and sustainable mitigation and adaptation options. Whilst different policy tools have been created to implement mitigation, adaptation and biodiversity, there is not, as yet, one single policy combining all three. It is vital that this integration occurs in order to avoid unwanted antagonisms and promote the synergies and thus the efficient use of resources.

Acknowledgements

This paper is based on work undertaken as part of the MACIS project (Minimisation of and adaptation to climate change: impacts on biodiversity), which was funded under the Sixth Framework Programme of the European Commission (Contract No. 044399).

References

[1] Adger W, Agrawala S, Mirza M, Conde C, O'Brien K, Pulhin J, Pulwarty R, Smit B and Takahashi K 2007 *Assessment of adaptation practices, options, constraints and capacity* ed Parry M L, Canziani O F, Palutikof J P, van der Linden P J and Hanson C E (Cambridge: Cambridge University Press) pp 717-43.

[2] Rosenzweig C, Casassa G, Karoly D J, Imeson A, Liu C, Menzel A, Rawlins S, Root T L, Seguin B and Tryjanowski P 2007 *Assessment of observed changes and responses in natural and managed systems* ed Parry M L, Canziani O F, Palutikof J P, van der Linden P J and Hanson C E (Cambridge: Cambridge University Press) pp 79-131.

[3] Pudas E, Leppälä M, Tolvanen A, Poikolainen J, Venäläinen a and Kubin E 2008 *Int. J. Biometeorol.* **52** 251-59

[4] Moreno-Rueda G, Pleguezuelos J and Alaminos E 2009 *Clim. Change.* **92** 235-42

[5] Kauserud H, Stige L C, Vik J O, Økland R H, Høiland K and Stenseth N C 2008 *Proc. Natl. Acad. Sci. USA.* **105** 3811-14

[6] Parmesan C 2007 *Global Change Biol.* **13** 1860-72

[7] Hegland S, Nielsen A, Lázaro A, Bjerknes A and Totland Ø 2009 *Ecol. Lett.* **12** 184-95

[8] MacDonald G, Kremenetski K and Beilman D 2008 *Philos. Trans. R. Soc. Lond., Ser. B: Biol. Sci.* **363** 2285-99

[9] Mitikka V, Heikkinen R, Luoto M, Araújo M, Saarinen K, Pöyry J and Fronzek S 2008 *Biodivers. Conserv.* **17** 623-41

[10] Beckage B, Osborne B, Gavin D, Pucko C, Siccama T and Perkins T 2008 *Proc. Natl. Acad. Sci. USA*. **105** 4197-202

[11] Kelly A and Goulden M 2008 *Proc. Natl. Acad. Sci. USA*. **105** 11823-26

[12] Devictor V, Julliard R, Couvet D and Jiguet F 2008 *Proc. R. Soc. Lond., Ser. B: Biol. Sci.*. **275** 2743-48

[13] Lal R 2008 *Philos. Trans. R. Soc. Lond., Ser. B: Biol. Sci.* **363** 815-30

[14] Sims R E H, Schock R N, Adegbululgbe A et al. 2007 *Energy supply* ed Metz B, Davidson O R, Bosch P R, Dave R and Meyer L A (Cambridge: Cambridge University Press) pp 251-322.

[15] Pimentel D and Patzek T A D 2006 *Bioscience*. **56** 875-875

[16] International Energy Agency 2006

[17] Canadell J G and Raupach M R 2008 *Science*. **320** 1456-57

[18] Goklany I 2007 *Mitig. Adapt. Strat. Global Change*. **12** 755-86

[19] Klein R J T, Huq S, Denton F, Downing T E, Richels R G, Robinson J B and Toth F L 2007 *Inter-relationships between adaptation and mitigation* ed Parry M L, Canziani O F, Palutikof J P, van der Linden P J and Hanson C E (Cambridge: Cambridge University Press) pp 745-77.

[20] Mata L J and Budhooram J 2007 *Mitig. Adapt. Strat. Global Change*. **12** 799-807

[21] Ayers J M and Huq S 2009 *Environ. Manage.* **43** 753-64

[22] Yohe G and Strzepek K 2007 *Mitig. Adapt. Strat. Global Change*. **12** 727-39

[23] Wilbanks T, Leiby P, Perlack R, Ensminger J and Wright S 2007 *Mitig. Adapt. Strat. Global Change*. **12** 713-25

[24] Berry P M (ed.) 2008 *Biodiversity in the Balance – Mitigation and Adaptation Conflicts and Synergies*. (Sofia: Pensoft Publishing) p 300

[25] Powlson D S, Riche A B and Shield I 2005 *Ann. Appl. Biol.* **146** 193-201

[26] Trieb F, Langniß O and Klaiß H 1997 *Sol. Energy*. **59** 89-99

[27] Copeman V A 1997 *J Chart. Inst. Water E* . **11** 431-36

[28] Dudgeon D, Arthington A H, Gessner M O et al. 2006 *Biol. Rev. Camb. Philos. Soc.* **81** 163-82

[29] Poff N L, Olden J D, Merritt D M and Pepin D M 2007 *Proc. Natl. Acad. Sci. USA*. **104** 5732-37

[30] Paterson J S, Araújo M B, Berry P M, Piper J and Rounsevell M D A 2008 *Conserv. Biol.* **22** 1352

[31] Tsoutsos T, Frantzeskaki N and Gekas V 2005 *Energ. Policy*. **33** 289-96

[32] Devereux C I, Denny M J H and Whittingham M J 2008 *J. Appl. Ecol.* **45** 1689-94

[33] Sovacool B K 2009 *Energ. Policy*. **37** 2241-48

[34] Jenner H A, Taylor C J L, Van Donk M and Khalanski M 1997 *Mar. Environ. Res.* **43** 279-93

[35] Greenwood M F D 2008 *Mar. Pollut. Bull.* **56** 723-39

[36] Carvalho I G, Cidu R, Fanfani L, Pitsch H, Beaucaire C and Zuddas P 2005 *Environ. Sci. Techno.* **39** 8646-52

[37] Thitakamol B, Veawab A and Aroonwilas A 2007 *Int. J. Greenhouse Gas Control*. **1** 318-42

[38] Zah R, Böni H, Gauch M, Hischier R, Lehmann M and Wäger P 2007 *Ökobilanz von Energieprodukten: Ökologische Bewertung von Biotreibstoffen* (Empa:St Gallen)

[39] Fitzherbert E B, Struebig M J, Morel A, Danielsen F, Bruhl C A, Donald P F and Phalan B

2008 *Trends Ecol. Evol.* **23** 538-45

[40] Eggers J, Troltzsch K, Falcucci A et al. 2009 *Global Change Biol. Bioenerg.* **1** 18-34

[41] Blanco-canqui H and Lal R 2009 *Soil Sci. Soc. Am. J.* **73** 418-26

[42] Trigo R M, Pozo-Vázquez D, Osborn T J, Castro-Díez Y, Gámiz-Fortis S and Esteban-Parra M J 2004 *Int. J. Climatol.* **24** 925-44

[43] Mason W L 2007 *Ibis.* **149** 41-52

[44] Gruenewald H, Brandt B, Schneider B, Bens O, Kendzia G and Hüttl R 2007 *Ecol. Eng.* **29** 319-28

[45] Roshetko J M, Lasco R D and Delos A, M.S. 2007 *Mitig. Adapt. Strat. Global Change.* **12** 219-42

[46] Tilman D, Hill J and Lehman C 2006 *Science.* **314** 1598-600

[47] Oberndorfer E, Lundholm J, Bass B, Coffman R R, Doshi H, Dunnett N, Gaffin S, Kohler M, Liu K K Y and Rowe B 2007 *Bioscience.* **57** 823-33

[48] Rowe R L, Street N R and Taylor G 2009 *Renewable Sustainable Energy Rev.* **13** 271-90

[49] Aronsson P G and Bergstrom L F 2001 *Biomass Bioenergy.* **21** 155-64

[50] French C J, Dickinson N M and Putwain P D 2006 *Environ. Poll.* **141** 387-95.

[51] Smeets E M W, Lewandowski I M and Faaij A P C 2009 *Renewable & Sustainable Energy Rev.* **13** 1230-45

[52] Hamin E M and Gurran N 2009 *Habitat Int.* **33** 238-45

[53] Forster P, Ramaswamy V, Artaxo P et al. 2007 *Changes in Atmospheric Constituents and in Radiative Forcing* ed Solomon S, Qin D, Manning M, Chen Z, Marquis M, Averyt K B, Tignor M and Miller H L (Cambridge: Cambridge University Press) pp 129-234.

[54] Pielke R, Prins G, Rayner S and Sarewitz D 2007 *Nature.* **445** 597-98

[55] Dalili N, Edrisy A and Carriveau R 2009 *Renewable & Sustainable Energy Rev.* **13** 428-38

Institute of Physics
Dirac House, Temple Back
Bristol BS1 6BE UK